高等职业教育软件技术专业系列教材

Java EE框架应用开发（SpringBoot+VueJS）

第2版

主　编　熊君丽　刘　鑫

参　编　王创鑫　扶卿妮

主　审　龙立功

机 械 工 业 出 版 社

本书由校企双元合作开发，以职业能力培养为本位，以企业常用的“权限管理系统”项目为载体，解析项目涉及的Java EE框架应用开发中的前端与后端知识技能点。本书基于工作任务与职业能力要求确定编写架构，结合职业岗位要求与国家职业标准，融入Java应用开发“1+X”职业技能等级证书的认证考核内容。

本书包括导学、12个工作单元及附录。工作单元1为搭建开发环境，工作单元2为生成数据库与约定接口规范，工作单元3为构建后端项目公共模块，工作单元4为实现用户和角色管理接口，工作单元5为实现登录认证和鉴权，工作单元6为实现菜单管理接口，工作单元7为构建前端项目，工作单元8为实现登录的前端功能，工作单元9为实现用户和角色管理的前端功能，工作单元10为实现菜单前端管理功能，工作单元11为打包部署项目，工作单元12为自动打包部署项目。

本书可作为各类职业院校软件技术及相关专业的教材，也可供软件开发爱好者参考使用。

为了方便读者学习，本书配有二维码微课视频，并在超星学习通建有在线开放课程和示范教学包。本书还配有教师授课用电子课件、源代码，可联系编辑（010-88379197）获取，或登录机工教育服务网（www.cmpedu.com）注册后免费下载。

图书在版编目（CIP）数据

Java EE框架应用开发 : SpringBoot+VueJS / 熊君丽，刘鑫主编. -- 2版. -- 北京 : 机械工业出版社，2025. 4. --（高等职业教育软件技术专业系列教材）.
ISBN 978-7-111-78472-2

Ⅰ. TP312. 8

中国国家版本馆CIP数据核字第20257XW023号

机械工业出版社（北京市百万庄大街22号　邮政编码100037）
策划编辑：徐梦然　　　　责任编辑：徐梦然
责任校对：李　杉　张亚楠　　封面设计：马精明
责任印制：张　博
北京建宏印刷有限公司印刷
2025年6月第2版第1次印刷
184mm×260mm・17.75印张・424千字
标准书号：ISBN 978-7-111-78472-2
定价：55.00元

电话服务
客服电话：010-88361066
　　　　　010-88379833
　　　　　010-68326294

网络服务
机　工　官　网：www.cmpbook.com
机　工　官　博：weibo.com/cmp1952
金　　书　　网：www.golden-book.com
机工教育服务网：www.cmpedu.com

前　言

本书以计算机软件人才需求、高等职业学校专业教学标准、“Java 应用开发”1+X 职业技能等级证书标准为基本依据，以提高学生的职业能力和职业素养为宗旨，坚持以职业能力培养为本位的课程设计原则，实现计算机软件应用型人才培养。

本书以企业典型项目产品为载体，以工作任务为讲述单元，以工作手册式为样板，对接企业最新技术和规范，践行教材改革。特点如下：

（1）素质教育元素融入　本书在动态修订过程中认真贯彻党的二十大精神，落实立德树人根本任务，通过将中华优秀传统文化等素质教育元素融入课程，用习近平新时代中国特色社会主义思想铸魂育人，实现价值引领，培养学生深度分析、大胆质疑、勇于创新、团结合作的精神和能力，引导学生形成积极正确的人生观、价值观和世界观。

（2）高阶性　本书针对有一定 Java 编程基础的学生，采用软件开发的新技术、新方法，开发一个完整的典型项目产品，将知识和能力有机融合，内容强调广度和深度，系统培养学生的整体设计与计划工作能力、解决复杂问题的综合能力。

（3）创新性　本书采用了 Java 企业应用开发的前沿技术和规范，体现与时俱进的时代性。本书配备了电子课件、源代码，另辅以教学云平台、代码仓库，增加了师生间的联系和互动，推进了现代信息技术与教学的深度融合，方便教师引导学生进行探究式与个性化学习。

（4）课证融通　本书有机融入 1+X 职业技能等级证书的技能点。在构思教材的工作单元和任务场景时，紧紧围绕该证书（中级）六大工作领域的职业技能等级要求，设计了“权限管理系统”项目作为载体，在任务实施过程中潜移默化地培养学生掌握证书所要求的职业技能。

（5）工作手册式　本书内容源自企业真实项目典型产品，参考了企业工作任务单。引领和提示学生明确工作任务的内容和质量标准；根据工作要求和技术条件制订合理的工作方案；基于工作任务的过程，提炼步骤和内容，进行学习和方案实施；对照工作标准对工作过程进行检查，对工作成果进行评价。能够帮助学生在学习的过程中迅速进入职业角色。

（6）实用性　本书配套了在线开放课程、示范教学包、云实训环境。学生和社会学习者能够通过自主使用资源实现不同起点的系统化、个性化学习，并实现一定的学习目标。教师可以针对不同的教学对象和课程要求，灵活组织教学内容，辅助实施课上课下、线上线下混合式教学。

本书面向企业 Java 全栈工程师岗位技能，包括搭建开发环境、实现后端功能、管理接口、实现前端和部署应用。以企业常用的“权限管理系统”项目为载体，解析项目涉及的知识技能点。该项目是一个综合性科目知识的总结，凝聚了很多课程的知识要素，学生在学习过程中会同时提高综合素质能力。

本书中涉及的后端技能主要涵盖数据表设计能力、项目构建能力、使用 Java 操作数据库能力、Restful 接口编写能力、JSON 数据交互能力、日志记录场景能力、登录场景解决能力、增删改查场景解决能力、树形菜单场景解决能力、权限认证授权能力、接口测试能力和数据监控场景能力。前端技能主要涵盖布局能力、首页布局能力、登录布局能力、增删改查页面布局能力、路由状态管理能力、后端交互能力、Mock 测试能力。本书教学课时建议如下：

工作单元	任务名称	课时
导学	项目介绍	2
工作单元 1 搭建开发环境	任务 1　搭建后端开发环境	1
	任务 2　搭建前端开发环境	1
工作单元 2 生成数据库与约定接口规范	任务 1　根据数据表设计生成数据库	1
	任务 2　根据 UI 设计编写后端接口文档	1
工作单元 3 构建后端项目公共模块	任务 1　使用 Spring Initializr 构建后端项目	1
	任务 2　使用 Spring Data JPA 构建数据访问层	8
	任务 3　导入常用工具类	2
	任务 4　封装统一接口响应的 HTTP 结果	2
工作单元 4 实现用户和角色管理接口	任务 1　实现用户列表接口	1
	任务 2　实现用户列表的查询与分页接口	2
	任务 3　实现用户新增接口	2
	任务 4　实现批处理删除用户接口	2
	任务 5　实现修改用户接口	2
	任务 6　实现获取用户信息与角色列表接口	2
工作单元 5 实现登录认证和鉴权	任务 1　实现基于 Spring Security 的权限控制功能	6
	任务 2　实现基于 JWT 的登录认证功能	6
	任务 3　实现基于 JWT 的访问鉴权功能	4
	任务 4　实现获取登录用户授权信息接口功能	1
工作单元 6 实现菜单管理接口	任务 1　实现菜单后端接口层	4
	任务 2　实现菜单后端业务层	3
	任务 3　实现菜单后端数据控制层	2
	任务 4　测试验证菜单后端接口	8
工作单元 7 构建前端项目	任务 1　初始化前端项目	4
	任务 2　裁剪前端项目结构	4
工作单元 8 实现登录的前端功能	任务 1　实现登录功能	6
	任务 2　实现菜单动态生成功能	6
工作单元 9 实现用户和角色管理的前端功能	任务 1　实现显示用户列表页面	5
	任务 2　实现用户列表分页	3
	任务 3　实现用户查询功能	4
	任务 4　实现用户新增功能	7
	任务 5　实现用户修改功能	5
	任务 6　实现用户批量删除功能	3
工作单元 10 实现菜单前端管理功能	任务 1　实现菜单列表与查询页面	3
	任务 2　实现菜单新增功能	7
	任务 3　实现菜单修改功能	3
	任务 4　实现菜单删除功能	3
工作单元 11 打包部署项目	任务 1　搭建云服务器环境	4
	任务 2　打包与部署项目后端	1
	任务 3　打包与部署项目前端	3
工作单元 12 自动打包部署项目	任务 1　安装 Docker 服务	3
	任务 2　自动发布后端项目到 Docker 容器	4
	任务 3　自动发布前端项目到 Docker 容器	5

本书建议一周安排 16 课时，学习之前需掌握一定的 SpringBoot 框架、Maven 构建工具、VueJS 前端框架等基础知识。

本书由广东科学技术职业学院熊君丽、珠海爱浦京软件股份有限公司刘鑫主编，珠海爱浦京软件股份有限公司王创鑫、广东科学技术职业学院扶卿妮参加编写，广东科学技术职业学院龙立功任主审。在本书的编写过程中，编者参考了大量相关技术材料，吸取了许多同仁的宝贵经验，在此对资料的提供者和支持、帮助过我们的同仁表示感谢。

限于编者的经验和水平，书中难免有疏漏之处，恳请各位读者批评指正。

编　者

二维码索引

（续）

序号	名称	图形	页码	序号	名称	图形	页码
17	工作单元 5-2　实现基于 JWT 的登录认证功能		108	25	工作单元 7-2　裁剪前端项目结构		155
18	工作单元 5-3　实现基于 JWT 的访问鉴权功能		117	26	工作单元 8-1　实现登录功能		162
19	工作单元 5-4　实现获取登录用户授权信息接口功能		123	27	工作单元 8-2　实现菜单动态生成功能		168
20	工作单元 6-1　实现菜单后端接口层		128	28	工作单元 9-1　实现显示用户列表页面		179
21	工作单元 6-2　实现菜单后端业务层		133	29	工作单元 9-2　实现用户列表分页		185
22	工作单元 6-3　实现菜单后端数据控制层		143	30	工作单元 9-3　实现用户查询功能		188
23	工作单元 6-4　测试验证菜单后端接口		146	31	工作单元 9-4　实现用户新增功能		193
24	工作单元 7-1　初始化前端项目		153	32	工作单元 9-5　实现用户修改功能		201

（续）

序号	名称	图形	页码	序号	名称	图形	页码
33	工作单元 9-6　实现用户批量删除功能		205	40	工作单元 11-3　打包与部署项目前端		238
34	工作单元 10-1　实现菜单列表与查询页面		210	41	工作单元 12-1　安装 Docker 服务		243
35	工作单元 10-2　实现菜单新增功能		216	42	工作单元 12-2　自动发布后端项目到 Docker 容器		245
36	工作单元 10-3　实现菜单修改功能		224	43	工作单元 12-3　自动发布前端项目到 Docker 容器		251
37	工作单元 10-4　实现菜单删除功能		227	44	在线开放课程		-
38	工作单元 11-1　搭建云服务器环境		230	45	代码仓库		-
39	工作单元 11-2　打包与部署项目后端		237	46	源代码		-

目　录

导学

项目介绍

职业能力

生活在信息化社会中，人们每天都在使用各种信息化系统，如电子商务系统、生产制造系统、管理信息系统等，其中权限管理系统几乎是所有信息化系统都会涉及的一个重要组成部分，也是信息化系统安全运行的基础，开发权限管理系统更是程序员的一项基本能力。

什么是权限管理系统？权限管理系统是对信息化系统的所有资源进行权限控制，它针对的对象是用户，其主要管控如下3个方面：

- 用户可以访问哪些页面。
- 用户可以进行哪些操作。
- 用户可以处理哪些数据。

本书中的Friday权限管理系统是基于角色的访问控制（Role-Based Access Control）模型设计，采用前后端分离架构实现的轻量级权限管理系统，其中前后端分离架构是基于SpringBoot、VueJS、Vue-Element-Admin等主流技术实现的，Friday权限管理系统之所以被称为轻量级，是因为设计时严格遵守“如无必要，勿增实体”的原则，让系统保持简单、高效。这样做有两个好处：首先它仅提供一个调用简单、可复用性高、满足一般需求的权限管理功能，便于理解和二次开发，可作为信息系统的脚手架项目；其次本系统没有无节制地引入各种技术框架，而是根据职业能力要求使用恰当的技术框架，降低学习和使用本系统的成本。

系统功能

权限管理系统的核心功能就是不同的用户具有不同的访问、操作、数据权限。本系统使用基于角色的访问控制模型（简称RBAC模型）进行设计，基础组成关系如图0-1所示。一组用户对应一个角色，同时该角色又对应一组权限，角色就是用户和权限的桥梁，通过角色来控制用户的访问权限。

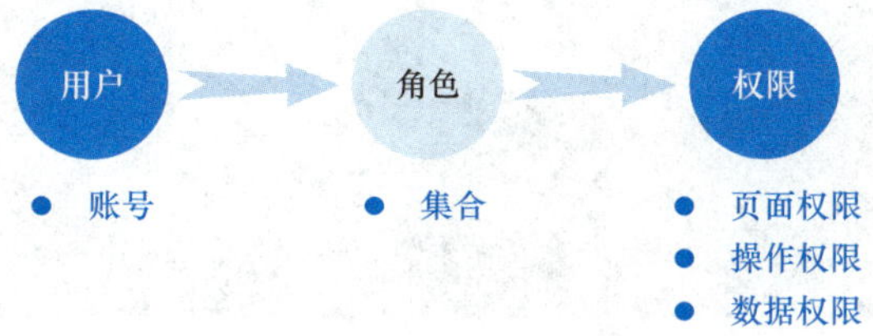

图0-1　RBAC模型基础组成关系

使用 RBAC 模型设计的权限系统，必然要管理用户、角色、菜单（权限）以及它们的关系，所以权限系统必须具备 3 个基础功能：用户管理、角色管理和菜单（权限）管理，图 0-2 列出了权限管理系统的基础功能模块结构图。

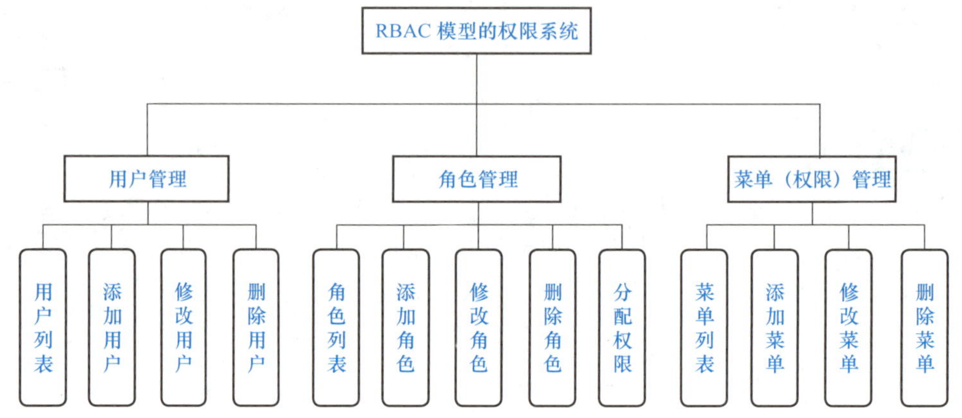

图 0-2　权限管理系统的基础功能模块结构图

下面详细介绍一下 Friday 权限管理系统的主要功能。

➢　登录功能：权限管理的核心就是认证和授权，用户通过登录功能完成认证授权，并且根据用户的授权信息在首页工作台中动态显示系统菜单，即控制用户可以访问哪些页面，如图 0-3 和图 0-4 所示。

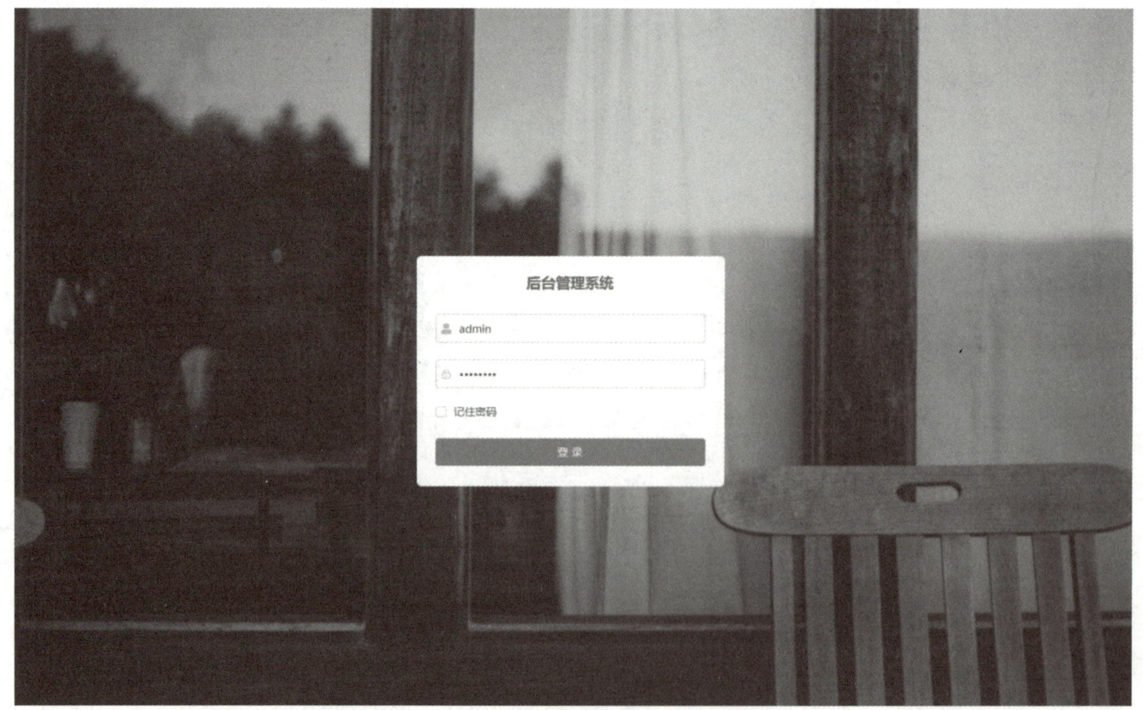

图 0-3　系统登录界面

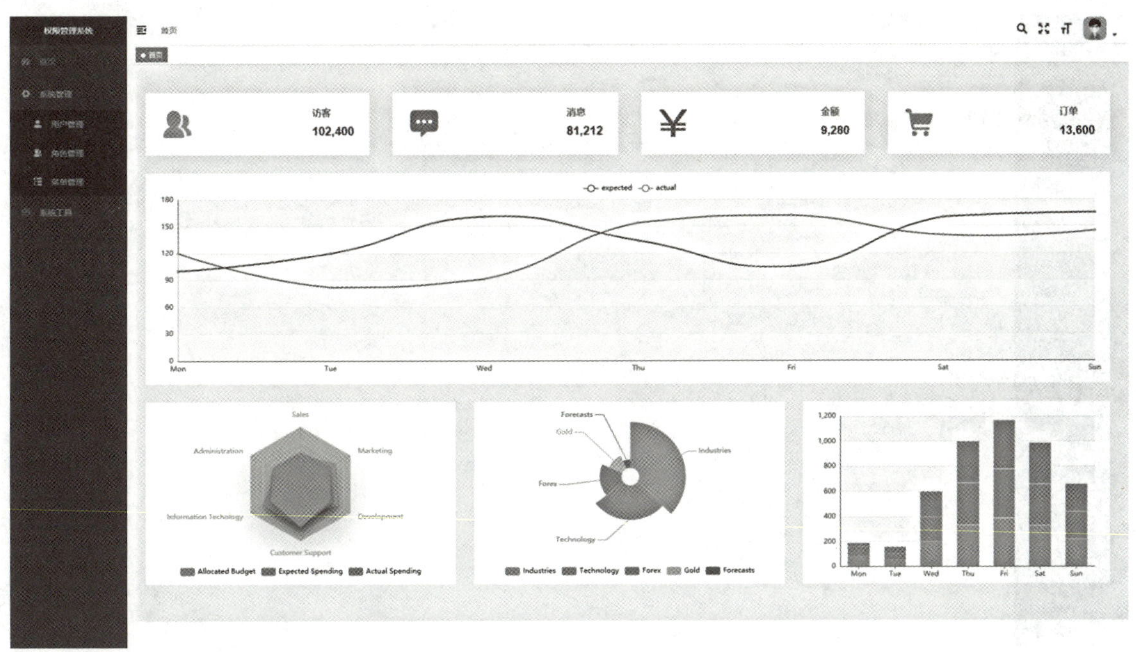

图 0-4　系统登录后的首页

➢ 用户管理：实现用户列表分页显示、查询用户、添加用户、修改用户、设置用户角色以及删除用户的功能，如图 0-5 所示。

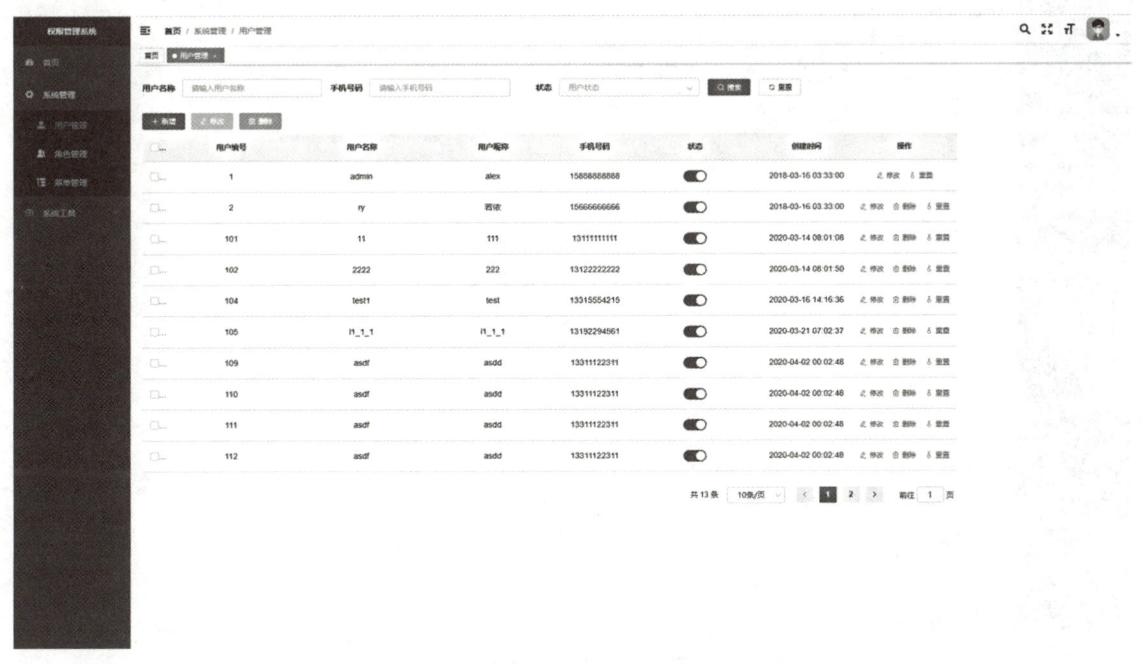

图 0-5　用户管理界面

➢ 角色管理：实现角色列表分页显示、查询角色、添加角色、修改角色、配置角色菜单权限以及删除角色的功能，如图 0-6 所示。

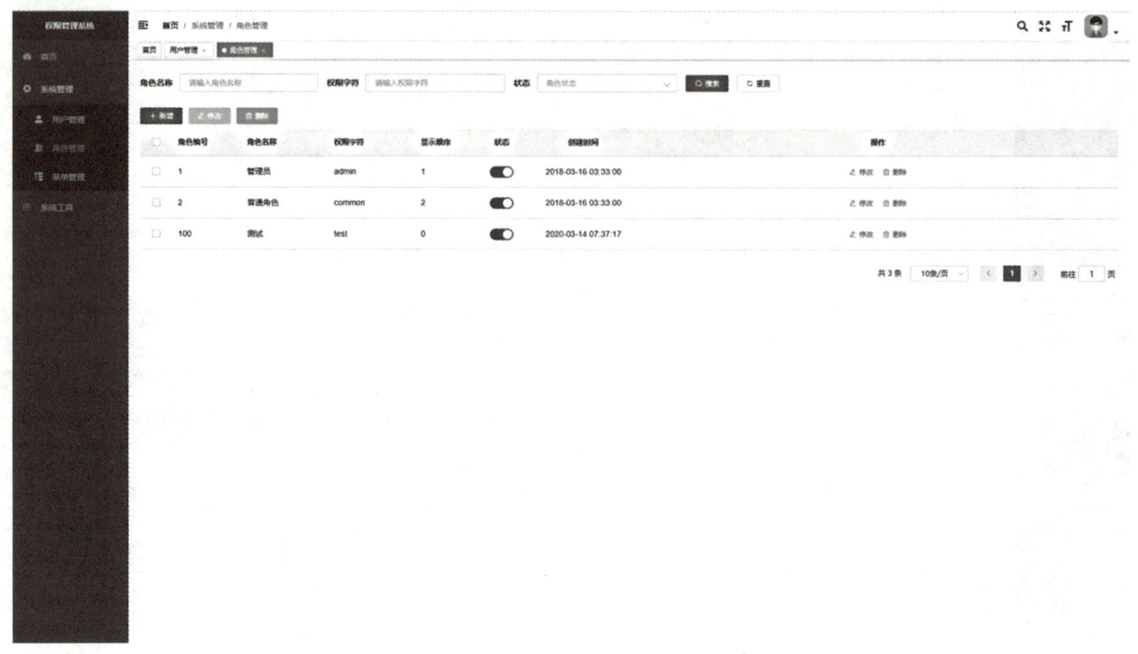

图 0-6　角色管理界面

➢　菜单管理：实现菜单列表、查询菜单、添加菜单、修改菜单以及删除菜单的功能，如图 0-7 所示。

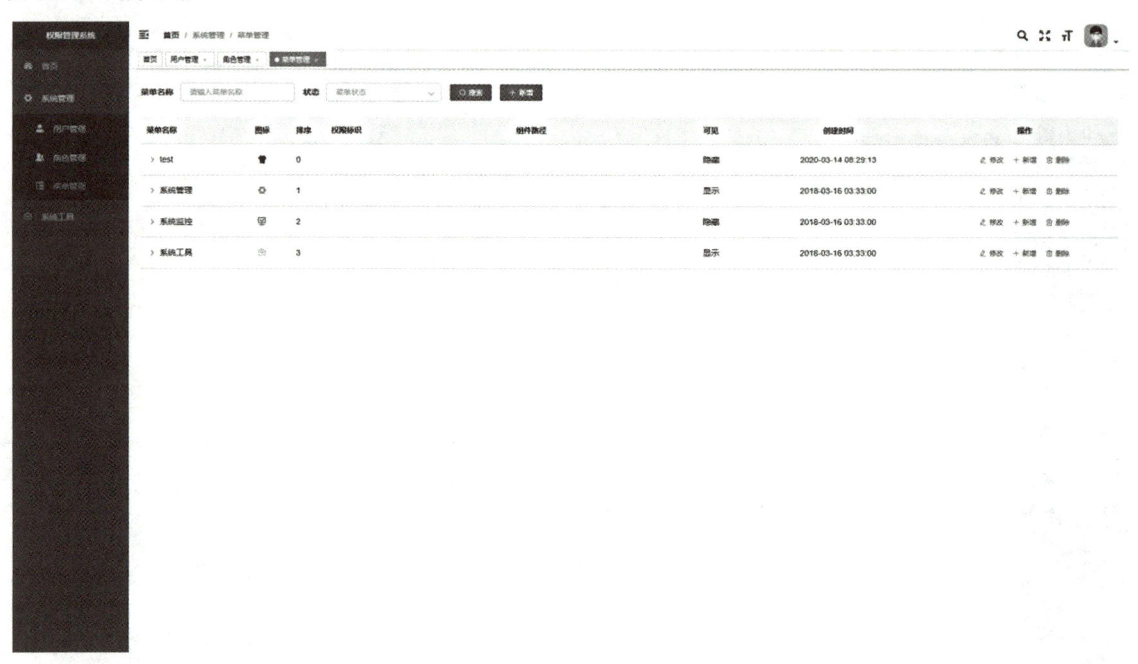

图 0-7　菜单管理界面

系统架构

Friday 权限管理系统采用前后端分离架构实现，在前后端分离的应用模式中，前端与后端独立部署，通过 JSON 格式进行数据交互，耦合度相对较低，系统结构图如图 0-8 所示。

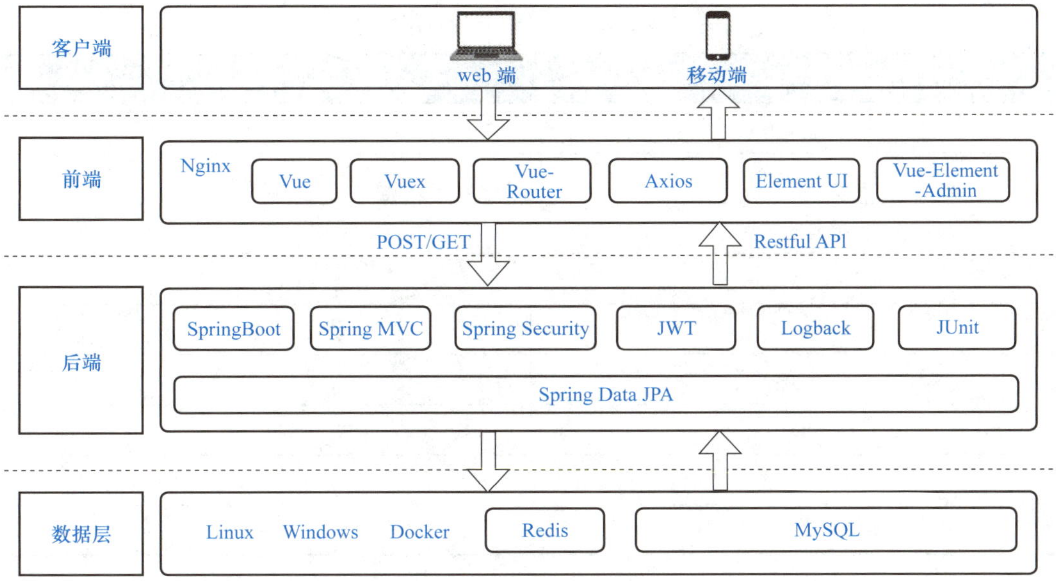

客户端	web 端　移动端
前端	Nginx　Vue　Vuex　Vue-Router　Axios　Element UI　Vue-Element-Admin
后端	SpringBoot　Spring MVC　Spring Security　JWT　Logback　JUnit Spring Data JPA
数据层	Linux　Windows　Docker　Redis　MySQL

（前端与后端之间：POST/GET，Restful API）

图 0-8　系统结构图

前端核心框架采用主流的 VueJS，页面展示使用基于 Element UI 封装的 Vue-Element-Admin 框架实现，页面跳转使用 Vue-Router 框架实现，前端与后端的数据交互通过 Axios 框架实现。

后端核心框架采用 SpringBoot + Spring MVC + Spring Data JPA + Spring Security 的 Spring“全家桶”技术框架实现，使用 SpringBoot 简化后端的初始搭建以及开发过程。使用 Spring MVC 实现后端 Restful 风格的接口（API），与前端通过 JSON 格式进行数据交互。使用 Spring Data JPA 实现对 MySQL 数据表进行操作，使用 Spring Security 实现对系统的认证和授权功能。

技术选型

前端技术选型见表 0-1。

表 0-1　前端软件包名称及版本号

软　件　包	版　　本
Vue	2.6.10
Vue-Router	3.0.2
Vuex	3.1.0
Axios	0.18.1
Element UI	2.11.1
Vue-Element-Admin	4.4.0
NPM	6.14.5

后端技术选型见表 0-2。

表 0-2　后端软件包名称及版本号

软件包	版本
SpringBoot	3.2.2
Spring Security	6.2.1
Spring Data JPA	3.2.2
Spring Data Redis	3.2.2
Lombok	1.18.30
Fastjson	1.2.56
Druid	1.1.10
JDK	21
MySQL	8.3.0

拓展阅读

《魏书》中记载了一个吐谷浑国王阿豺的故事。阿豺有子二十人，及老，临终谓子曰："汝等各奉吾一只箭，折之地下。"俄而命母弟慕利延曰："汝取一只箭折之。"慕利延折之。又曰："汝取十九只箭折之。"延不能折。阿豺曰："汝曹知否？单者易折，众则难摧，戮力一心，然后社稷可固也。"言终而死。由这个故事可以看出，团结合作的力量是强大的。

团结合作是事业成功的基础。依靠团结合作的力量，把个人的愿望和团队的目标结合起来，可以超越个体的局限，产生"1+1>2"的效果。计算机项目的设计与实施，离不开团队的团结合作。擅长美术的做美工，擅长数据的设计数据库，擅长写程序的编写代码。团队成员充分发挥个体的优势，分工明确，团结合作，从而顺利交付项目。

工作单元 1

搭建开发环境

职业能力

本书以权限管理系统作为项目载体，采用 SpringBoot+VueJS 的前后端分离技术驱动分解工作任务。本工作单元主要完成项目前后端开发环境的搭建，最终希望学生达成如下职业能力目标：

1．熟练掌握后端开发环境包括 Redis、Postman、IDEA 软件的安装和配置

2．熟练掌握前端开发环境包括 VS Code、Node.js 软件的安装

任务情景

由于本书采用 SpringBoot+VueJS 的前后端分离技术，需要在开发环境安装数据库、前后端的集成开发环境及接口测试等软件。基于上述要求，约定安装环境条件与约束如下：

- 操作系统 Windows 10。
- 软件安装包存放路径 C:\friday\package。
- 服务安装目录 C:\friday\tool。
- 代码存放 C:friday\friday。
- 软件安装路径 C:\friday\tool。
- 浏览器 Chrome（正式版，64 位）。

前置知识

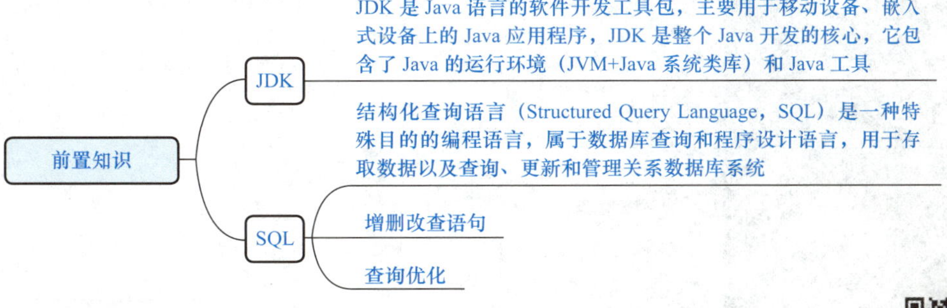

任务 1 搭建后端开发环境

微课 1-1 搭建后端开发环境

后端开发环境总共要安装 5 个软件：JDK21 作为 Java 运行环境，MySQL8 作为数据库

服务器，Redis 是一个完全开源、遵守 BSD 协议、高性能的 key-value 数据库，Postman 作为后端应用的测试工具，IntelliJ IDEA 则是后端代码的集成开发平台。

任务实施

步骤 1 安装 JDK21

访问网址：http://www.oracle.com/technetwork/java/javase/downloads/，下载 Windows 64 版本的 JDK21。这里就不赘述安装过程了。

步骤 2 安装 MySQL8

1）访问网址：https://dev.mysql.com/downloads/mysql/，下载安装包，进入安装程序。

2）选择安装类型为“Developer Default”，单击“Next”按钮继续，如图 1-1 所示。

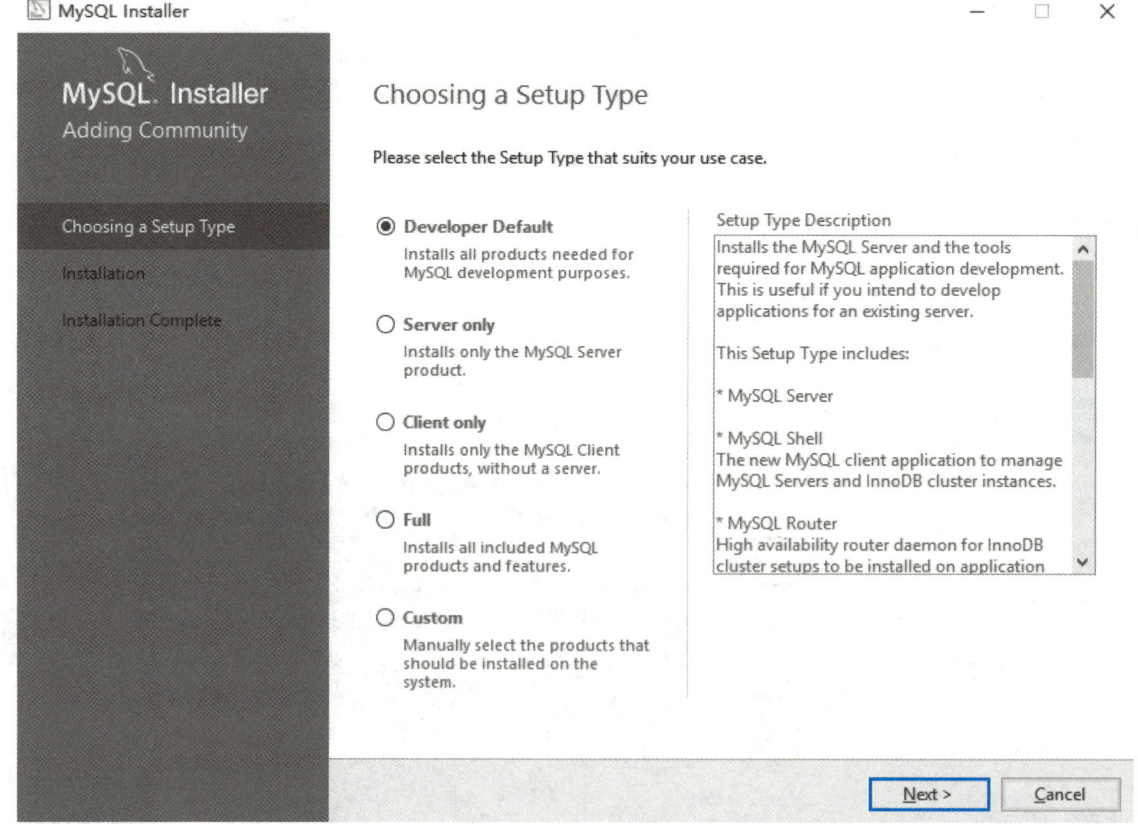

图 1-1 选择安装类型

3）检查软件依赖，单击“Next”按钮继续，如图 1-2 所示。

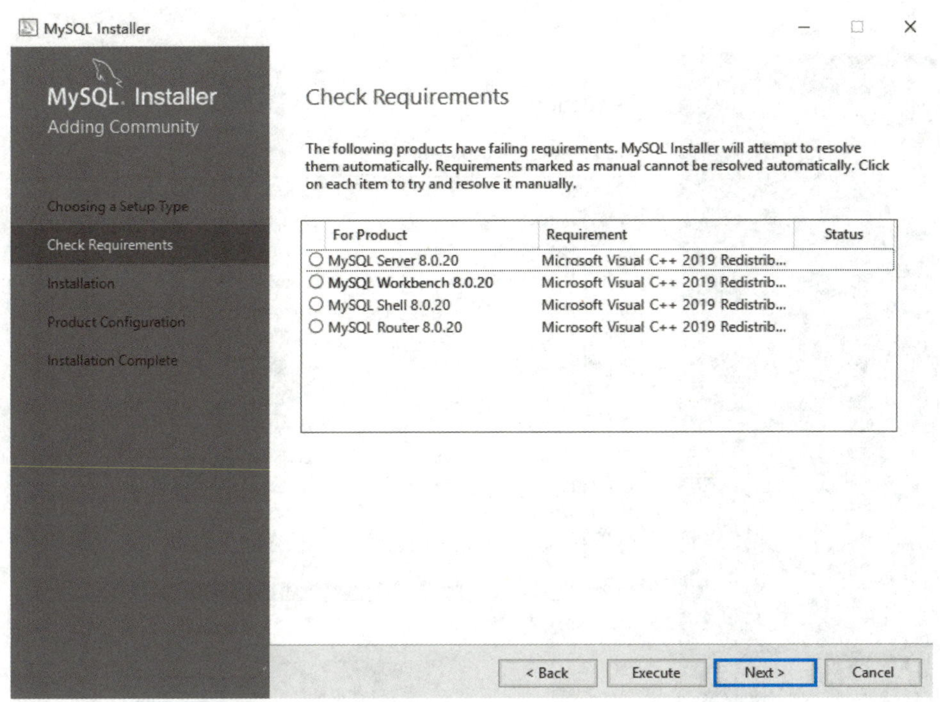

图 1-2 检查软件依赖

4）执行安装，单击“Execute”按钮执行安装，如图 1-3 所示。

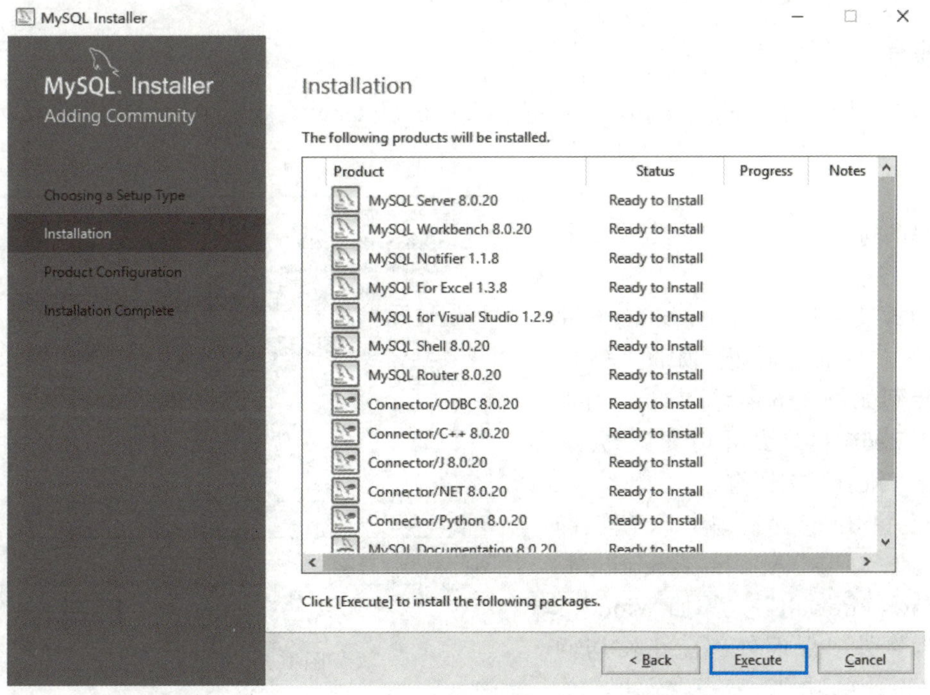

图 1-3 执行安装

5）安装完毕，单击“Next”按钮继续，如图 1-4 所示。

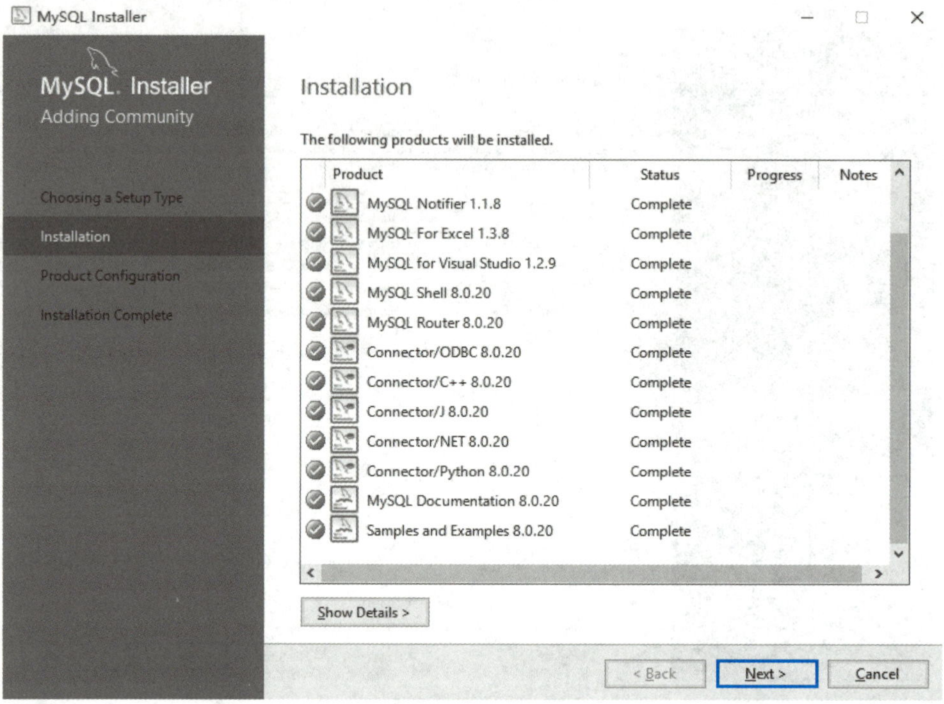

图 1-4　安装完毕

6）对于后面的界面，一直单击“Next”按钮即可，最后单击“Finish”按钮完成安装。

步骤 3　安装 Redis6

1）访问网址：download.redis.io/release/redis-6.0.9.tar.gz，下载 Redis6 安装包。

2）进入安装界面，单击“Next”按钮继续。

3）勾选同意协议条款，单击“Next”按钮继续安装。

4）更改安装目录为：C:\friday\tool\redis\，单击“OK”按钮，如图 1-5 所示；跳转到如图 1-6 所示界面，勾选方框添加 Redis 目录到环境变量 PATH 中，单击“Next”按钮继续。

5）设置 Redis 运行端口（默认即可），同时勾选“Add an exception to the Windows Firewall”，添加 Windows 防火墙为例外，保证外部能正常访问 Redis 服务，单击“Next”按钮，如图 1-7 所示。

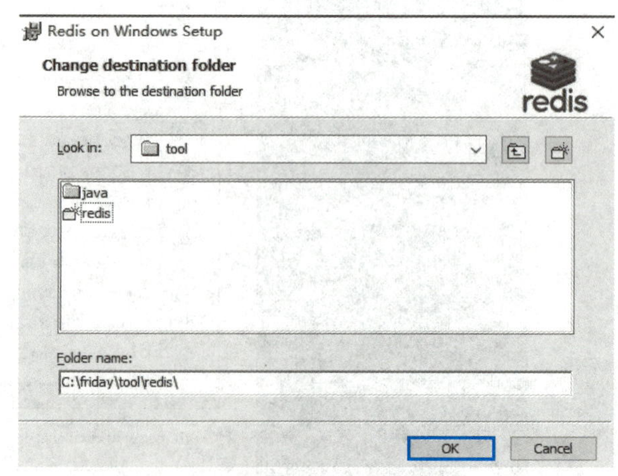

图 1-5　更改安装目录

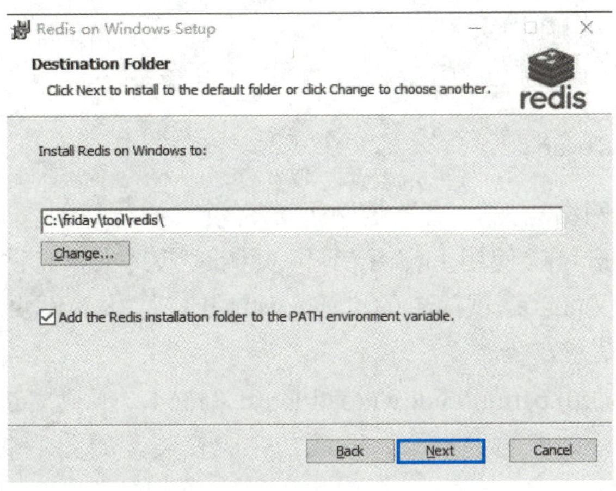

图 1-6　添加 Redis 目录到环境变量

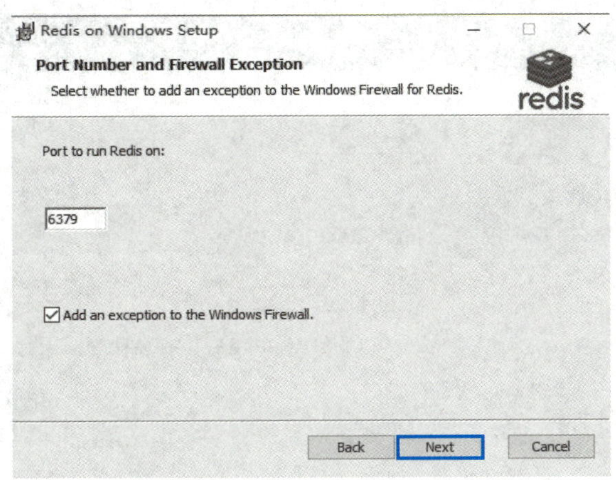

图 1-7　设置 Redis 运行端口

6）设置 Redis 最大占用内存，可根据实际情况设置，一般情况下设置为物理内存的 1/4，单击“Next”按钮，如图 1-8 所示。

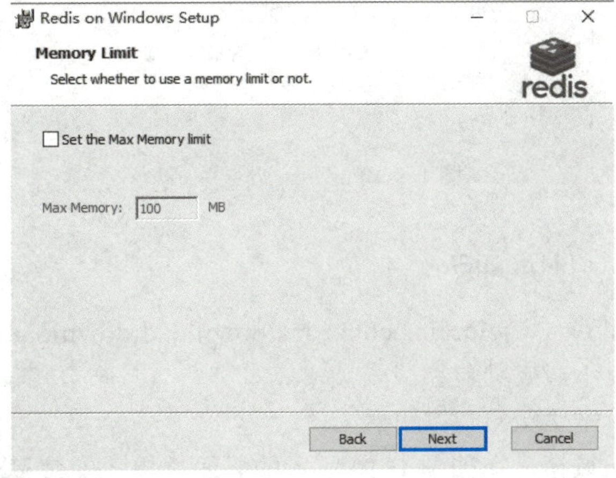

图 1-8　设置 Redis 最大占用内存

7）单击“Install”按钮开始安装。

8）单击“Finish”按钮完成安装。

步骤 4 安装 Postman

用户在开发、调试网络程序，或者是网页 B/S 模式的程序时，需要使用一些方法来跟踪网页请求，例如，用户可以使用 Firebug 等网页调试工具来跟踪网页请求。Postman 作为一款网页调试工具，不仅可以调试简单的 CSS、HTML、脚本等网页基本信息，还可以发送几乎所有类型的 HTTP 请求。

1）访问网址：https://dl.pstmn.io/download/latest/win64，下载安装包完成安装。

2）打开 Postman，注册账号并登录，即成功完成安装，如图 1-9 所示。

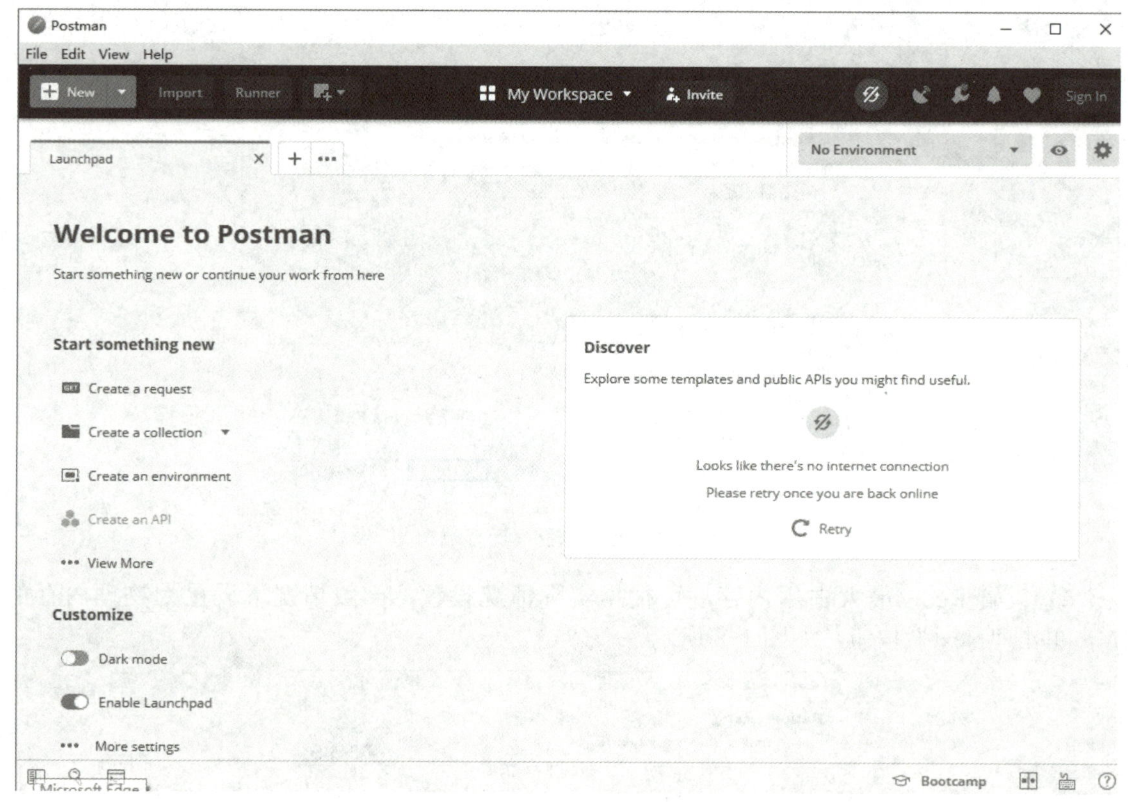

图 1-9　Postman 安装完成

步骤 5 安装 IntelliJ IDEA

1）访问网址：https://www.jetbrains.com/idea/download/download-thanks.html?platform=windowsZip&code=IIC，下载安装包完成安装。

2）在欢迎界面，单击“Next”按钮继续。

3）单击“Browse”按钮，在如图 1-10 所示的界面，更改安装目录为：C:\ProgramFiles\JetBrains\IntelliJ IDEA 2020.1，单击“Next”按钮继续。

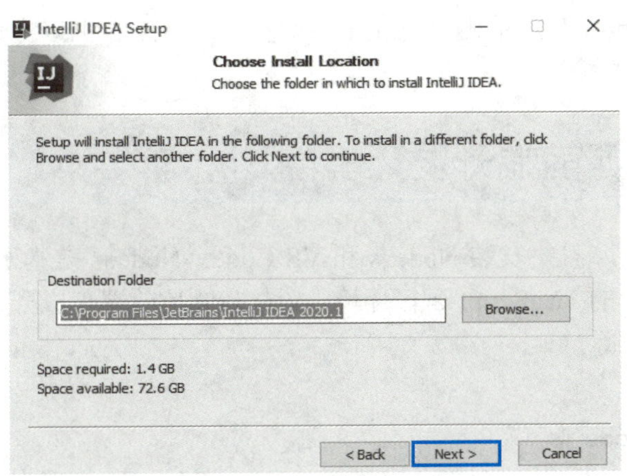

图 1-10 更改安装目录

4）在安装选项中勾选“Add launchers dir to the PATH”（添加环境变量），单击“Next”按钮继续，如图 1-11 所示。

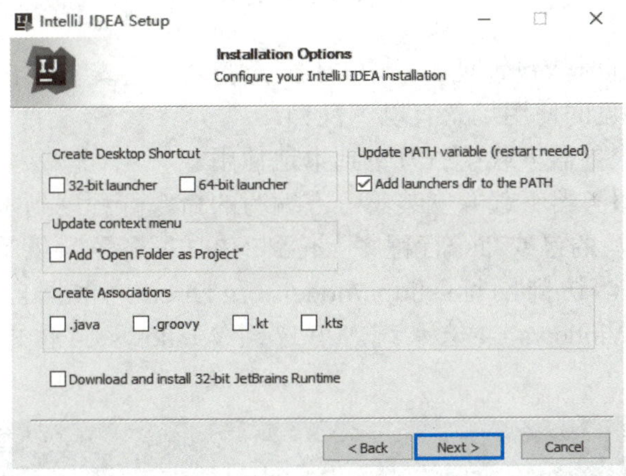

图 1-11 添加环境变量

5）创建“开始”菜单快捷方式，单击“Install”按钮开始安装，如图 1-12 所示。

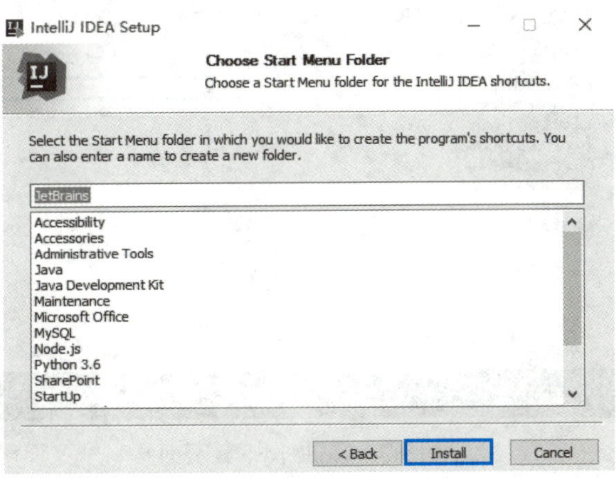

图 1-12 开始安装 IDEA

6）选择“Reboot now”，单击“Finish”按钮，重启计算机以完成安装。

任务 2 搭建前端开发环境

微课 1-2 搭建前端开发环境

搭建前端开发环境，需要安装 Node.js 和 VS Code。Node.js 是基于 Chrome V8 引擎的开源、跨平台的 JavaScript 运行环境，VS Code 是前端集成开发环境。

任务实施

步骤 1 安装 Node.js

Node.js 是本项目在前端所依赖的重要环境，根据 Node.js 的版本不同，项目的运行结果也不尽相同，故特此说明。

出于稳定性和兼容性的考虑，本项目推荐 Node.js 为 10 或以上的版本。如果 Node.js 版本低于 10，建议升级到 10 或者更高的版本。本项目在新版本（v14.4.0）的 Node.js 上可以正常运行。

在安装 Node.js 的同时，包管理工具 NPM 也会随之一起安装，它能解决 Node.js 在代码部署上的很多问题，常见的使用场景有以下几种：

- 用户从 NPM 服务器下载第三方包到本地使用。
- 用户从 NPM 服务器下载并安装第三方编写的命令行程序到本地使用。
- 用户将自己编写的包或命令行程序上传到 NPM 服务器，供第三方使用。

1）Node.js 的获取。访问网址：http://nodejs.org/zh-cn/，根据自己的系统版本，下载对应的文件，例如，Windows 64 位系统，可选用 Windows 64 位的 msi 安装包文件，如图 1-13 所示。

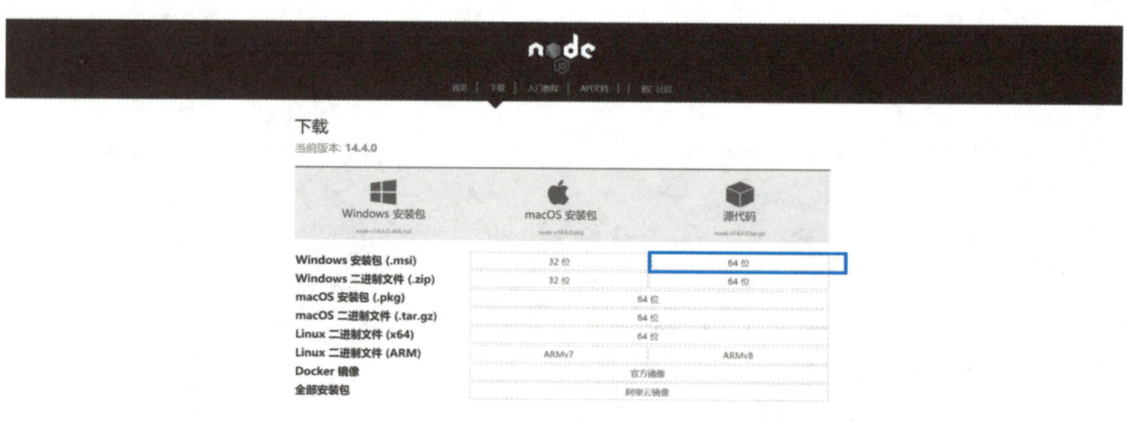

图 1-13 Node.js 下载页面

2）找到下载好的 Node.js 安装包，双击并运行安装程序。单击“Next”按钮进入安装流程。

3）单击“Next”按钮进入协议确认页面，确认过协议内容后，勾选“I accept the terms in the License Agreement”选项以接受协议内容，然后单击“Next”按钮继续。

4）选择 Node.js 的安装位置，默认位置是“C:\Program Files\Node.js\”，修改安装路径为“C:\friday\tool\node.js\”，如图 1-14 所示。

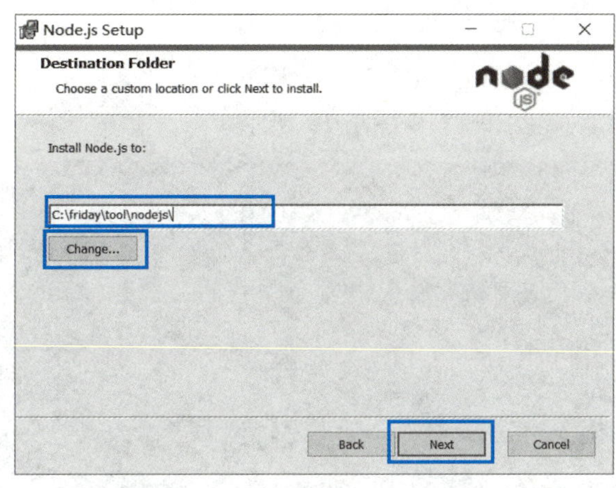

图 1-14　修改 Node.js 安装路径

5）选择自定义的安装选项，这里一般保持默认选项就可以了。但是，如果需要自定义一些安装选项，可以在这里进行设置。设置完成后，单击“Next”按钮进入下一步。单击“Next”按钮进入程序安装过程并出现进度条，等进度条走完，说明安装就完成了，单击“Finish”按钮完成安装过程。

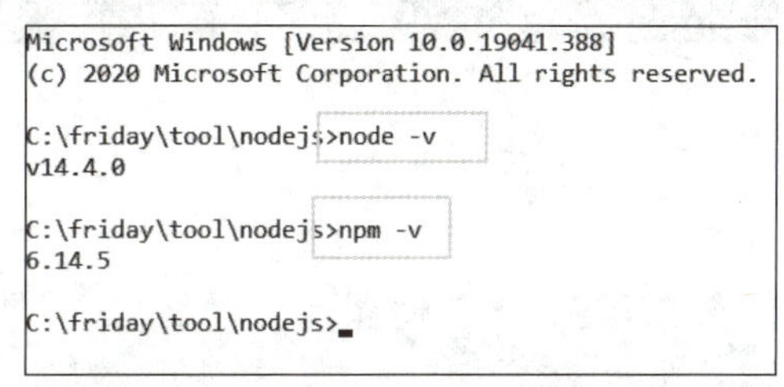

图 1-15　检测 Node.js 是否安装成功

6）Node.js 的测试和基本介绍。按 <Win+R> 组合键打开“运行”窗口，在输入框中输入“cmd”打开命令行界面。在命令行中输入“node -v”并按 <Enter> 键以检查 Node.js 的版本号信息。在命令行中输入“npm -v”并按 <Enter> 键以检查 npm 的版本号信息，如图 1-15 所示。如果两个命令都正常输出版本号，则说明 Node.js 安装成功。

7）配置 Node.js。软件安装好后，可以在安装目录下双击运行或者在命令行界面使用。但是，如果想要在安装目录之外的地方使用 Node.js 时，则需要在环境变量里配置 Node.js 的安装目录。系统可以通过环境变量，找到 Node.js 的安装目录，进而运行想要执行的命令。

正常情况下，安装程序会自动把安装目录添加到环境变量中，如图 1-16 所示。

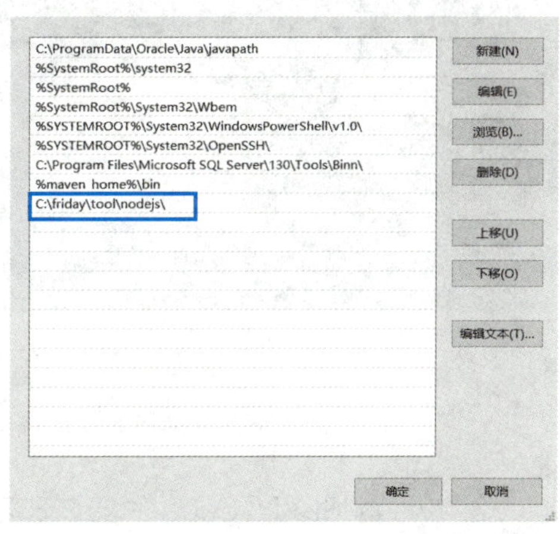

图 1-16　配置 Node.js 环境变量

步骤 2 安装 Visual Studio Code（VS Code）

Visual Studio Code 是 Microsoft 公司在 2015 年 4 月 30 日 Build 开发者大会上正式发布，运行于 Mac OS X、Windows 和 Linux 之上，针对编写现代 Web 和云应用的跨平台源代码编辑器。

1. Visual Studio Code 安装包的获取

访问网址：https://code.visualstudio.com/ 下载程序安装包。找到首页的下载按钮，根据计算机系统下载对应的版本，如图 1-17 和图 1-18 所示。

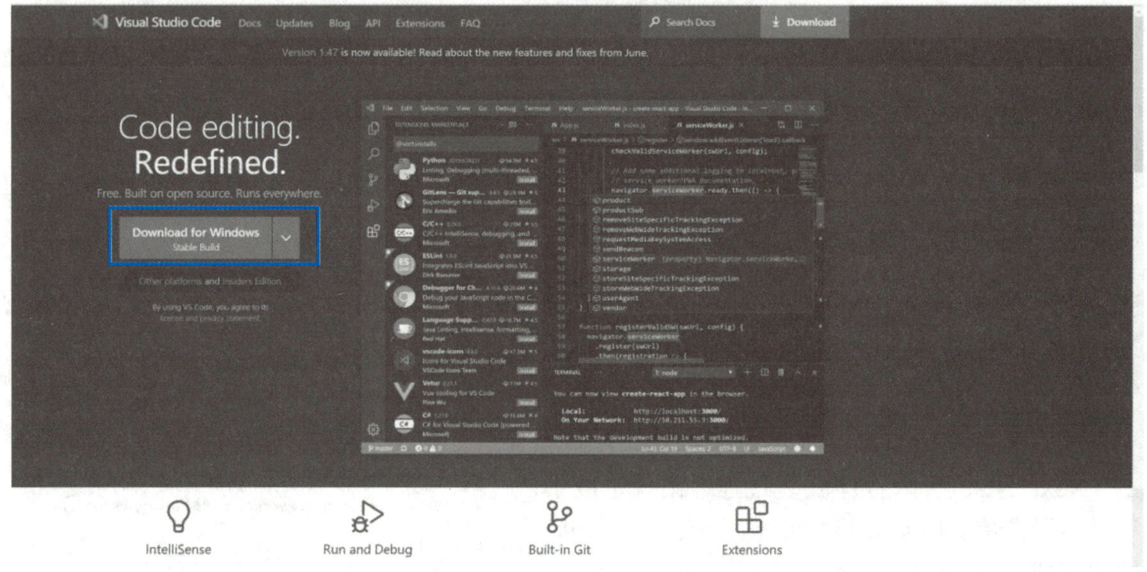

图 1-17　VS Code 下载页面

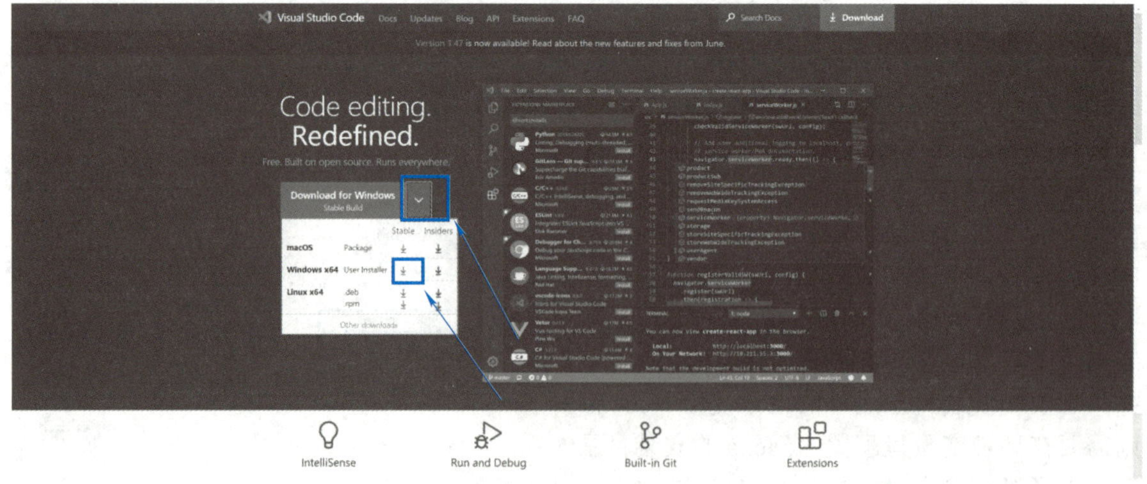

图 1-18　下载 VS Code 程序安装包

2. Visual Studio Code 的安装和基本使用

下载完成后，双击运行程序安装包，进入“许可协议”页面，如图 1-19 所示。确认过内容后，选择“我接受协议”，然后单击“下一步”按钮进入安装设置环节。

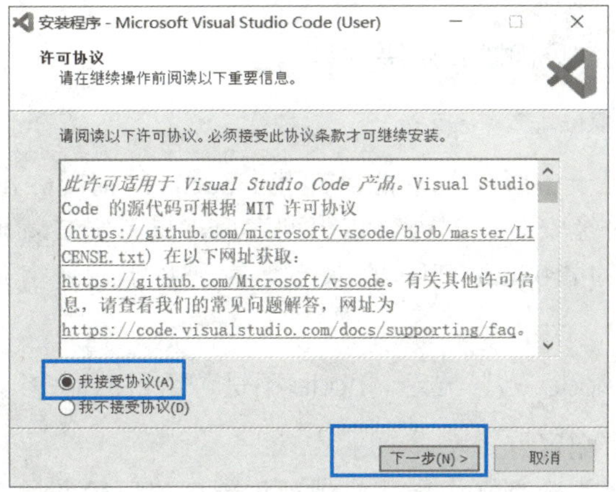

图 1-19　“许可协议”页面

在安装设置环节，如果有特殊的安装需求，可以进行相应的设置。如果没有特殊需求，可以直接单击“下一步”按钮跳过。

设置好之后，单击“安装”按钮进入安装过程，进度条走完即安装完成，单击“完成”按钮，退出安装流程并启动 Visual Studio Code。

步骤 3　VS Code 工作界面介绍

VS Code 工作界面主要分为 5 个区域，分别是活动栏、侧边栏、编辑栏、面板栏和状态栏，如图 1-20 所示。

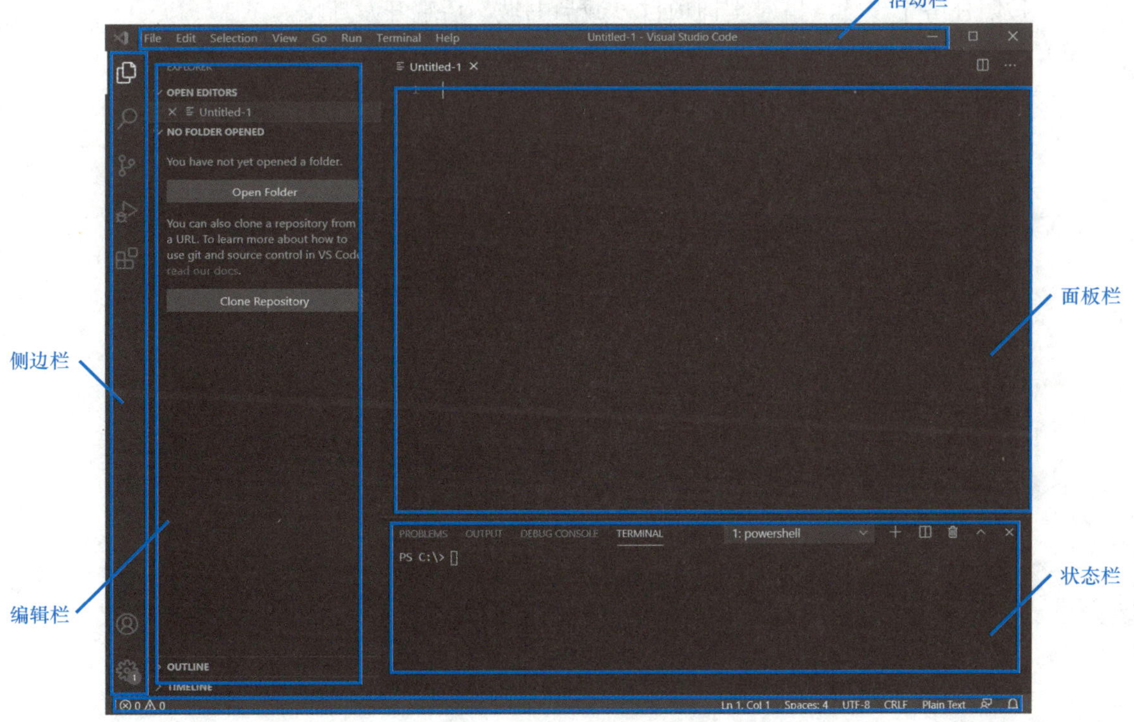

图 1-20　VS Code 工作界面

知识小结

安装配置 Node.js 的常见问题和解决方案

1. 安装 Node.js 失败

原因：这种情况常见于曾经安装，然后卸载 Node.js 的情况。出现这种情况一般是卸载不彻底导致的，只要把曾经安装过的内容删除彻底后，重新安装就可以。

解决方案：在 C 盘的 Program Files 目录下，找到并删除名字开头为 Node.js 和 npm 的文件夹，重新安装即可。

2. 在命令行运行“node -v”显示“node 不是可运行的命令”

原因：环境变量没有配置好。

解决方案：对照步骤 1 检查操作步骤，然后重新配置环境变量，重启 cmd 运行命令。

单元小结

搭建开发环境是程序员的基本能力，构建前后端分离技术的开发环境会比较复杂。约定安装环境，准备与之相适应的软件版本等工作是搭建开发环境的前提工作，十分重要。根据软件系统开发，掌握 Java 应用软件开发与运行基础环境的搭建，是 1+X 职业技能等级证书（Java 应用开发）初级的职业技能要求。通过完成本工作单元的任务和后面的实战强化，熟练掌握开发和运行环境搭建的职业技能。

实战强化

1）按照任务 1 步骤，搭建实训项目“诚品书城”后端开发环境。

2）按照任务 2 步骤，搭建实训项目“诚品书城”前端开发环境。

工作单元 2

生成数据库与约定接口规范

职业能力

本工作单元主要是生成项目需要的数据库表与编写项目接口规范文档，最终希望学生达成如下职业能力目标：

1．熟练掌握创建数据库、数据表以及导入数据

2．熟练掌握基于 UI 设计模型编写接口规范文档

任务情景

在前后端分离开发的项目中，研发人员开始编码之前，需要完成图 2-1 中的 3 个任务：

➢ 设计 UI 页面，用于展示系统操作的界面。

➢ 设计数据表，用于存储系统的业务数据。

➢ 编写接口服务文档，以便于有一个统一的文件进行前后端分离开发任务。

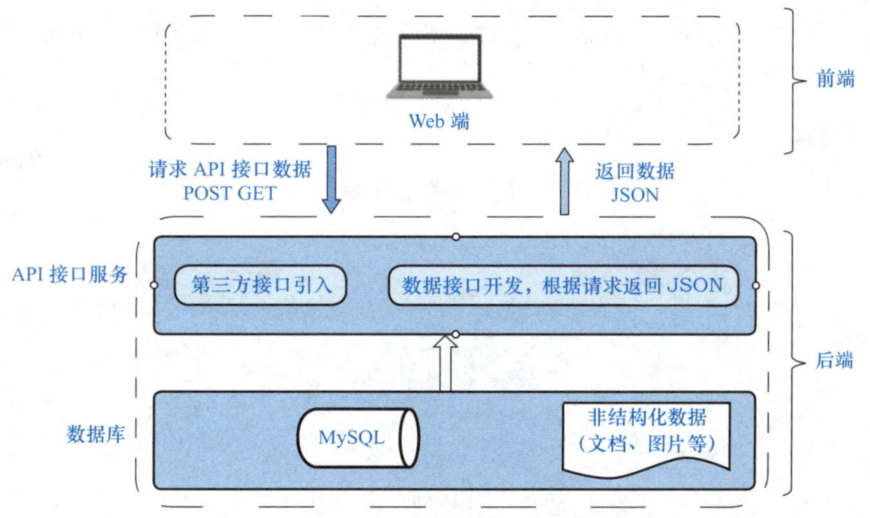

图 2-1　前后端分离开发的项目

在设计人员完成设计 UI 页面和设计数据表的任务之后，研发人员需要根据数据表生成数据库，以及根据 UI 设计编写统一的后端接口文档，研发人员开始编码之前，将面临如下 2 个问题：

1）如何创建数据库与数据表以及导入初始数据？

2）在前后端分离开发项目中如何根据 UI 设计编写后端接口文档？

基于上述 2 个问题，本工作单元的具体任务如下：

1）根据数据表设计生成数据库。
2）根据 UI 设计编写后端接口文档。

前置知识

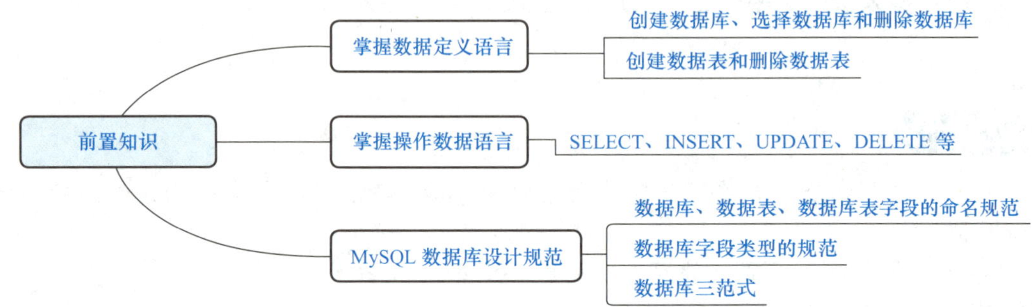

任务 1 根据数据表设计生成数据库

微课 2-1　根据数据表设计生成数据库

任务实施

步骤 1　理解权限管理系统的数据表

根据 RBAC 原则设计了 Friday 权限管理系统的实体关系图，如图 2-2 所示，然后根据该实体关系图在 MySQL 数据库中创建数据表。

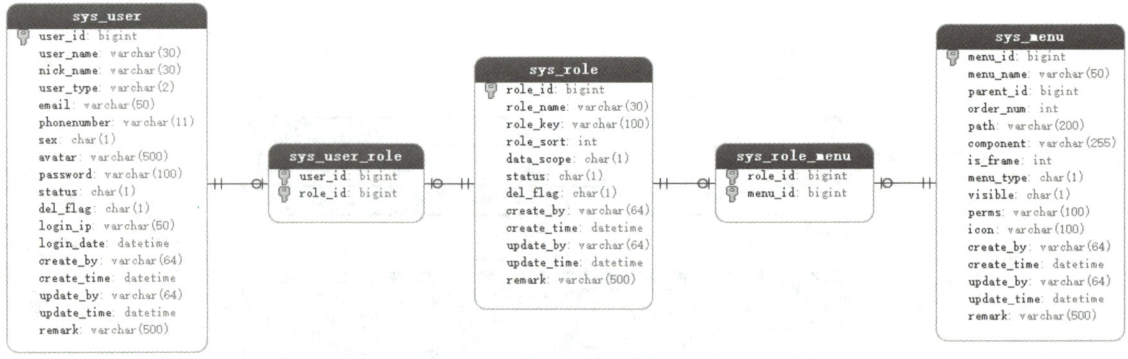

图 2-2　Friday 权限管理系统的实体关系图

Friday 权限管理系统的数据表描述见表 2-1。

表 2-1　Friday 权限管理系统数据表

表　名	表　描　述
sys_user	用户信息表
sys_role	角色信息表
sys_user_role	用户和角色关联表
sys_menu	菜单权限表
sys_role_menu	角色和菜单关联表

Friday 权限管理系统的数据表详细说明如下。

1. sys_user 用户信息表（表 2-2）

表 2-2　sys_user 用户信息表

字　段　名	字 段 描 述	字段数据类型	备　　注
user_id	用户 ID	bigint	主键，自增
user_name	用户账号	varchar	非空
nick_name	用户昵称	varchar	非空
user_type	用户类型	varchar	00：系统用户
email	用户邮箱	varchar	
phonenumber	手机号码	varchar	
sex	用户性别	char	0：男；1：女；2：未知
avatar	头像	varchar	
password	密码	varchar	
status	账号状态	char	0：正常；1：停用
del_flag	删除标识	char	0：存在；2：删除
login_ip	最后登录 IP	varchar	
login_date	最后登录时间	datetime	
create_by	创建者	varchar	
create_time	创建时间	datetime	
update_by	更新者	varchar	
update_time	更新时间	datetime	
remark	备注	varchar	

2. sys_role 角色信息表（表 2-3）

表 2-3　sys_role 角色信息表

字　段　名	字 段 描 述	字段数据类型	备　　注
role_id	角色 ID	bigint	主键，自增
role_name	角色名称	varchar	非空
role_key	角色权限字符串	varchar	非空
role_sort	显示顺序	int	非空
data_scope	数据范围	char	1：全部数据权限；2：自定数据权限；3：本部门数据权限；4：本部门及以下数据权限
status	账号状态	char	非空，0：正常；1：停用
del_flag	删除标识	char	0：存在；2：删除
create_by	创建者	varchar	
create_time	创建时间	datetime	
update_by	更新者	varchar	
update_time	更新时间	datetime	
remark	备注	varchar	

3. sys_user_role 用户和角色关联表（表 2-4）

表 2-4　sys_user_role 用户和角色关联表

字 段 名	字 段 描 述	字段数据类型	备　注
user_id	用户 ID	bigint	主键，非空
role_id	角色 ID	bigint	主键，非空

4. sys_menu 菜单权限表（表 2-5）

表 2-5　sys_menu 菜单权限表

字 段 名	字 段 描 述	字段数据类型	备　注
menu_id	菜单 ID	bigint	主键，自增
menu_name	菜单名称	varchar	非空
parent_id	父菜单 ID	bigint	
order_num	显示顺序	int	
path	路由地址	varchar	
component	组件路径	varchar	
is_frame	是否为外链	int	0：是；1：否
menu_type	菜单类型	char	M：目录；C：菜单；F：按钮
visible	菜单状态	char	0：显示；1：隐藏
perms	权限标识	varchar	
icon	菜单图标	varchar	
create_by	创建者	varchar	
create_time	创建时间	datetime	
update_by	更新者	varchar	
update_time	更新时间	datetime	
remark	备注	varchar	

5. sys_role_menu 角色和菜单关联表（表 2-6）

表 2-6　sys_role_menu 角色和菜单关联表

字 段 名	字 段 描 述	字段数据类型	备　注
role_id	角色 ID	bigint	主键，非空
menu_id	菜单 ID	bigint	主键，非空

步骤 2　创建数据库

1）打开 MySQL Workbench 客户端，连接 MySQL 服务端，如图 2-3 所示。

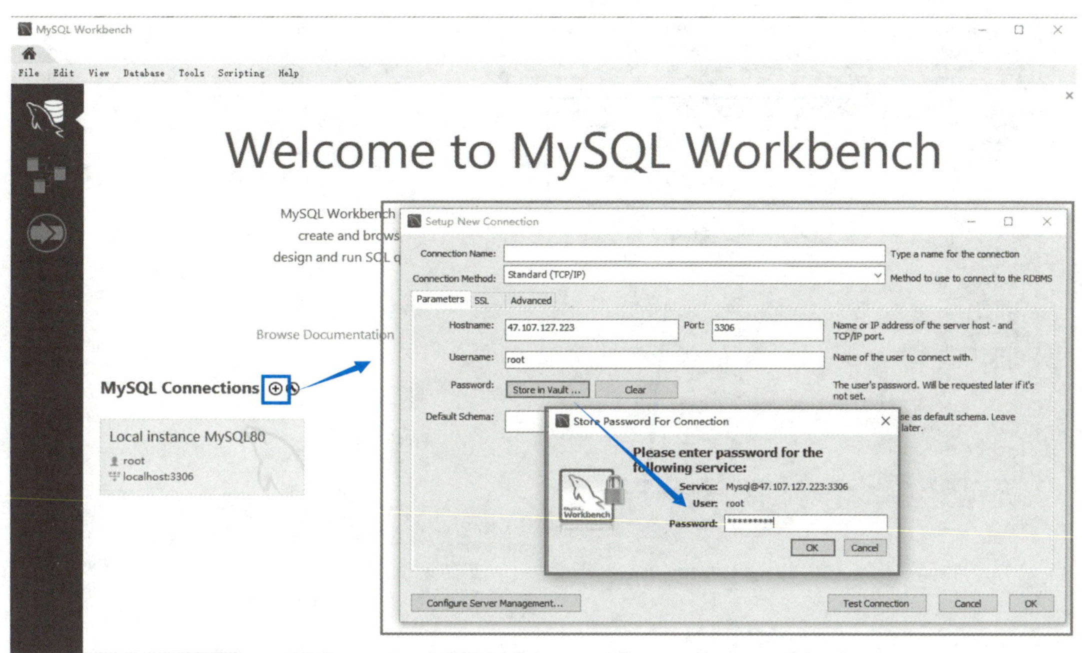

图 2-3　连接 MySQL 服务端

连接成功后，界面如图 2-4 所示。

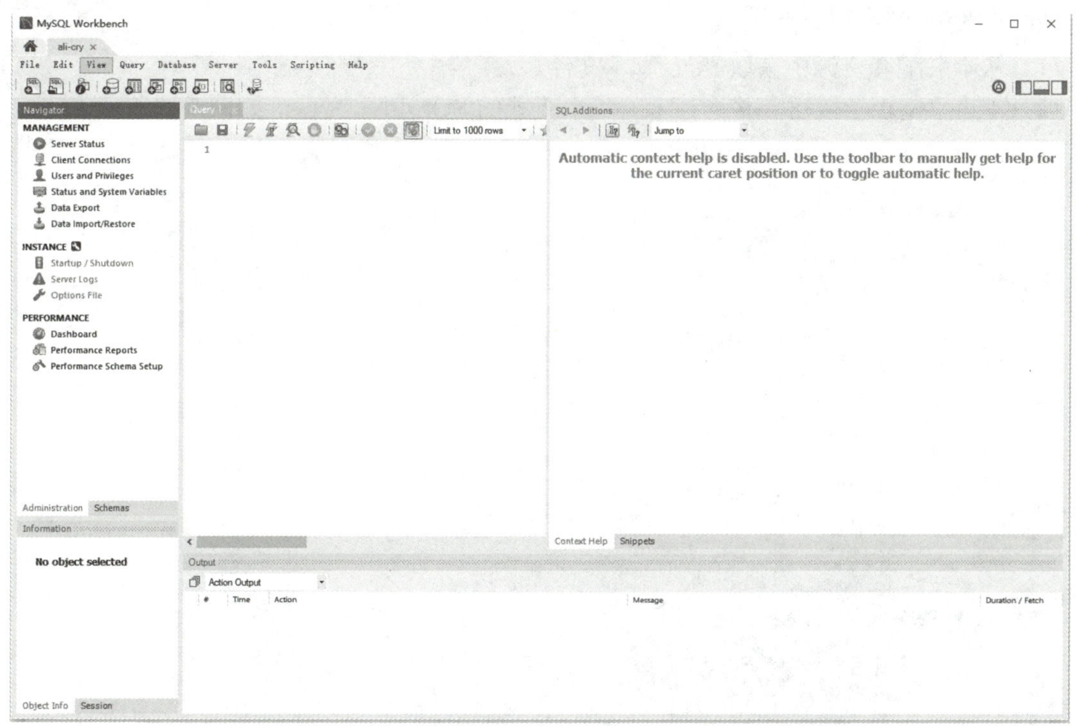

图 2-4　连接成功

2）创建一个名为“friday”的 Schema，如图 2-5 所示，选择“Charset/Collation”的值为“utf8”和“utf8_general_ci”，单击“Apply”按钮创建 Schema。

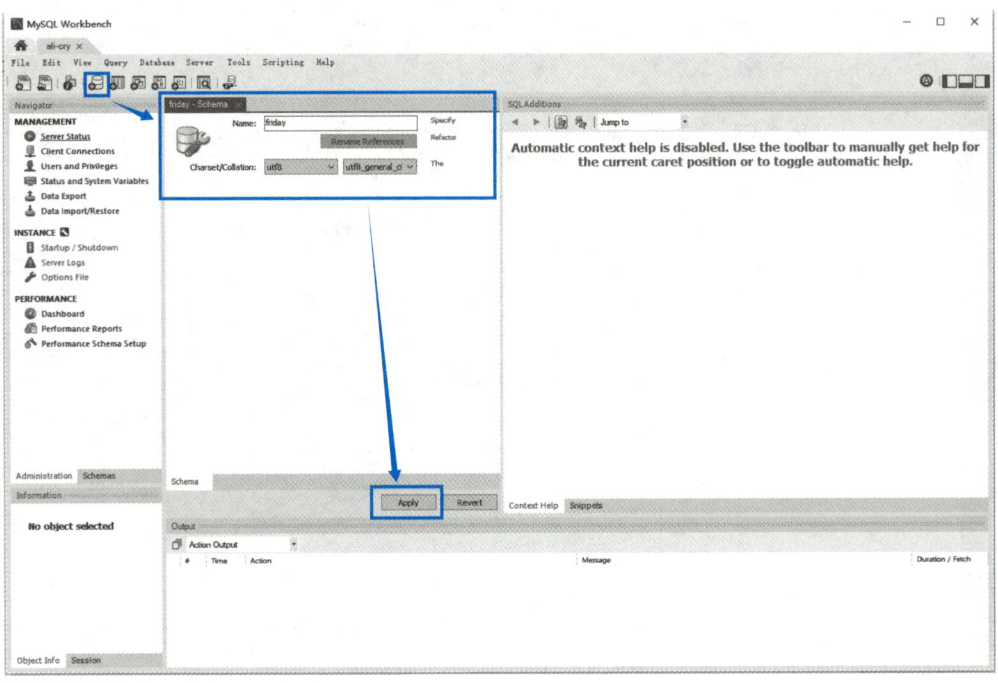

图 2-5　创建 Schema

步骤 3　导入数据表与初始数据

1）从本书配套资源中获取 SQL 脚本文件。

2）单击“SQL”按钮，并选择上一步中下载的 SQL 脚本文件，如图 2-6 所示。

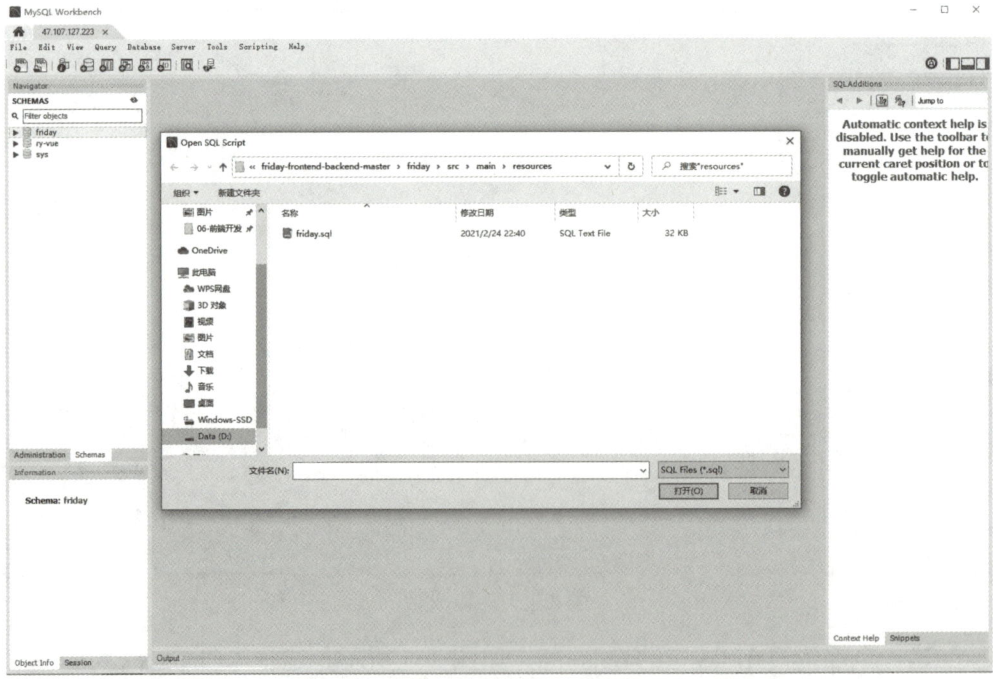

图 2-6　选择下载的 SQL 脚本文件

3）单击“黄色闪电”按钮，执行导入的 SQL 脚本文件，如图 2-7 所示。

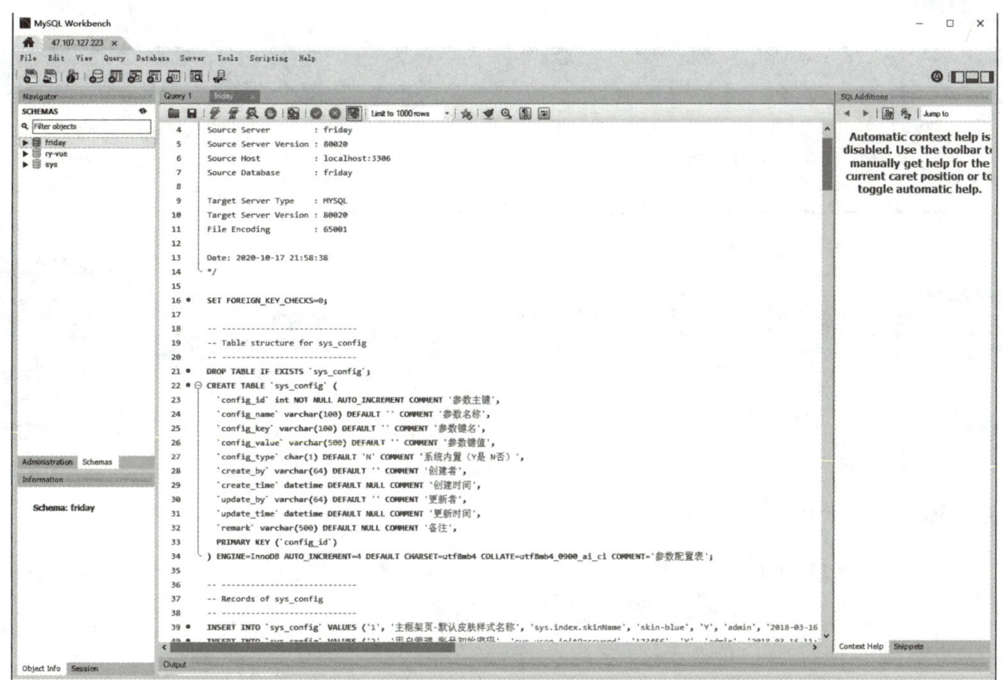

图 2-7　执行导入的 SQL 脚本文件

4）验证数据表是否创建成功，Tables 中显示 friday 数据表创建成功，如图 2-8 所示。

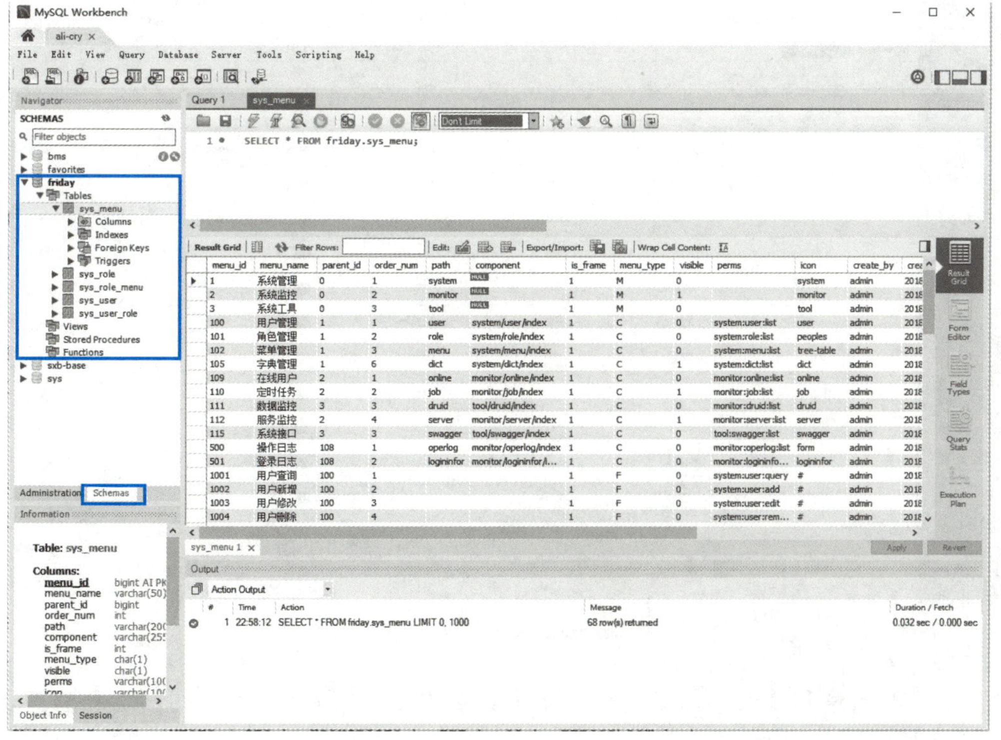

图 2-8　friday 数据表创建成功

任务评价

技　能　点	知　识　点	自我评价（不熟悉 / 基本掌握 / 熟练掌握 / 灵活运用）
创建数据库	MySQL 创建数据库	
导入数据表		

任务 2　根据 UI 设计编写后端接口文档

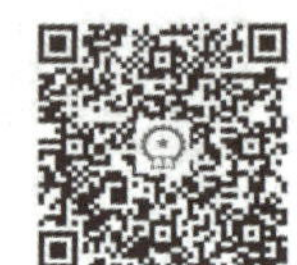

微课 2-2　根据 UI 设计编写后端接口文档

任务实施

步骤 1　设计用户管理功能的后端接口

1）分析用户管理页面，如图 2-9 所示，用户管理页面需要如下 4 个后台接口功能：

- 数据列表区的获取用户列表数据的接口功能。
- 功能区的保存用户信息到数据库的接口功能。
- 功能区的更新用户信息到数据库的接口功能。
- 功能区的删除用户信息的接口功能。

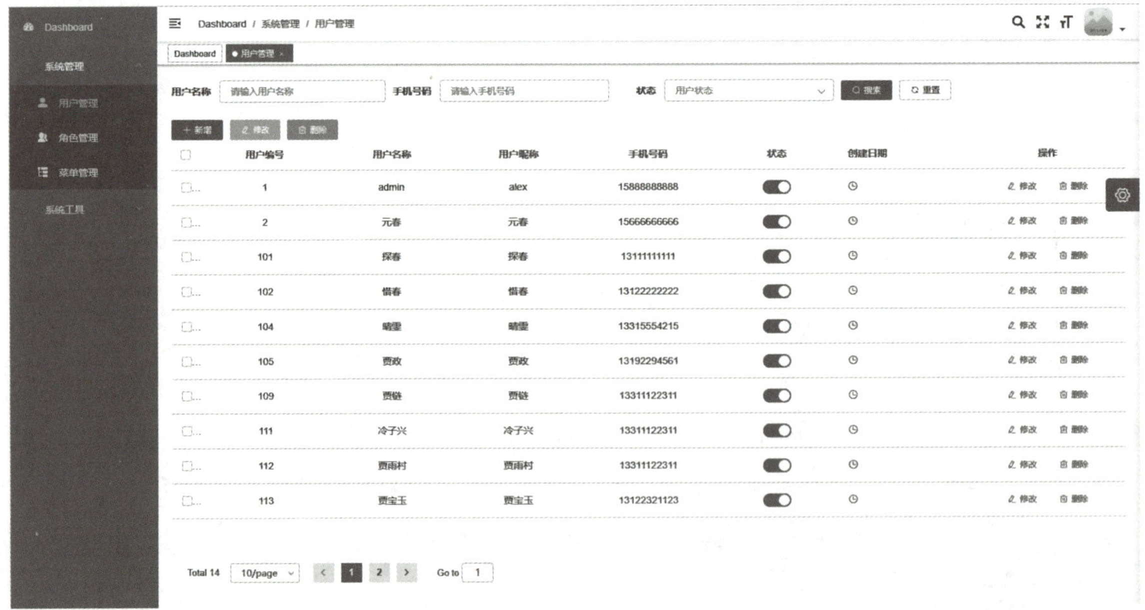

图 2-9　用户管理页面

2）用户的新增和修改功能共用一个页面，如图 2-10 所示，用户的新增和修改页面需要如下 2 个后台接口功能：

➢ 新增、修改页面需要获取角色列表的接口功能。
➢ 修改页面需要根据用户编号获取用户信息的接口功能。

图 2-10　用户的新增和修改页面

3）根据识别出的接口功能设计 UI 页面与后端接口的对应关系，见表 2-7。

表 2-7　用户的 UI 页面与后端接口对应关系表

接 口 功 能	HTTP 请求方式	URL 地址	支 持 格 式
获取用户列表	GET	/system/user/list	JSON
新增用户数据	POST	/system/user	JSON
修改用户数据	PUT	/system/user	JSON
删除用户数据	DELETE	/system/user/{userIds}	JSON
根据用户编号获取用户信息与角色列表	GET	/system/user/{userId}, /system/user/	JSON

步骤 2　设计角色管理功能的后端接口

1）分析角色管理页面，如图 2-11 所示，角色管理页面中需要如下 4 个后台接口功能：
➢ 数据列表区的获取角色列表数据的接口功能。
➢ 功能区的保存角色信息到数据库的接口功能。
➢ 功能区的更新角色信息到数据库的接口功能。
➢ 功能区的删除角色信息的接口功能。

图 2-11　角色管理页面

2）角色的新增和修改功能共用一个页面，如图 2-12 所示，角色的新增和修改页面需要如下 3 个后台接口功能：

➢ 新增页面需要获取树状菜单列表的接口功能。
➢ 修改页面需要根据角色编号获取详细信息的接口功能。
➢ 修改页面需要根据角色编号获取树状菜单列表的接口功能。

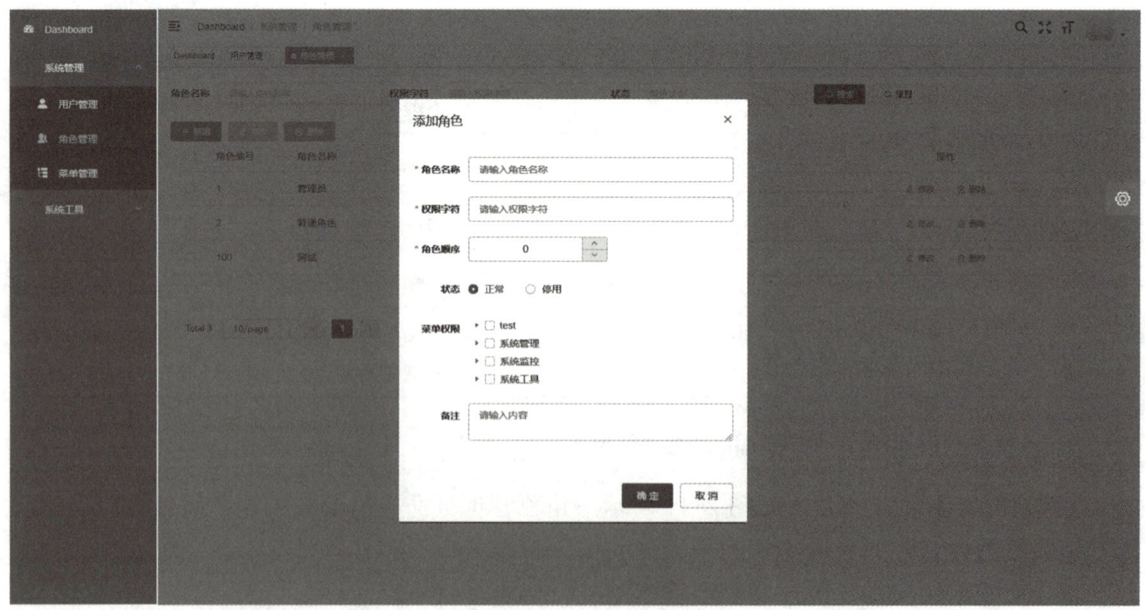

图 2-12　角色的新增和修改页面

3）根据识别出的接口功能设计 UI 页面与后端接口的对应关系，见表 2-8。

表 2-8　角色的 UI 页面与后端接口对应关系表

接口功能	HTTP 请求方式	URL 地址	支持格式
获取角色列表	GET	/system/role/list	JSON
新增角色数据	POST	/system/role	JSON
修改角色数据	PUT	/system/role	JSON
删除角色数据	DELETE	/system/role/{roleIds}	JSON
获取树状菜单列表	GET	/system/menu/treeselect	JSON
根据角色编号获取详细信息	GET	/system/role/{roleId}	JSON
根据角色编号获取树状菜单列表	GET	/system/menu/roleMenuTreeselect/{roleIds}	JSON
修改角色状态	PUT	/system/role/changeStatus	JSON

步骤 3　设计菜单管理功能的后端接口

1）分析菜单管理页面，如图 2-13 所示，菜单管理页面中需要如下 4 个后台接口功能：

- 数据列表区的获取菜单列表数据的接口功能。
- 功能区的保存菜单信息到数据库的接口功能。
- 功能区的更新菜单信息到数据库的接口功能。
- 功能区的删除菜单信息的接口功能。

2）菜单的新增和修改功能共用一个页面，如图 2-14 所示，菜单的新增和修改页面需要的后台接口功能如下：

- 根据菜单编号获取详细信息的接口功能。

图 2-13　菜单管理页面

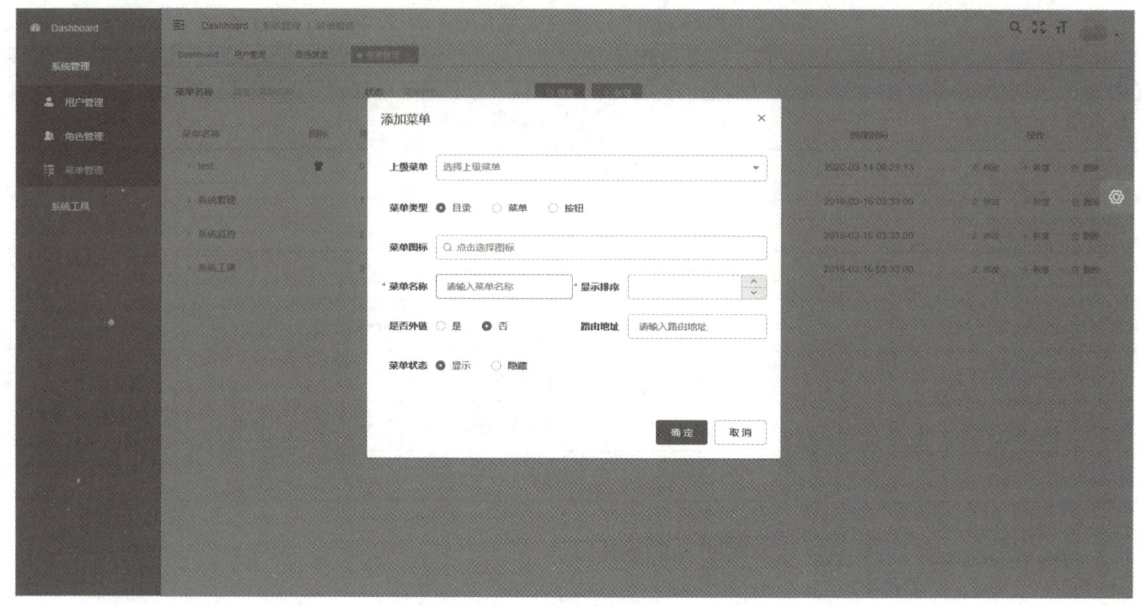

图 2-14 菜单的新增、修改页面

3）根据识别出的接口功能设计 UI 页面与后端接口的对应关系，见表 2-9。

表 2-9 菜单的 UI 页面与后端接口对应关系表

接 口 功 能	HTTP 请求方式	URL 地址	支 持 格 式
获取树状菜单列表	GET	/system/menu/list	JSON
新增菜单数据	POST	/system/menu	JSON
修改菜单数据	PUT	/system/menu	JSON
删除菜单数据	DELETE	/system/menu/{menuIds}	JSON
根据菜单编号获取详细信息	GET	/system/menu/{menuId}	JSON

步骤 4 设计登录授权的后端接口

1）分析登录页面，如图 2-15 所示，登录页面中需要的后台接口功能如下：

➢ 用户登录验证的接口功能。

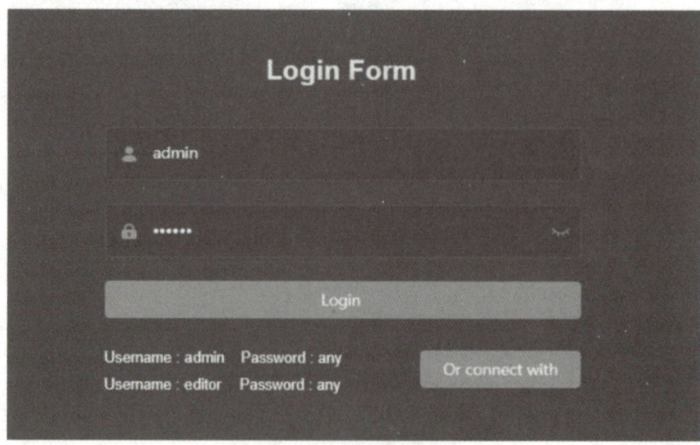

图 2-15 登录页面

2）登录成功后的首页，如图 2-16 所示，首页页面需要如下 2 个后台接口功能：

➢ 获取登录用户授权信息的接口功能。
➢ 获取菜单路由信息的接口功能。

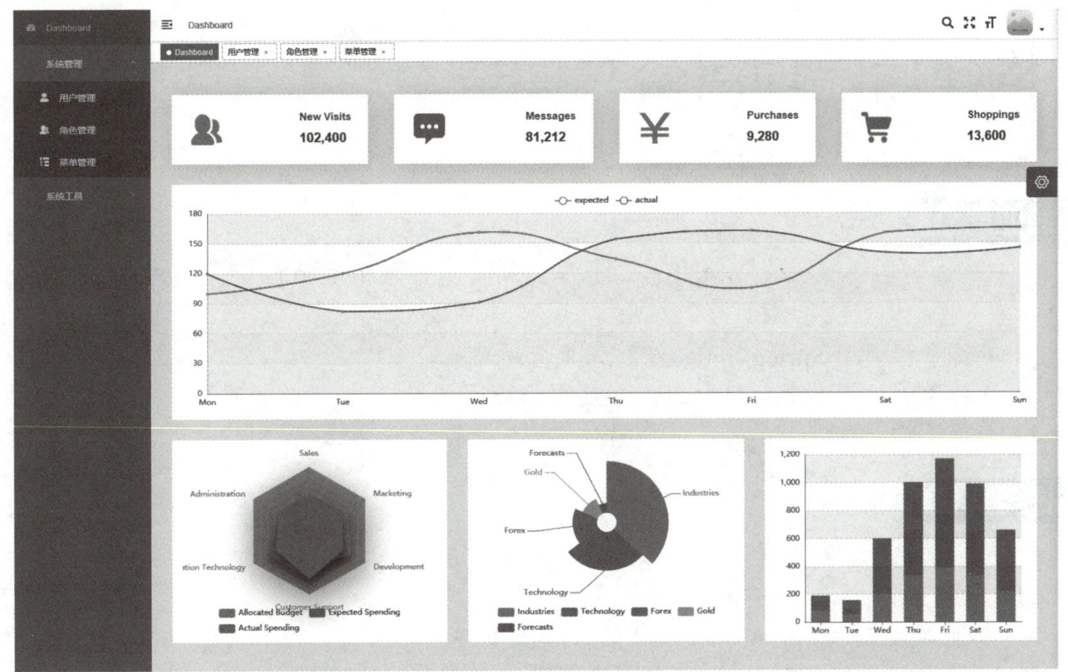

图 2-16　登录成功后的首页

3）设计 UI 页面与后端接口对应关系，见表 2-10。

表 2-10　登录授权的 UI 页面与后端接口对应关系表

接 口 功 能	HTTP 请求方式	URL 地址	支 持 格 式
登录验证	POST	/login	JSON
获取登录用户授权信息	GET	/getInfo	JSON
获取菜单路由信息	GET	/getRouters	JSON

完成了后端接口文档之后，研发人员就可以根据该接口文档进行后端项目的开发了。

任务评价

技 能 点	知 识 点	自我评价（不熟悉 / 基本掌握 / 熟练掌握 / 灵活运用）
编写接口文档	HTTP 基础知识	

单元小结

基于数据库设计生成数据库以及根据 UI 设计编写后端接口文档，是前后端分离架构项目开发前期的重要环节。通过完成本工作单元的任务和后面的实战强化，熟练掌握系统生成数据库与编写后端接口文档的职业技能。

实战强化

1）按照任务 1 步骤，生成实训项目“诚品书城”的数据库。
2）按照任务 2 步骤，编写实训项目“诚品书城”的接口设计文档。

工作单元 3

构建后端项目公共模块

职业能力

本工作单元主要是基于 SpringBoot 框架构建后端项目公共模块，最终希望学生达成如下职业能力目标：

1．熟练掌握使用 Spring Initializr 初始化后端项目

2．熟练掌握使用 Spring Data JPA 构建数据访问层

3．熟练掌握导入常用工具类以及编写后端项目的公共模块

任务情景

在前后端分离架构的项目中，构建后端服务需要解决下面 3 个问题（如图 3-1 所示的①、②、③位置）：

➢ 如何构建基于 SpringBoot 框架的后端服务？

➢ 在后端服务中，如何操作 MySQL 数据库的数据表？

➢ 在后端项目中，如何把数据表中得到的数据返回到前端页面？

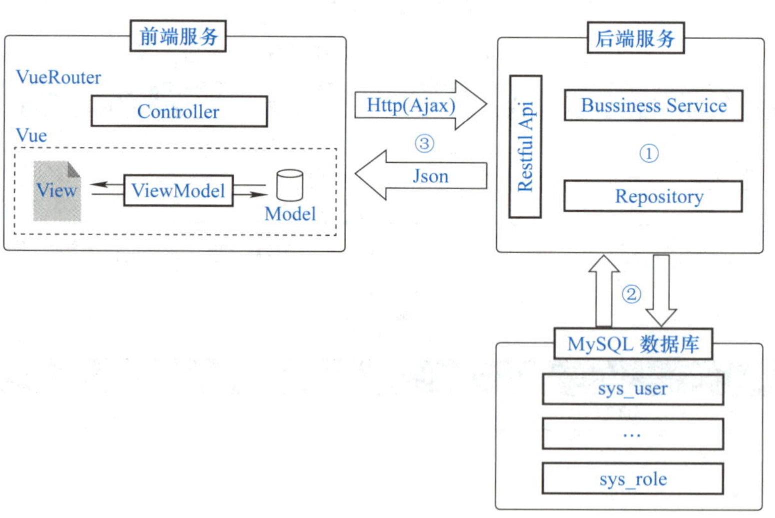

图 3-1　前后端分离架构

基于上述 3 个问题，本工作单元的具体任务如下：

➢ 使用 Spring Initializr 构建后端项目。

➢ 使用 Spring Data JPA 构建数据访问层。

➢ 封装统一接口响应的 HTTP 结果。

前置知识

为顺利完成本项目的相关功能，需要掌握如下知识和技术：

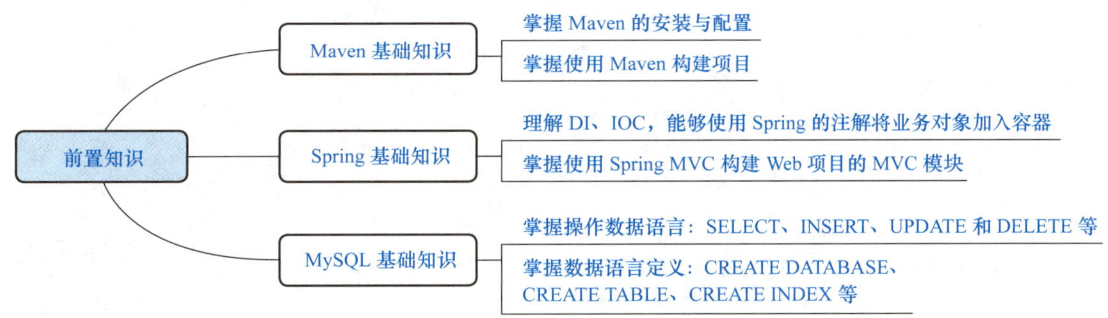

任务 1　使用 Spring Initializr 构建后端项目

微课 3-1　使用 Spring Initializr 构建后端项目

任务分析

本任务的目的是在 IntelliJ IDEA 下使用 Spring Initializr 构建具备 Web 响应能力和 MySQL 数据库操作能力的 SpringBoot 项目，也是本书所要完成的权限管理系统后端服务的基础。

任务实施

步骤 1　使用 Spring Initializr 构建项目

1）打开 IntelliJ IDEA，选择“File”→“New”→“Project...”，如图 3-2 所示。

2）在左侧导航菜单中选择“Spring Initializr”选项，右侧界面“Project SDK”选项选择“1.8”，单击“Next”按钮，如图 3-3 所示。

3）在“Spring Initializr Project Settings”中，“Group”（组织）填写“edu.friday”，“Artifact”（成品名）填写“friday”，其他保持默认不变，单击“Next”按钮，如图 3-4 所示。

4）选择项目的依赖模块，选中“Developer Tools”里面的“Spring Boot DevTools”和“Lombok”来提升开发效率，如图 3-5 所示。

5）选择“Web”里的“Spring Web”模块来引入 Spring MVC 框架，使项目具备 Web 能力，如图 3-6 所示。

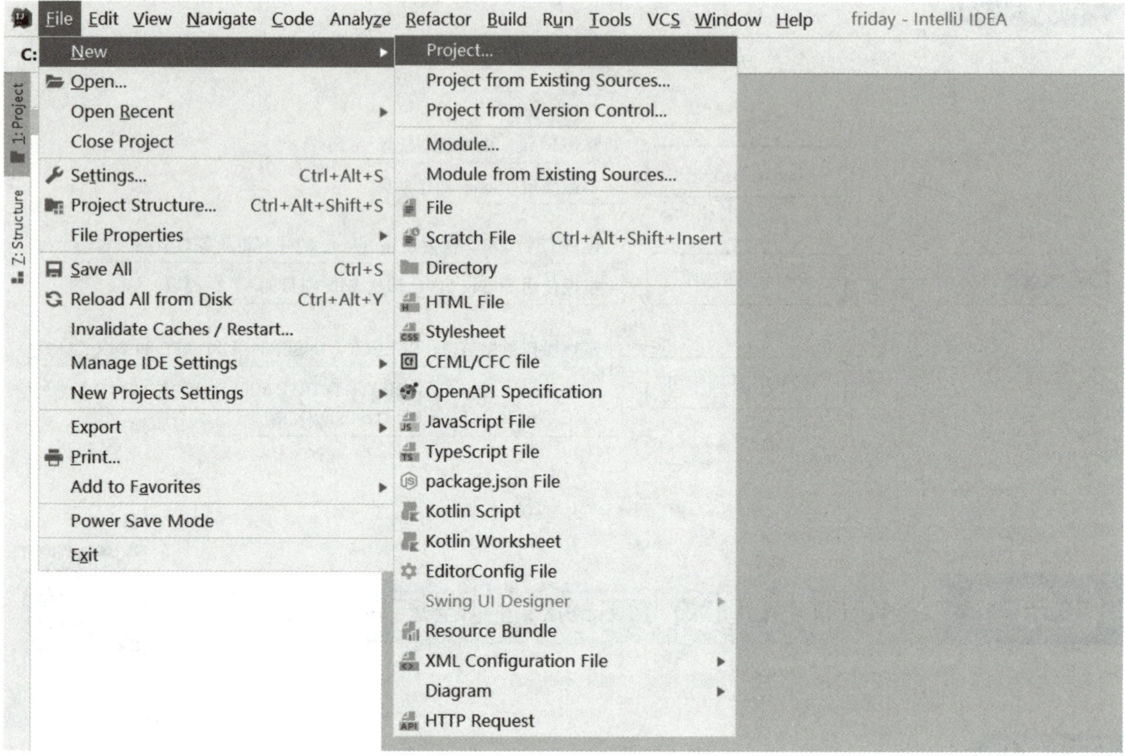

图 3-2 “IntelliJ IDEA”窗口

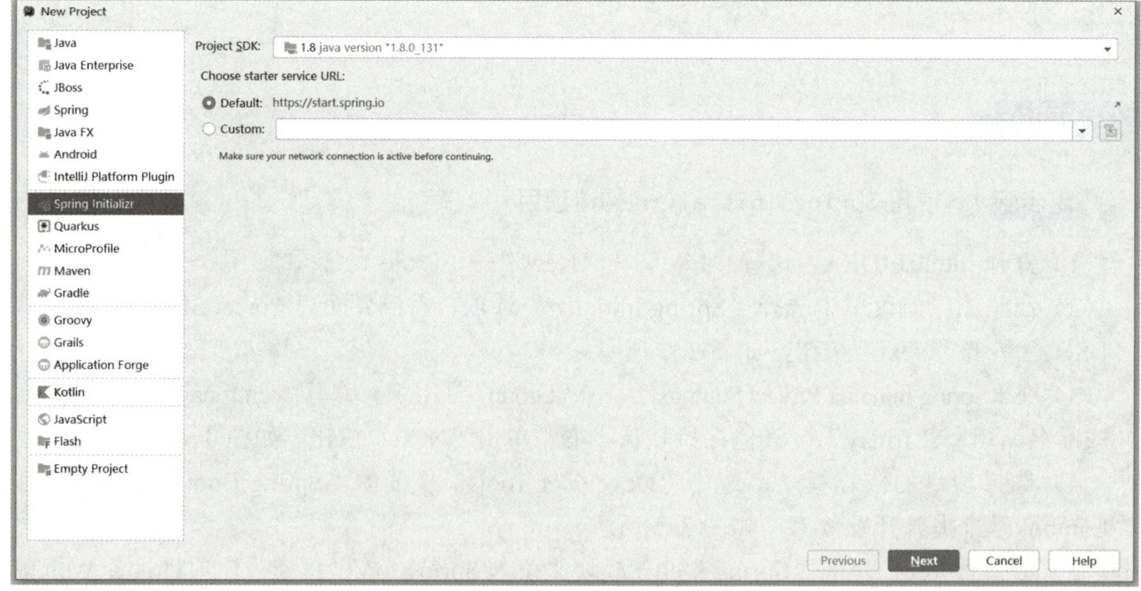

图 3-3 “Spring Initializr”选项

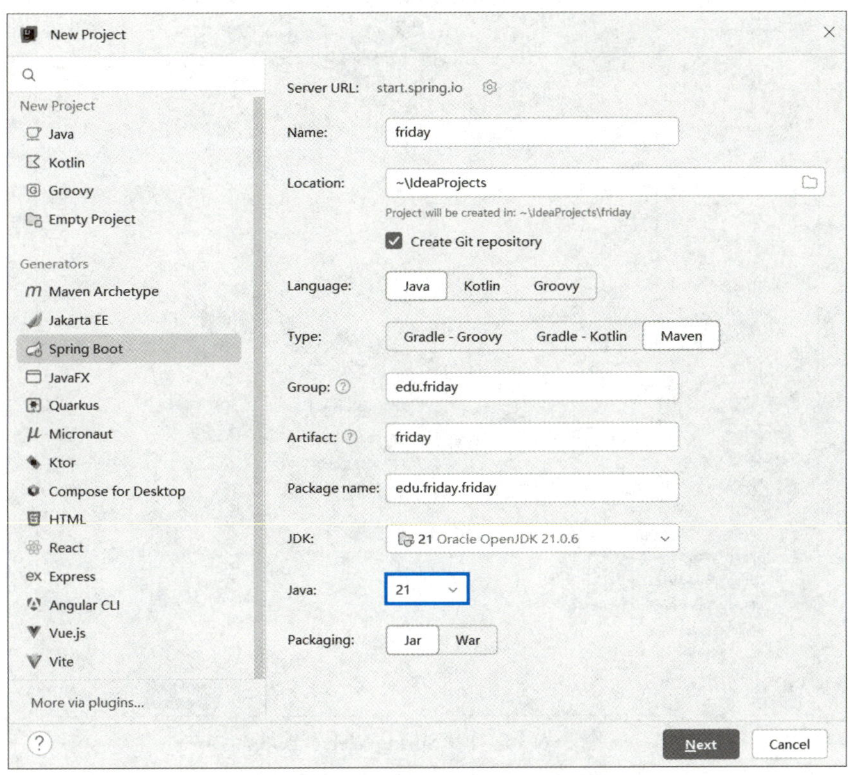

图 3-4　Spring Initializr Project Settings

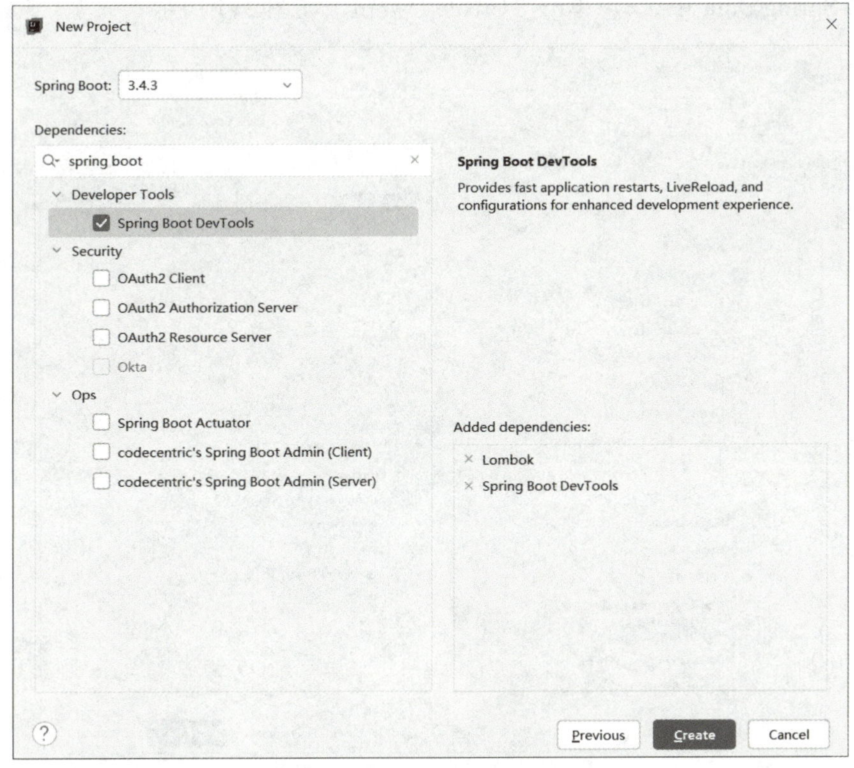

图 3-5　选择项目的依赖模块

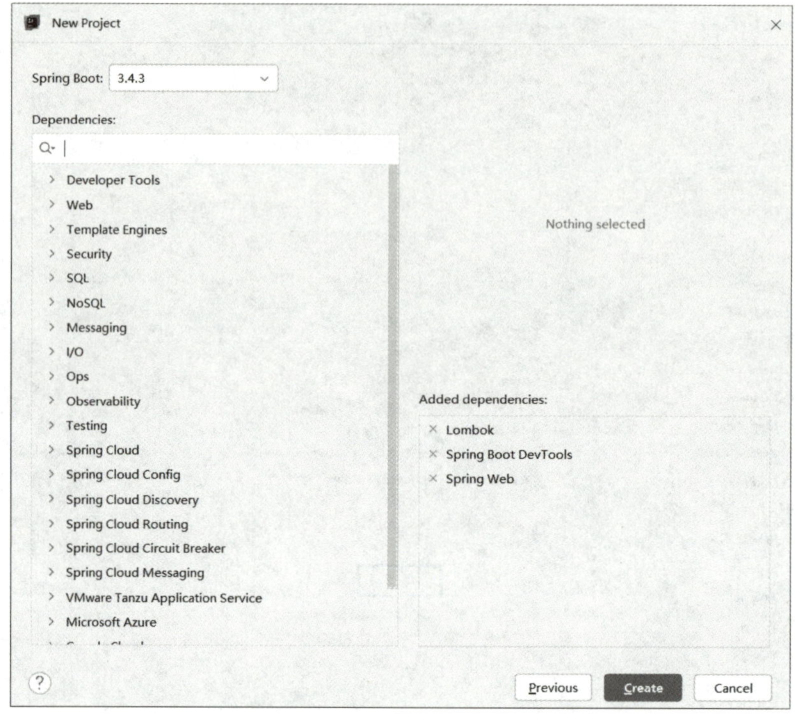

图 3-6　选择“Spring Web”模块

6）项目需要具备 MySQL 数据库的操作能力，所以这里选择“SQL”里面的“MySQL Driver”和“Spring Data JPA”，单击“Create”按钮，如图 3-7 所示。

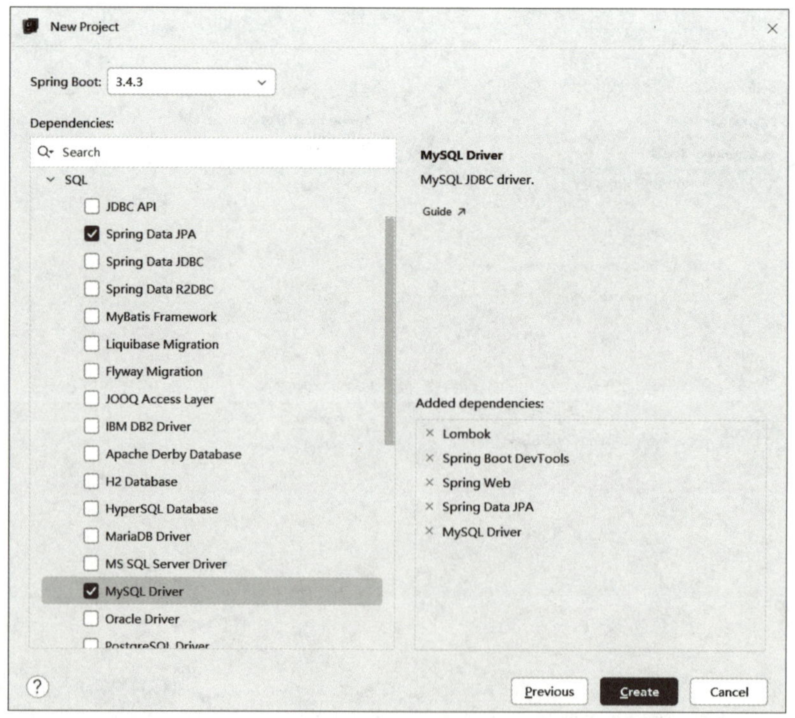

图 3-7　选择“MySQL Driver”和“Spring Data JPA”

在操作过程中，选择依赖模块时要注意图3-7中所示的细节，必须确保选择了“Spring Boot DevTools”“Lombok”“Spring Web”“Spring Data JPA”和“MySQL Driver”这5个模块。

7）正在下载模板构建项目，如图3-8所示。等下载完毕，项目即创建完成。

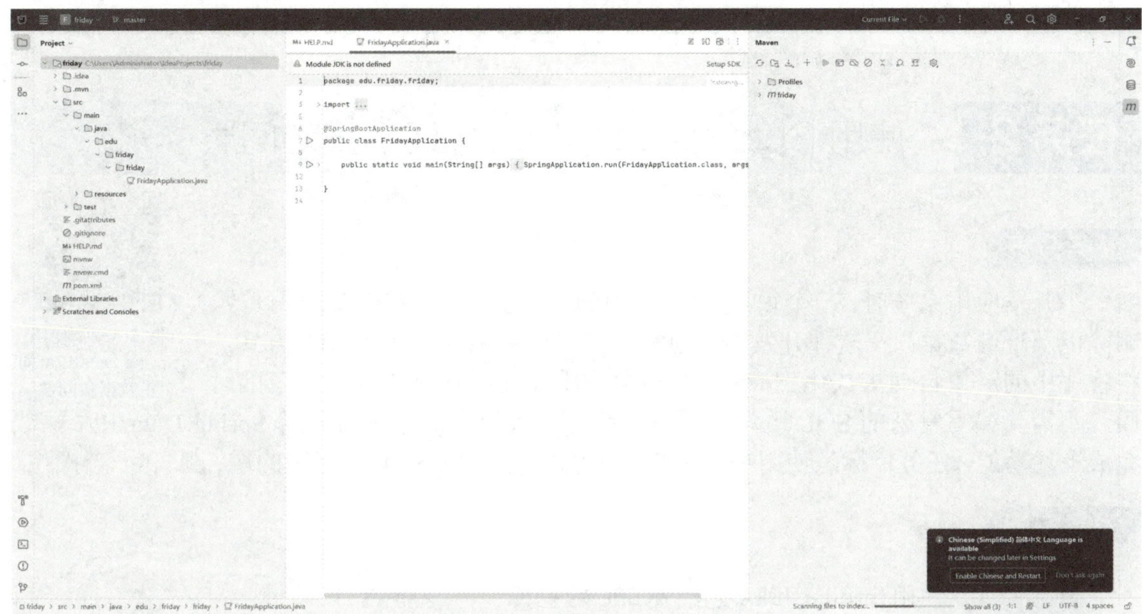

图3-8 下载模板构建项目

步骤2 安装Lombok插件

如果IDEA还没有安装Lombok，则可以根据提示直接选择安装，如图3-9所示。

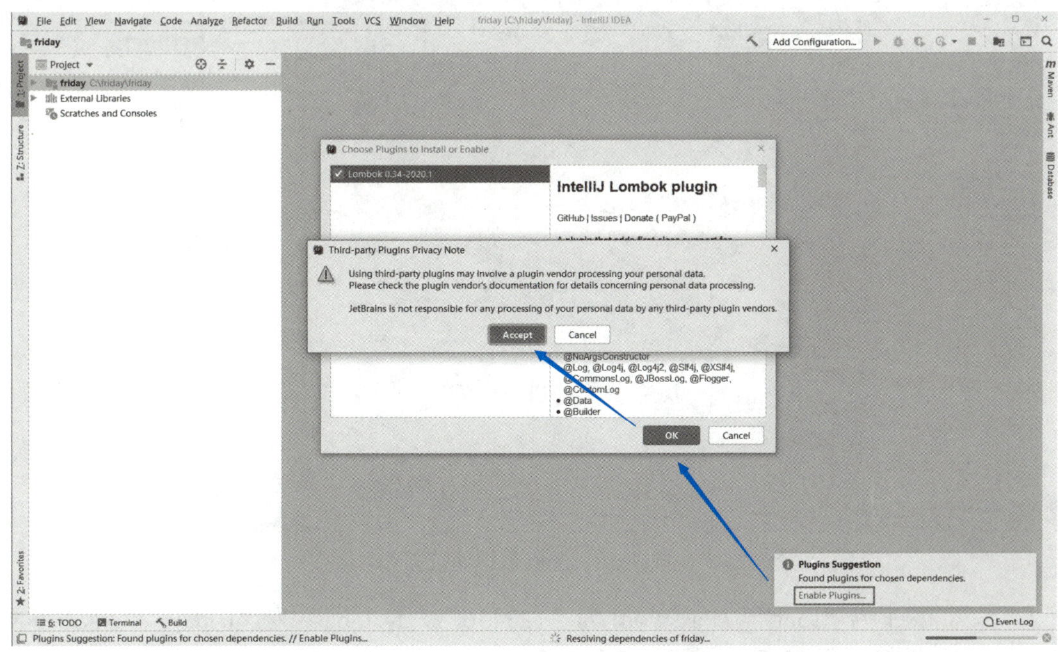

图3-9 安装Lombok插件

任务评价

技　能　点	知　识　点	自我评价（不熟悉 / 基本掌握 / 熟练掌握 / 灵活运用）
使用 Spring Initializr 构建项目	SpringBoot 基础知识	

任务 2　使用 Spring Data JPA 构建数据访问层

微课 3-2　使用 Spring Data JPA 构建数据访问层

任务分析

当开发应用系统时，一个很重要的基础模块就是在数据库中新增、修改、删除以及查询数据，在软件开发中，通常通过对象关系映射（ORM）方式，将程序中的对象自动持久化到 MySQL 数据库中，这样在操作业务对象的时候，不需要编写复杂的 SQL 语句，开发效率可大大提高。本任务采用 Spring Data JPA 框架完成上述功能，任务目标是使用 Spring Data JPA 实现对 MySQL 数据表的增、删、改、查操作。

任务实施

步骤 1　使用 IntelliJ IDEA 连接 MySQL 数据库

1）配置 MySQL 数据连接，单击 IntelliJ IDEA 右侧的“Database”，选中“Data Source”→“MySQL”选项，如图 3-10 所示。

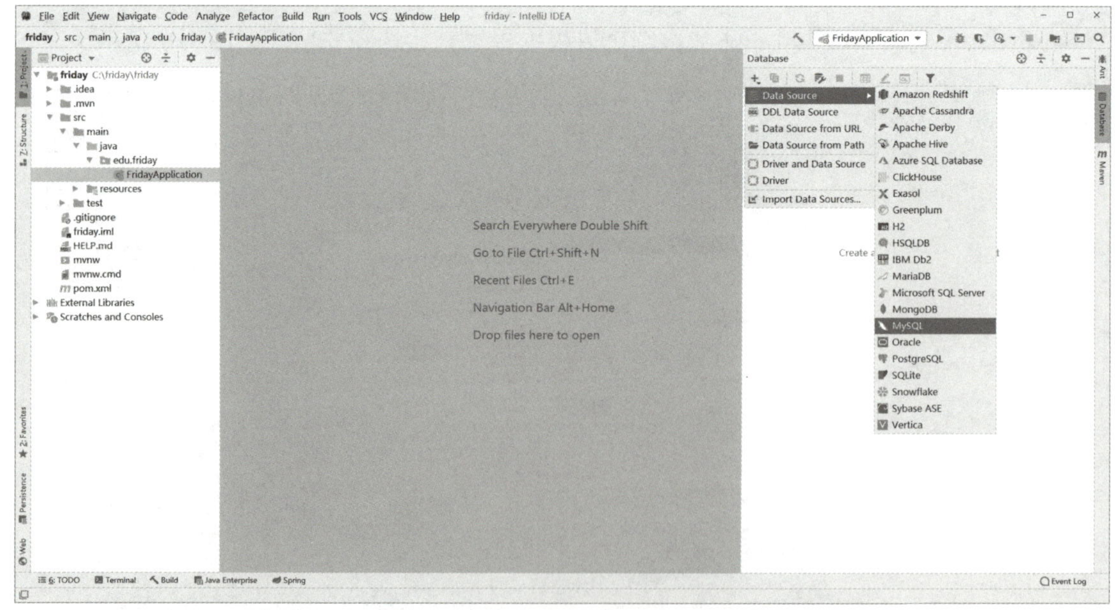

图 3-10　配置数据库连接

2）输入 User 为“root”，Password 为“123456”，Database 为“friday”，单击下面的“Download”链接下载 MySQL 驱动，MySQL 驱动安装完成后，单击“Test Connection”按

钮，如图 3-11 所示。

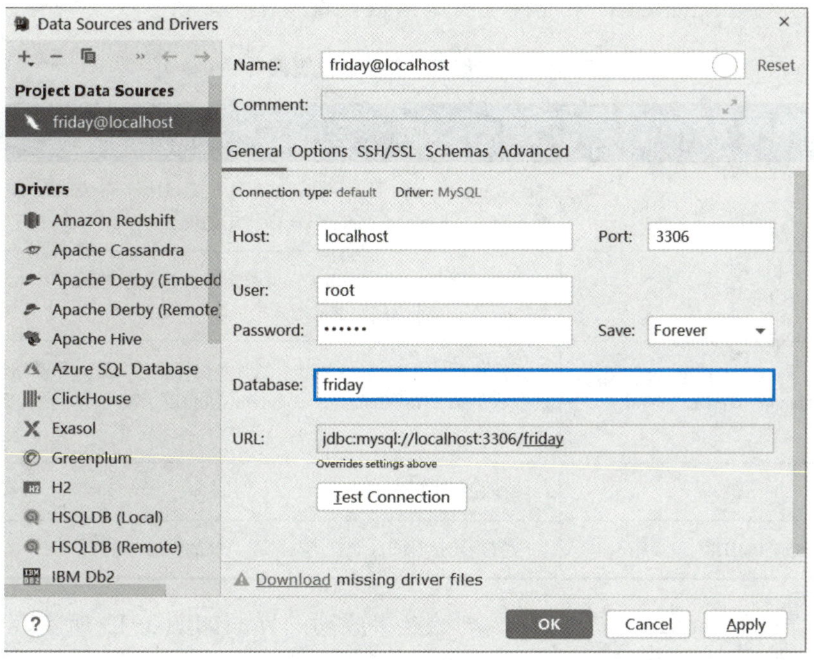

图 3-11　测试连接

3）测试连接时，会提示“Server returns innvalid timezone”，这是因为驱动包用的是 mysql-connector-java-8.0.11.jar，相应的新版驱动连接 URL 也有所改动，输入 URL：“jdbc:mysql://localhost:3306/friday?useUnicode=true&characterEncoding=utf8&zeroDateTimeBehavior=convertToNull&useSSL=true&serverTimezone=GMT%2B8”，如图 3-12 所示。

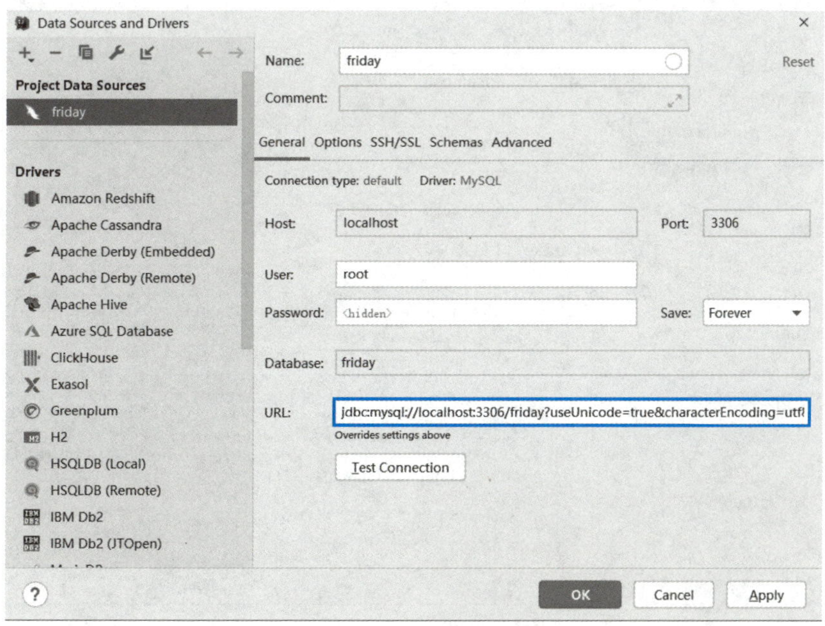

图 3-12　输入 URL

表 3-1 中介绍了上面 URL 中连接字符串参数设置及作用。

表 3-1　连接字符串参数设置及作用

参数设置	参数作用
useUnicode=true	useUnicode=true 是默认值，如果指定 useUnicode=false，那么数据库就会使用服务端变量 character-set-server 所指定的值
characterEncoding=utf8	设置字符集编码为 utf8，告诉数据库此次连接传输 UTF8 数据，解决了读取数据库产生的数据乱码的问题
zeroDateTimeBehavior=convertToNull	在操作值为 0 的 timestamp 类型时不能正确地处理，而是默认抛出一个异常，设置 zeroDateTimeBehavior=convertToNull 后，策略就是将日期转换成 Null 值，而不抛出异常
useSSL=true	指定使用 SSL 连接
serverTimezone=GMT%2B8	设置为北京时间（北京所在的东八区的区时）

4）单击“Test Connection”按钮，测试连接成功的界面如图 3-13 所示，修改 Name 为“friday”，单击“OK”按钮。

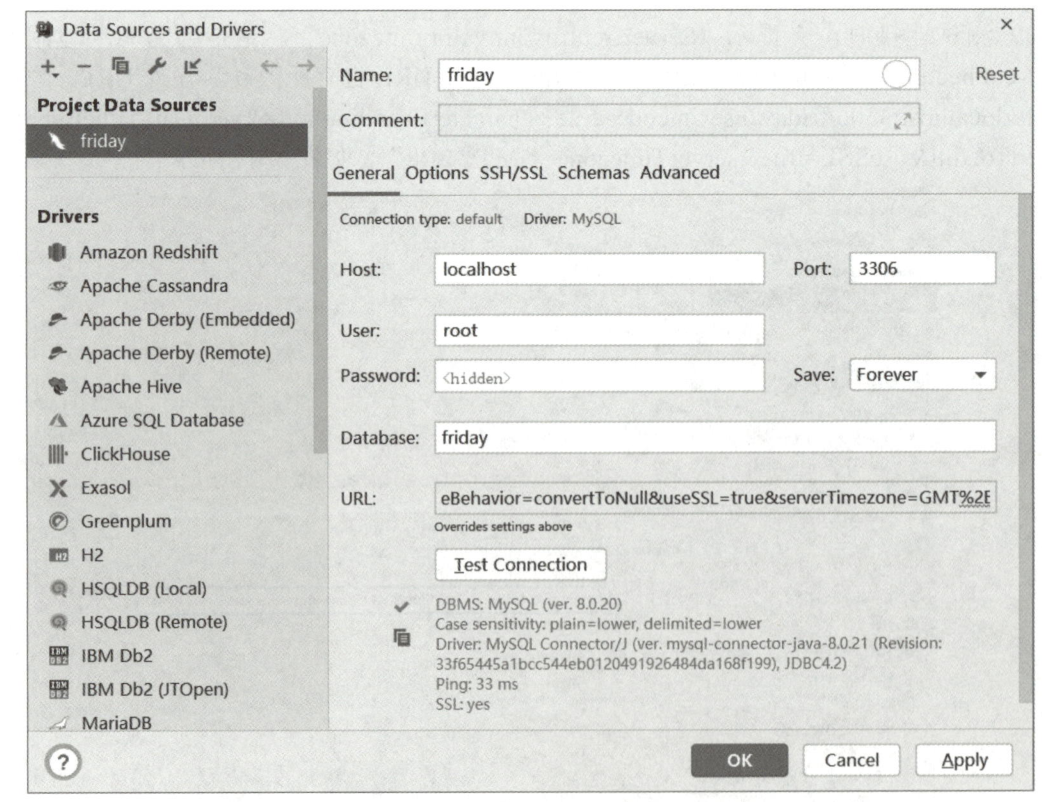

图 3-13　测试连接成功

5）选择使用“friday”，然后单击“schemas”→“friday”选项查看数据表，如图 3-14 所示。

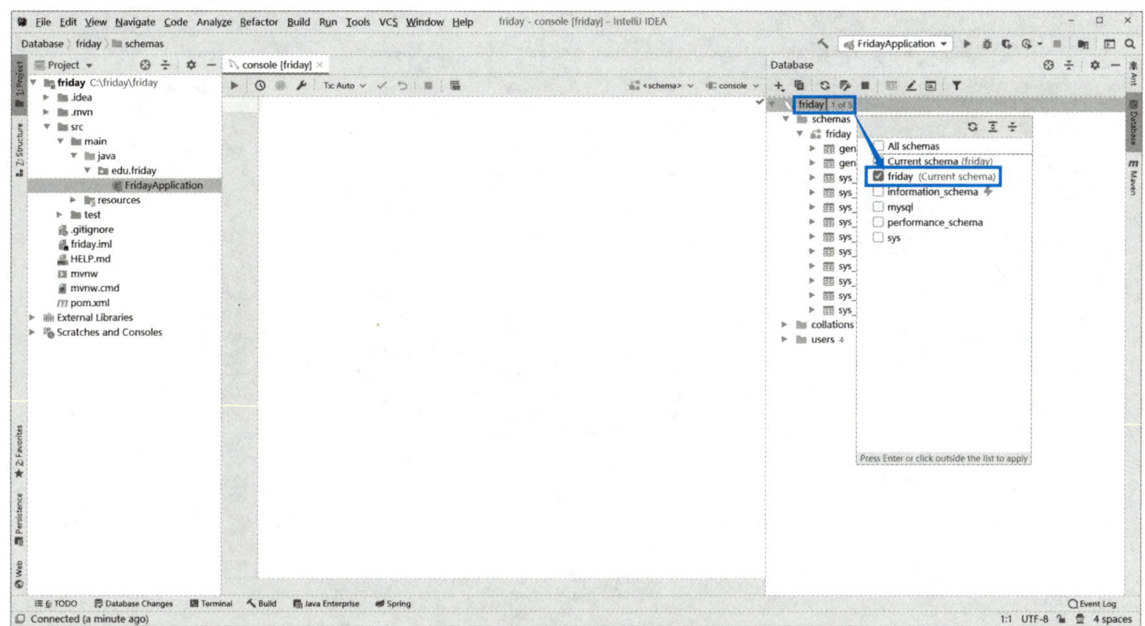

图 3-14　查看数据表

6）配置数据库完成后，如果出现如图 3-15 所示的界面，说明数据库配置成功。

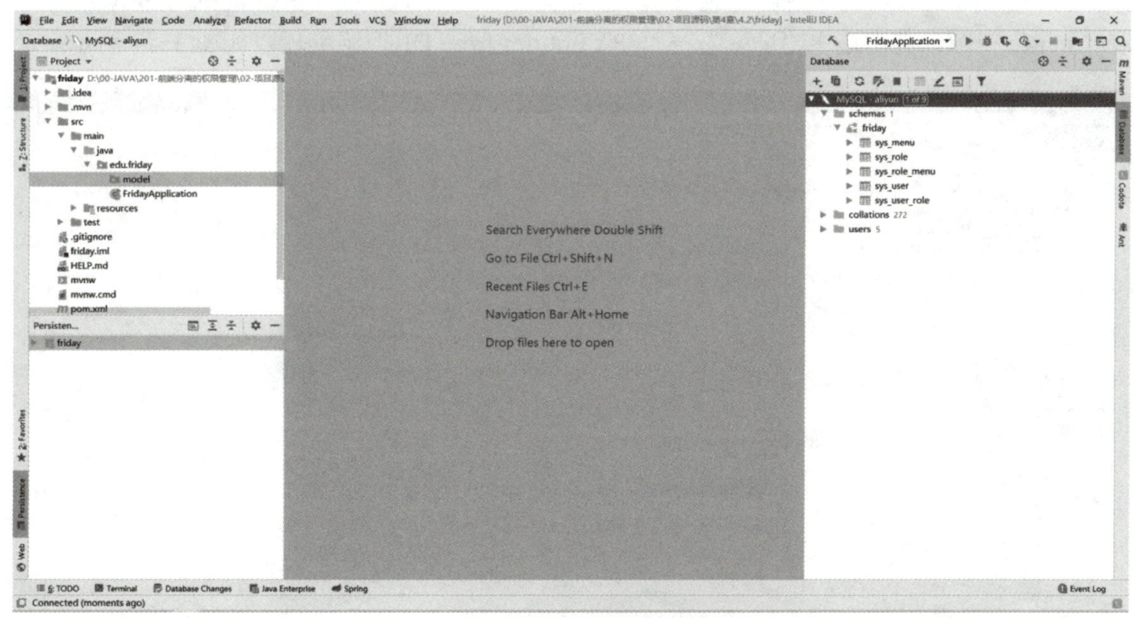

图 3-15　数据库配置成功

7）双击“sys_user”表，就可以对数据进行常规的操作了，如图 3-16 所示。至此数据库配置就全部完成。

	user_id	user_name	nick_name	user_type	email	phonenumber	sex	avatar
1	1	admin	alex	00	alex@163.com	15888888888	1	https://\
2	2	元春	元春	00	yr@qq.com	15666666666	1	https://\
3	100	迎春	迎春	00	test@qq.com	13192294560	0	https://\
4	101	探春	探春	00	11@qqw.bb	13111111111	0	https://\
5	102	惜春	惜春	00	22@Qq.mm	13122222222	0	https://\
6	103	秦可卿	秦可卿	00	22@qq.com	13222332222	0	
7	104	晴雯	晴雯	00	test1@qq.com	13315554215	0	
8	105	贾政	贾政	00	l1_1_1@qq.com	13192294561	1	
9	109	贾链	贾链	00	asd@q.com	13311122311	<null>	<null>
10	111	冷子兴	冷子兴	00	asd@q.com	13311122311	<null>	<null>
11	112	贾雨村	贾雨村	00	asd@q.com	13311122311	<null>	<null>
12	113	贾宝玉	贾宝玉	00	asdasd@qq.com	13122321123	<null>	<null>
13	114	王熙凤	王熙凤	00	asdasda@qq.com	13412311222	<null>	https://\
14	115	妙玉	妙玉	00	del_flag@qq.com	15312112311	<null>	https://\
15	124	234	2413213	00	2@qq.com	13523423523	<null>	<null>
16	125	甄士隐	甄士隐	00	a@qq.com	13512312312	<null>	<null>
17	126	123	asd	00	a@qq.com	13512312312	<null>	<null>
18	127	123123	11	00	11@qq.com	13512312312	<null>	<null>
19	128	alex123123	222	00	222@qq.com	13512321312	<null>	<null>
20	134	10112	10113	<null>	11@qq.com	13545679122	0	<null>
21	135	Alex	alex test	<null>	<null>	<null>	<null>	<null>
22	136	Alex	alex test	<null>	<null>	<null>	<null>	<null>

图 3-16　双击“sys_user”表对数据进行常规操作

步骤 2　使用 Persistence 插件生成实体对象

1）创建包“edu.friday.model”，作为实体对象的生成位置。

2）使用 Persistence 插件生成数据表对应的实体对象，单击 IntelliJ IDEA 左侧的 Persistence 视图，右击选中“New”→“JPA Entities from DB”菜单项，如图 3-17 所示。

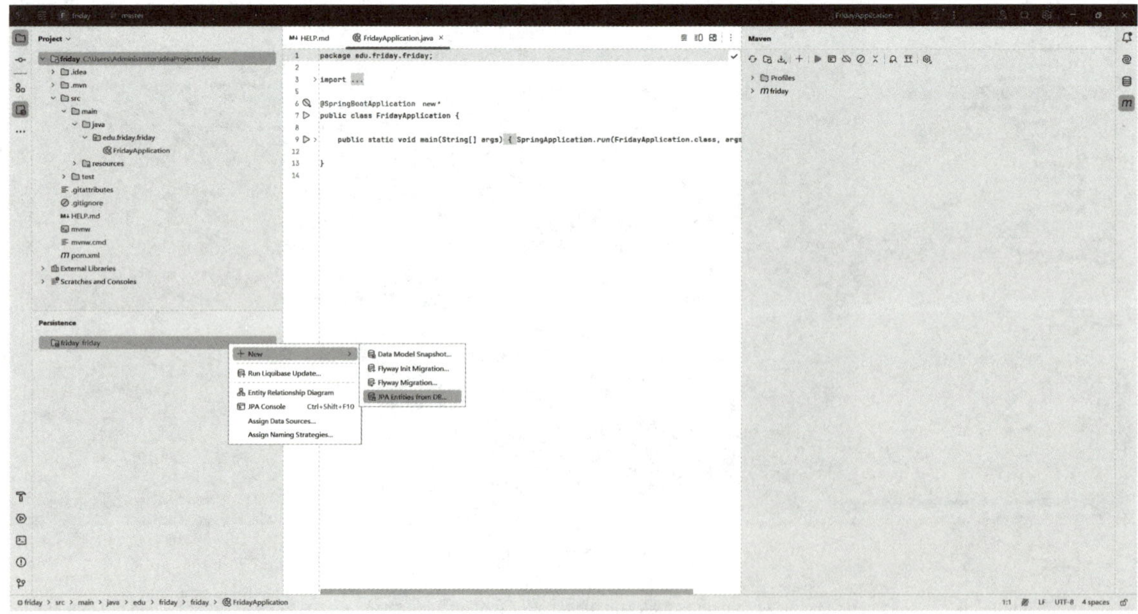

图 3-17　生成数据表对应的实体对象

3）数据库连接。选择配置好数据库 friday 的数据源，“Source root”选择“src/main/java”，选项“Migrate default values”打勾，“Target package”选择项目的“model”，选择全部的 friday 数据表，单击“OK”按钮，如图 3-18 所示。

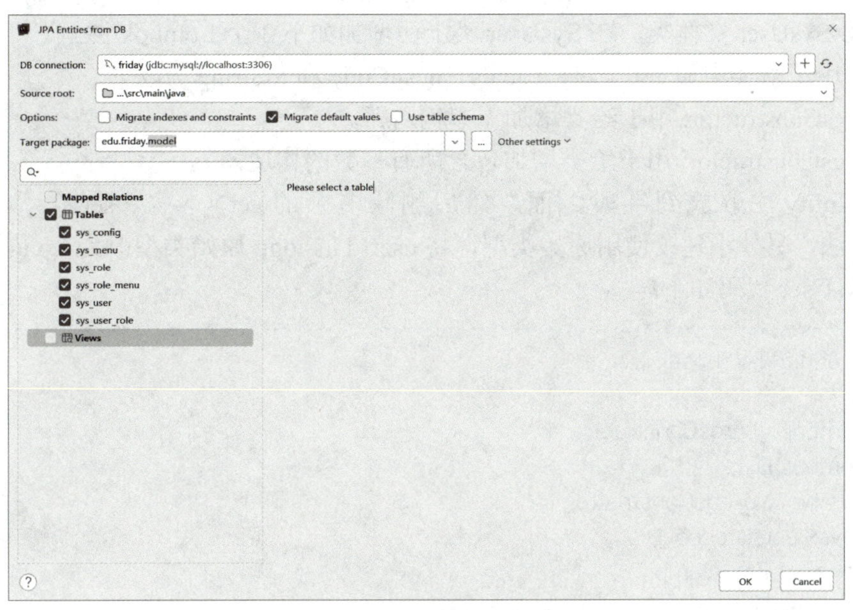

图 3-18 配置实体类的生成规则

4）在包“edu.friday.model”中成功生成数据表对应的实体对象，如图 3-19 所示，后续会通过操作这些实体对象来实现数据表的增、删、改、查操作。

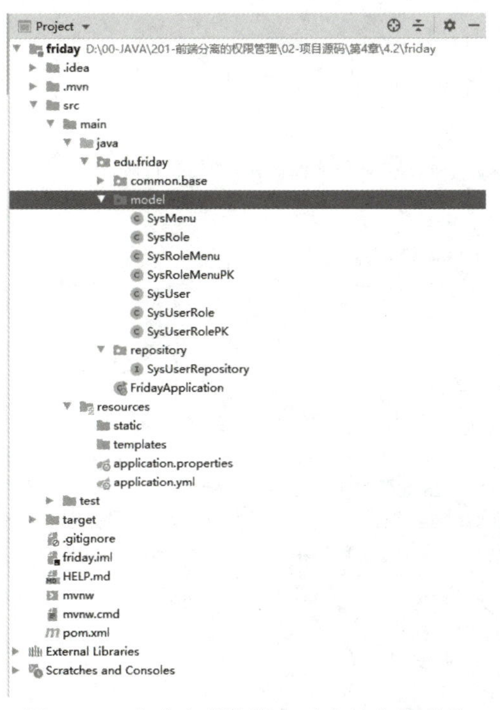

图 3-19 成功生成数据表对应的实体对象

步骤 3 引入 Lombok 插件简化实体类代码

生成的实体类中有大量 get××× 和 set××× 的代码片段，导致实体类比较冗长，可以使用 Lombok 注解来消除 Java 类中的大量冗长代码。

1）改写 SysUser 实体类，在 SysUser 类上面增加如下 3 个 Lombok 注解。

@Data: 用于给类增加 get、set、equals、hashCode 和 toString 方法。

@NoArgsConstructor: 用于给类增加无参构造器。

@AllArgsConstructor: 用于给类增加包含所有参数的构造器。

2）将 Entity 注解放到字段上面，然后删除多余的 get×××、set×××、equals 和 hashCode 方法，另外不建议使用原始类型，将 userId 的 long 原始类型改为 Long 封装类型，精简后的 SysUser 代码如下：

```
package edu.friday.model;

import lombok.AllArgsConstructor;
import lombok.Data;
import lombok.NoArgsConstructor;
import javax.persistence.*;
import java.sql.Timestamp;
import java.util.Objects;

@Entity
@Table(name = "sys_user", schema = "friday")
@Data
@NoArgsConstructor
@AllArgsConstructor
public class SysUser {
    @Id
    @GeneratedValue(strategy = GenerationType.IDENTITY)
    @Column(name = "user_id")
    private Long userId;

    @Basic
    @Column(name = "user_name")
    private String userName;

    @Basic
    @Column(name = "nick_name")
    private String nickName;

    @Basic
    @Column(name = "user_type")
    private String userType;
```

```
@Basic
@Column(name = "email")
private String email;

@Basic
@Column(name = "phonenumber")
private String phonenumber;

@Basic
@Column(name = "sex")
private String sex;

@Basic
@Column(name = "avatar")
private String avatar;

@Basic
@Column(name = "password")
private String password;

@Basic
@Column(name = "status")
private String status;

@Basic
@Column(name = "del_flag")
private String delFlag;

@Basic
@Column(name = "login_ip")
private String loginIp;

@Basic
@Column(name = "login_date")
private Timestamp loginDate;

@Basic
@Column(name = "create_by")
private String createBy;

@Basic
@Column(name = "create_time")
private Timestamp createTime;

@Basic
@Column(name = "update_by")
```

```
    private String updateBy;

    @Basic
    @Column(name = "update_time")
    private Timestamp updateTime;

    @Basic
    @Column(name = "remark")
    private String remark;
}
```

表 3-2 列出了 SysUser 代码中使用的 Entity 注解以及它们的作用。

表 3-2　Entity 注解说明表

注解名称	注解作用
@Entity	定义对象将会成为被 JPA 管理的实体，映射到指定的数据库表
@Table	用于指定数据库的表名
@Id	定义属性为数据库的主键，一个实体里面必须有一个
@GeneratedValue(strategy = GenerationType.IDENTITY)	主键生成策略，GenerationType.IDENTITY 表示采用数据库 ID 自增长，一般用于 MySQL 数据库
@Basic	表示属性是到数据库表的字段的映射。如果实体的字段上没有任何注解，默认为 @Basic
@Column(name = "user_id")	定义该属性对应数据库中的列名

步骤 4　编写 SysUserRepository 接口操作 SysUser 实体类

1）在包“edu.friday.repository”中创建 SysUserRepository 接口，继承 JpaRepository<SysUser, Long> 可实现对 SysUser 对象的操作，代码如下：

```
package edu.friday.repository;
import org.springframework.data.jpa.repository.JpaRepository;
import java.edu.friday.model.SysUser;
@Repository
public interface SysUserReporsitory extends JpaRepository<SysUser, Long> {
}
```

2）单击继承的接口“JpaRepository”，按 <Ctrl+F12> 组合键查看方法，可以看到 JpaRepository 已经实现了新增、删除和查询的方法，不需要额外写任何的代码，如图 3-20 所示。

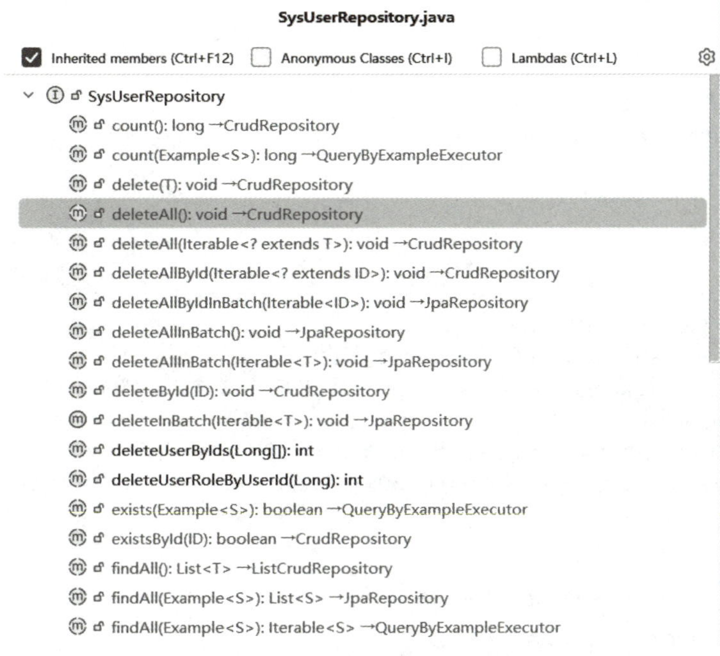

图 3-20　JpaRepository 类

步骤 5　配置数据源连接池

1）在 pom.xml 文件中添加 alibaba druid 数据连接池的依赖。

```
<dependencies>
    ...
    <dependency>
        <groupId>com.alibaba</groupId>
        <artifactId>druid-spring-boot-starter</artifactId>
        <version>1.1.10</version>
    </dependency>
    <dependency>
        <groupId>org.apache.commons</groupId>
        <artifactId>commons-pool2</artifactId>
    </dependency>
</dependencies>
```

2）修改 resources 下的 application.yml 文件，添加如下配置参数。

```
spring:
  datasource:
    driver-class-name: com.mysql.cj.jdbc.Driver
    url: jdbc:mysql://localhost:3306/friday?useUnicode=true&characterEncoding=utf8&zeroDateTimeBehavior=convertToNull&useSSL=true&serverTimezone=GMT%2B8
    username: root
    password: 123456
    platform: mysql
    type: com.alibaba.druid.pool.DruidDataSource
```

```
      initialSize: 1
      minIdle: 3
      maxActive: 20
      maxWait: 60000
      timeBetweenEvictionRunsMillis: 60000
      minEvictableIdleTimeMillis: 30000
      validationQuery: select 'x'
      testWhileIdle: true
      testOnBorrow: false
      testOnReturn: false
      poolPreparedStatements: true
      maxPoolPreparedStatementPerConnectionSize: 20
      # 配置监控统计拦截的 filters，去掉后监控界面 SQL 无法统计，“wall”用于防火墙
      filters: stat,wall,slf4j
      # 通过 connectProperties 属性来打开 mergeSql 功能；
      connectionProperties: druid.stat.mergeSql=true;druid.stat.slowSqlMillis=5000

    jpa:
      show-sql: true
      properties:
        hibernate.format_sql: true
        hibernate.enable_lazy_load_no_trans: true
```

步骤 6 编写 JUnit 代码测试增、删、改、查数据操作

1）在 FridayApplicationTests 类中编写测试代码，测试 SysUserRepository 接口的增、删、改、查方法。

```
package edu.friday;

import edu.friday.model.SysUser;
import edu.friday.repository.SysUserRepository;
import org.junit.jupiter.api.Test;
import org.springframework.beans.factory.annotation.Autowired;
import org.springframework.boot.test.context.SpringBootTest;

@SpringBootTest
class FridayApplicationTests {

    @Autowired
    SysUserRepository userRepository;

    @Test
    void contextLoads() {
    }

    @Test
    void testSaveUser() {
        SysUser user = new SysUser();
        user.setUserName("Alex");
```

```
            user.setNickName("alex test");
            userRepository.save(user);
        }

        @Test
        void testUpdateUser() {
            SysUser user = userRepository.getOne(new Long(110));
            user.setUserName("Alex 1");
            user.setNickName("alex test");
            userRepository.save(user);
        }

        @Test
        void testFindUsers() {
            System.out.println(userRepository.findAll());
        }

        @Test
        void testDeleteUser() {
            userRepository.deleteById(new Long(110));
        }
    }
```

2）执行“testSaveUser()”测试代码，看数据是否正常保存到数据库中，如图 3-21 所示，单击要测试方法的运行三角，选择“Run'testSaveUser()'”。

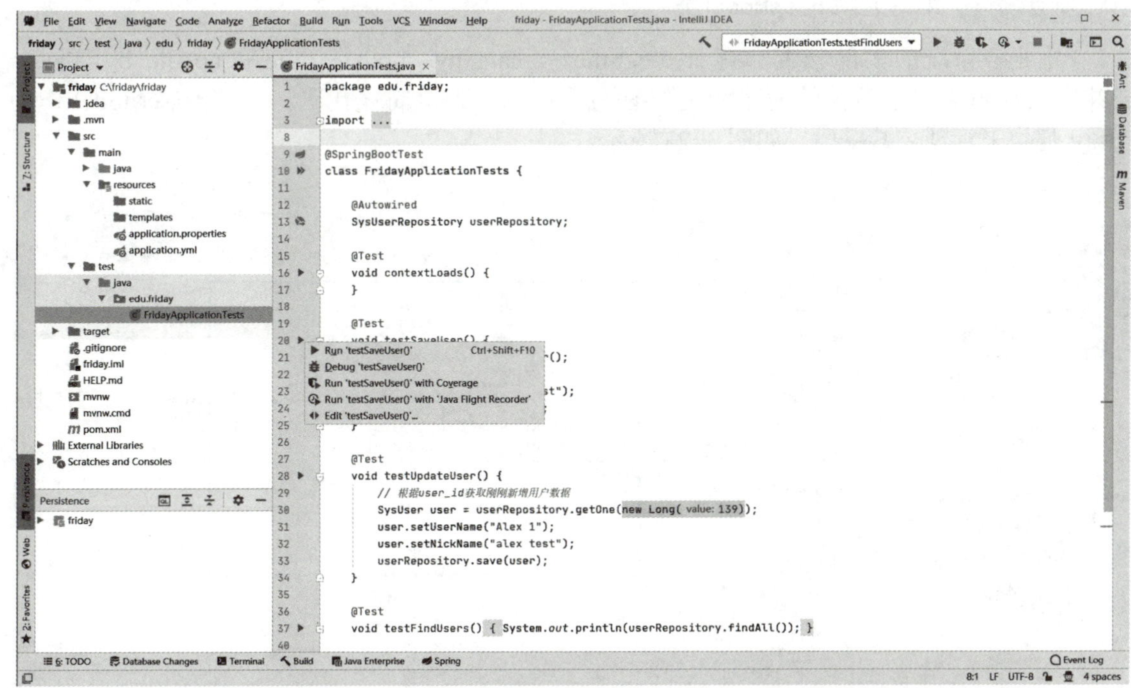

图 3-21　测试运行

3）运行之后，需要检查 2 个地方，如图 3-22 所示，①的位置是绿色，代表测试通过，双击“sys_user”数据表，检查②的位置，数据是否保存到“sys_user”数据表中。

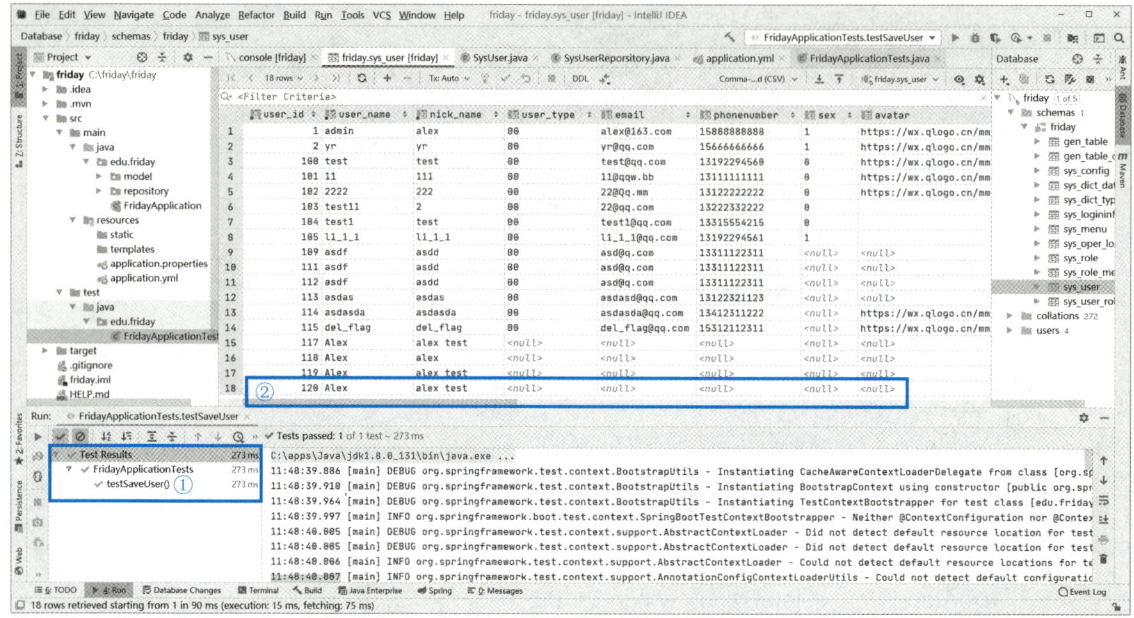

图 3-22　运行结果

SysUserRepository 接口其他方法的测试参照上面“testSaveUser()”方法进行，要确保 SysUserRepository 接口的方法全部通过测试。

步骤 7　编写通用 BaseModel 类

1）业务系统的大部分表都有 create_time、create_by、update_time 和 update_by 这四个公共字段，用来记录数据的创建和更新时间等信息，为了简化代码，编写 BaseModel 作为 Entity 基类来管理公共字段，如图 3-23 所示。

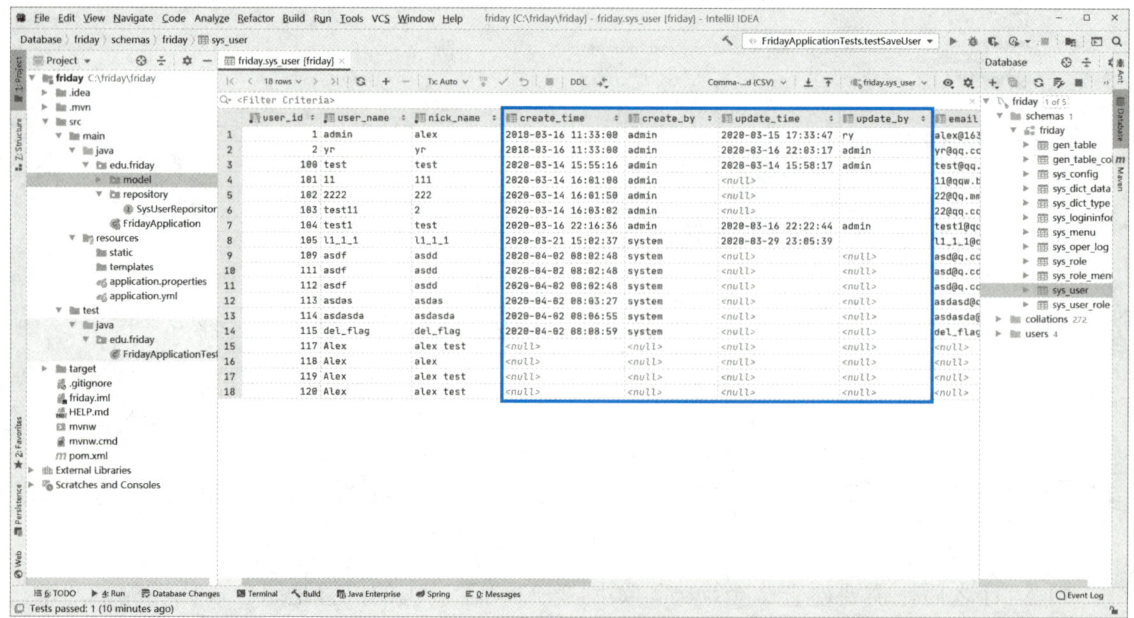

图 3-23　公共字段

2）创建“edu.friday.common.base”包，在包内新增 BaseModel 类，增加公共基础字段，代码如下。

```
@MappedSuperclass
@Data
@NoArgsConstructor
@AllArgsConstructor
public class BaseModel implements Serializable {
    private static final long serialVersionUID = 1L;
    private String createBy;
    @Column(updatable = false)
    @JsonFormat(pattern = "yyyy-MM-dd HH:mm:ss")
    private Date createTime;
    private String updateBy;
    @JsonFormat(pattern = "yyyy-MM-dd HH:mm:ss")
    private Date updateTime;
    private String remark;
    @PrePersist
    protected void onCreate() {
        createTime = new Date();
    }
    @PreUpdate
    protected void onUpdate() {
        updateTime = new Date();
    }
}
```

BaseModel 类源码中新增注解的作用见表 3-3。

表 3-3 注解作用

@MappedSuperclass	标注为 @MappedSuperclass 的类不是一个完整的实体类，它不会映射到数据库表，但是它的属性都会映射到其子类的数据库字段中
@JsonFormat(pattern = "yyyy-MM-dd HH:mm:ss")	为时间格式化注解，当读取日期的时候，会格式化成“yyyy-MM-dd HH:mm:ss”这样格式的中文时间
@PrePersist	save 之前被调用，帮助在持久化之前自动填充实体属性。在 friday 项目中用来在保存数据之前，自动给 createTime 赋值，这样就不用在业务代码中进行赋值操作
@PreUpdate	update 之前被调用，帮助在持久化之前自动填充实体属性。在 friday 项目中用来在保存数据之前，自动给 updateTime 赋值，这样就不用在业务代码中进行赋值操作

3）修改 SysUser 继承 BaseModel 基类。

```
public class SysUser extends BaseModel {
    @Id
    @GeneratedValue(strategy = GenerationType.IDENTITY)
    @Column(name = "user_id")
    private Long userId;

    @Basic
    @Column(name = "user_name")
```

```
    private String userName;

    @Basic
    @Column(name = "nick_name")
    private String nickName;

    @Basic
    @Column(name = "user_type")
    private String userType;

    @Basic
    @Column(name = "email")
    private String email;

    @Basic
    @Column(name = "phonenumber")
    private String phonenumber;

    @Basic
    @Column(name = "sex")
    private String sex;

    @Basic
    @Column(name = "avatar")
    private String avatar;

    @Basic
    @Column(name = "password")
    private String password;

    @Basic
    @Column(name = "status")
    private String status;

    @Basic
    @Column(name = "del_flag")
    private String delFlag;

    @Basic
    @Column(name = "login_ip")
    private String loginIp;

    @Basic
    @Column(name = "login_date")
    private Timestamp loginDate;
}
```

4）对修改的代码使用 JUnit 执行单元测试，执行“testSaveUser”测试方法，如图 3-24 所示，可以看到数据正常保存，而且 create_time 也保存了当前创建时间。

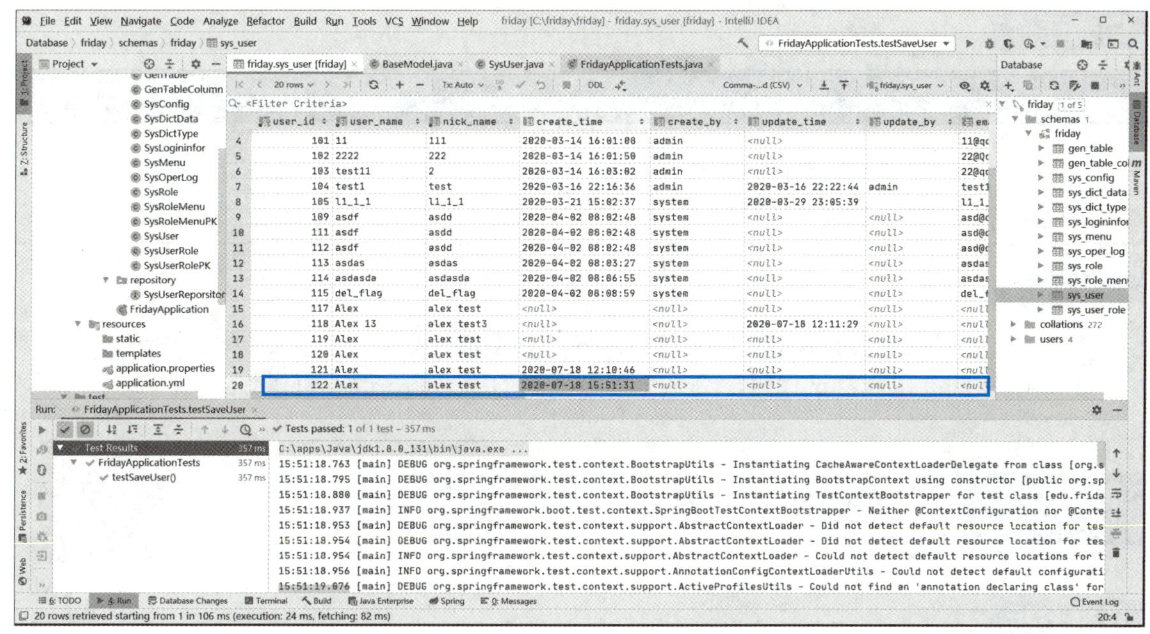

图 3-24　新增用户数据的测试结果

5）执行“testUpdateUser”测试方法，更新 user_id 为 122 的数据，如图 3-25 所示，可以看到数据正常保存，而且 update_time 也保存了当前更新时间。

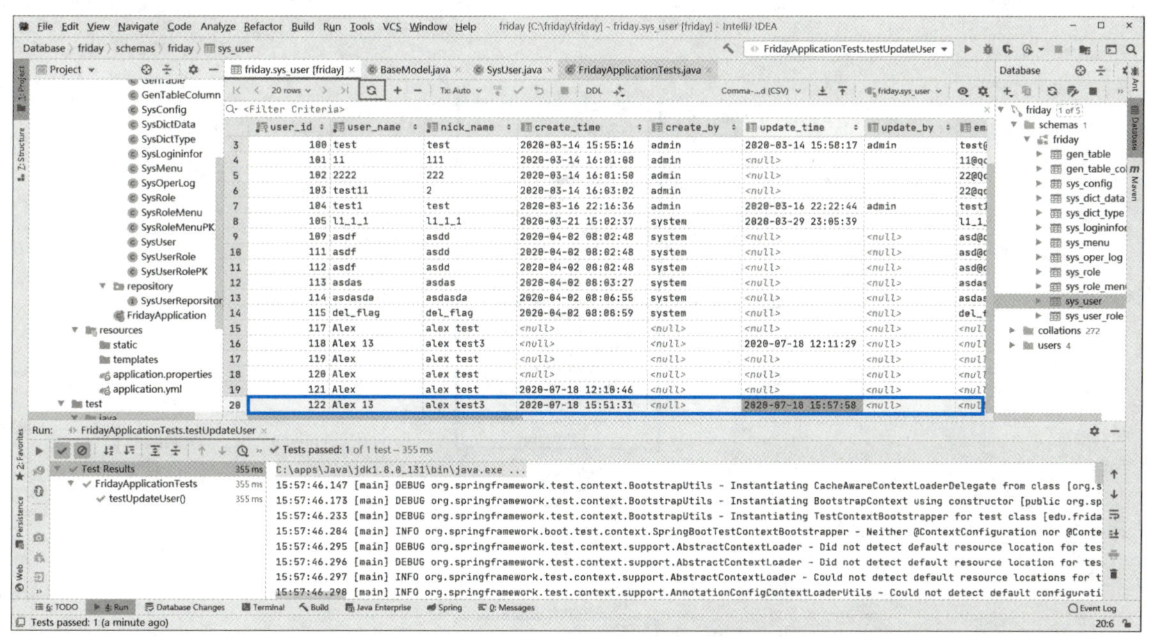

图 3-25　更新用户数据的测试结果

至此，就完成了使用 Spring Data JPA 对 MySQL 数据表的增、删、改、查操作的全部任务。

任务评价

技 能 点	知 识 点	自我评价（不熟悉 / 基本掌握 / 熟练掌握 / 灵活运用）
使用 Spring Data JPA 操作 MySQL 数据库	Spring Data JPA 常用注解；JUnit 常用注解	
使用 JUnit 编写测试代码		

拓展练习

1）完成全部数据表的 Repository 接口，使其具备增、删、改、查数据表的能力。

2）编写 JUnit 单元测试代码，测试全部数据表的增、删、改、查操作。

任务 3 导入常用工具类

微课 3-3 导入常用工具类

任务分析

实现项目业务功能前，通常会引入一些设计好的工具类，这些工具类的作用是把一个通用的功能包装在一个类中，让它可以重复使用，避免重复代码，提高开发效率，如字符串工具类、异常处理类等。图 3-26 所示的工具类目录图，是本项目中使用到的所有工具类。

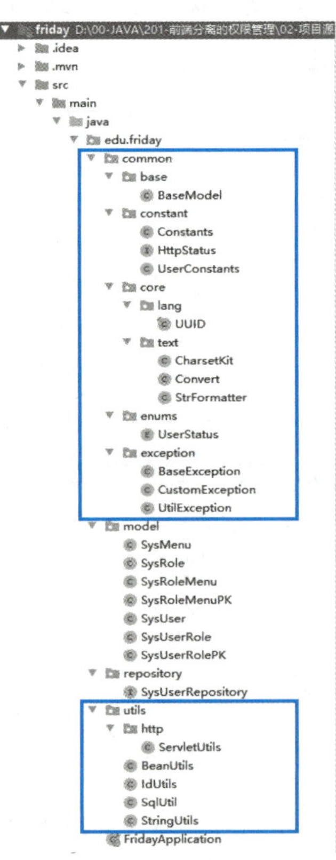

图 3-26 工具类目录图

任务实施

步骤 1　导入常用工具类到项目

参考图 3-26 中所示的工具类目录，将配套资源中的工具类源码复制到项目中的 src 目录。

步骤 2　熟悉常用工具类的用处

工具类在项目中的作用见表 3-4。

表 3-4　工具类作用

工具类名	工具类的作用
HttpStatus -HTTP 状态码接口	客户端向服务器发送异步请求后，服务器会返回状态码给客户端。HTTP 状态码接口就是描述服务器返回给客户端的所有状态码代码和内容
UserConstants - 用户常量类	用户常量类 UserConstants 定义了诸如系统用户名、状态等系统默认常量值
CharsetKit - 字符集工具类	字符集工具类 CharsetKit 定义了各种字符集常量以及字符集转换方法
Convert - 类型转换器	类型转换器 Convert 类提供将对象转换字符串、字符等类型的方法
StrFormatter - 字符串格式化	字符串格式化 StrFormatter 类负责将占位符 {} 按照顺序替换为参数
CustomException - 异常处理类	异常处理类 CustomException 定义异常代码和信息
BeanUtils - 实体工具类	实体工具类 BeanUtils 里支持 3 个方法：转换实体类的 List 对象，检查实体类的空置字段，忽略空值 Null（不包括空字符串）复制对象值
SqlUtil -SQL 操作工具类	SQL 操作工具类 SqlUtil 支持 3 个方法：检查字符（防止注入绕过），验证 order by 语法是否符合规范，生成批量插入的预编译语句
StringUtils - 字符串工具类	字符串工具类 StringUtils 支持的功能有：判断参数是否为空，判断集合是否为空或非空，判断对象数组是否为空或者非空，判断字符串是否为空或者非空等
IdUtils -ID 生成器工具类	用来生成唯一主键 ID
UUID - 唯一识别码工具类	用来提供生成通用唯一识别码的方法

任务 4　封装统一接口响应的 HTTP 结果

微课 3-4　封装统一接口响应的 HTTP 结果

任务分析

在前后端分离架构的项目中，基本的工作流程就是前端服务发起一个 HTTP 请求，后端服务处理业务逻辑，然后返回结果给前端服务。但是，每个开发人员都习惯于根据自己的需要来确定响应的数据结构，导致后续的维护工作较难开展。本任务就是封装统一接口响应的 HTTP 结果，解决由于不同的开发人员使用不同的数据结构响应前端服务，而产生的规范不统一的问题。

封装统一接口响应的 HTTP 结果的数据结构及参数含义见表 3-5。

表 3-5　HTTP 结果的数据结构及参数含义

数据结构	参数含义
code	对响应结果进一步细化，200 表示请求成功，400 表示用户操作导致的异常，500 表示系统异常，999 表示其他异常
message	友好的提示信息，或者请求结果提示信息
data	通常用于查询数据请求，成功之后将查询数据响应给前端

任务实施

步骤 1　构建统一接口响应结果类

1）构建增、删、改、查场景下的响应结果 RestResult。

- 当请求结果成功的情况下，使用 RestResult.success() 构建返回结果给前端。
- 当查询请求需要返回业务数据时，在请求成功的情况下，使用 RestResult.success(data) 构建返回结果给前端。
- 在请求结果失败的情况下，使用 RestResult.error() 构建返回结果给前端。

```
package edu.friday.common.result;

import edu.friday.common.constant.HttpStatus;
import edu.friday.utils.StringUtils;
import java.util.HashMap;

/**
 * 操作消息提醒
 */
public class RestResult extends HashMap<String, Object> {
    private static final long serialVersionUID = 1L;

    /**
     * 状态码
     */
    public static final String CODE_TAG = "code";

    /**
     * 返回内容
     */
    public static final String MSG_TAG = "msg";

    /**
     * 数据对象
     */
    public static final String DATA_TAG = "data";
```

```
/**
 * 初始化一个新创建的 RestResult 对象，使其表示一个空消息。
 */
public RestResult() {
}

/**
 * 初始化一个新创建的 RestResult 对象
 *
 * @param code 状态码
 * @param msg  返回内容
 */
public RestResult(int code, String msg) {
    super.put(CODE_TAG, code);
    super.put(MSG_TAG, msg);
}

/**
 * 初始化一个新创建的 RestResult 对象
 *
 * @param code 状态码
 * @param msg  返回内容
 * @param data 数据对象
 */
public RestResult(int code, String msg, Object data) {
    super.put(CODE_TAG, code);
    super.put(MSG_TAG, msg);
    if (StringUtils.isNotNull(data)) {
      super.put(DATA_TAG, data);
    }
}

/**
 * 返回成功消息
 *
 * @return 成功消息
 */
public static RestResult success() {
    return RestResult.success(" 操作成功 ");
}

/**
 * 返回成功数据
 *
 * @return 成功消息
 */
public static RestResult success(Object data) {
    return RestResult.success(" 操作成功 " , data);
}
```

```
/**
 * 返回成功消息
 *
 * @param msg 返回内容
 * @return 成功消息
 */
public static RestResult success(String msg) {
      return RestResult.success(msg, null);
}

/**
 * 返回成功消息
 *
 * @param msg  返回内容
 * @param data 数据对象
 * @return 成功消息
 */
public static RestResult success(String msg, Object data) {
      return new RestResult(HttpStatus.SUCCESS, msg, data);
}

/**
 * 返回错误消息
 *
 * @return
 */
public static RestResult error() {
      return RestResult.error(" 操作失败 ");
}

/**
 * 返回错误消息
 *
 * @param msg 返回内容
 * @return 警告消息
 */
public static RestResult error(String msg) {
 return RestResult.error(msg, null);
}

/**
 * 返回错误消息
 *
 * @param msg  返回内容
 * @param data 数据对象
 * @return 警告消息
 */
public static RestResult error(String msg, Object data) {
      return new RestResult(HttpStatus.ERROR, msg, data);
```

```
    }

    /**
     * 返回错误消息
     *
     * @param code 状态码
     * @param msg  返回内容
     * @return 警告消息
     */
    public static RestResult error(int code, String msg) {
        return new RestResult(code, msg, null);
    }
}
```

2）构建分页数据表格的响应结果 TableDataInfo。

```
package edu.friday.common.result;

import edu.friday.common.constant.HttpStatus;
import java.io.Serializable;
import java.util.List;

/**
 * 表格分页数据对象
 */
public class TableDataInfo implements Serializable {
    private static final long serialVersionUID = 1L;

    /**
     * 总记录数
     */
    private long total;

    /**
     * 列表数据
     */
    private List<?> rows;

    /**
     * 消息状态码
     */
    private int code;

    /**
     * 消息内容
     */
    private int msg;

    /**
     * 表格数据对象
```

```
     */
    public TableDataInfo() {
    }

    /**
     * 分页
     *
     * @param list 列表数据
     * @param total 总记录数
     */
    public TableDataInfo(List<?> list, long total) {
        this.rows = list;
        this.total = total;
    }

    /**
     * 成功的返回方法
     */
    public static TableDataInfo success(List<?> list, long total) {
        TableDataInfo tableDataInfo = new TableDataInfo(list, total);
        tableDataInfo.setCode(HttpStatus.SUCCESS);
        return tableDataInfo;
    }

    /**
     * 失败的返回方法
     */
    public static TableDataInfo failure(int msg) {
        TableDataInfo tableDataInfo = new TableDataInfo();
        tableDataInfo.setCode(HttpStatus.ERROR);
        tableDataInfo.setMsg(msg);
        return tableDataInfo;
    }

    public long getTotal() {
        return total;
    }

    public void setTotal(long total) {
        this.total = total;
    }

    public List<?> getRows() {
        return rows;
    }

    public void setRows(List<?> rows) {
        this.rows = rows;
    }
```

```
    public int getCode() {
        return code;
    }

    public void setCode(int code) {
        this.code = code;
    }

    public int getMsg() {
        return msg;
    }

    public void setMsg(int msg) {
        this.msg = msg;
    }
}
```

步骤 2 构建 Web 层通用数据处理类

1）定义 Web 层通用数据处理 BaseController，用于控制返回响应结果为成功还是失败。

```
package edu.friday.common.base;

import edu.friday.common.result.RestResult;

/**
 * Web 层通用数据处理
 */
public class BaseController {
    /**
     * 响应返回结果
     *
     * @param rows 影响行数
     * @return 操作结果
     */
    protected RestResult toAjax(int rows) {
        return rows > 0 ? RestResult.success() : RestResult.error();
    }
}
```

2）编写通用响应结果的代码如下。

```
package edu.friday.controller;

import edu.friday.common.base.BaseController;
import edu.friday.common.result.RestResult;
import edu.friday.common.result.TableDataInfo;
import edu.friday.model.SysUser;
import org.springframework.web.bind.annotation.*;
import java.util.ArrayList;
```

```
@RestController
@RequestMapping("/demo")
public class DemoController extends BaseController {

    /**
     * 新增
     */
    @PostMapping
    public RestResult add() {
        boolean flag = true;
        return toAjax(flag ? 1 : 0);
    }

    /**
     * 修改
     */
    @PutMapping
    public RestResult edit() {
        boolean flag = true;
        return toAjax(flag ? 1 : 0);
  }

    /**
     * 删除
     */
    @DeleteMapping()
    public RestResult remove() {
        boolean flag = true;
        return toAjax(flag ? 1 : 0);
    }

    /**
     * 根据用户编号获取详细信息
     */
    @GetMapping(value = {"/", "/{userId}"})
    public RestResult getInfo(@PathVariable(value = "userId", required = false) Long userId) {
        RestResult ajax = RestResult.success();
        ajax.put("user", new SysUser());
        return ajax;
    }

    @GetMapping("/list")
    public TableDataInfo list() {
        return TableDataInfo.success(new ArrayList<>(), 100);
    }
}
```

3）验证响应结果，在浏览器 URL 地址栏中输入“http://localhost:8080/demo/list”，返

回统一接口的响应结果，如图 3-27 所示。

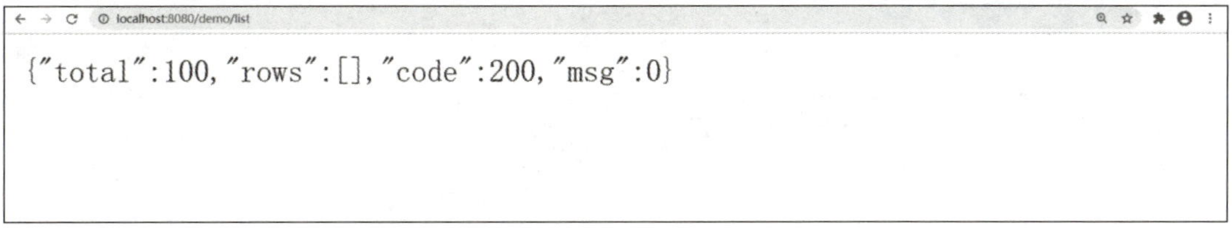

```
← → C ⓘ localhost:8080/demo/list
{"total":100,"rows":[],"code":200,"msg":0}
```

图 3-27　返回统一接口的响应结果

任务评价

技　能　点	知　识　点	自我评价（不熟悉 / 基本掌握 / 熟练掌握 / 灵活运用）
编写统一接口响应结果类	HTTP 状态码； Spring MVC 请求响应注解	
编写 Web 层通用数据处理类		

单元小结

在前后端分离架构的项目中，其设计的核心思想是前端项目通过调用后端的 API 接口对数据库进行操作，并使用 JSON 数据进行交互。基于 Spring Initializr 初始化后端项目、使用 Spring Data JPA 构建数据访问层以及封装统一接口响应的 HTTP 结果，是项目构建的重要环节。通过完成本工作单元的任务和后面的实战强化，熟练掌握构建后端项目公共模块的职业技能。

实战强化

1）按照任务 1 步骤，使用 Spring Initializr 构建实训项目“诚品书城”后端。

2）按照任务 2 步骤，构建实训项目“诚品书城”后端的数据访问层。

3）按照任务 3 步骤，导入常用工具类到实训项目“诚品书城”前端。

4）按照任务 4 步骤，实现实训项目“诚品书城”后端的统一接口响应的 HTTP 结果。

工作单元 4

实现用户和角色管理接口

职业能力

本工作单元主要完成用户和角色管理的接口功能，最终希望学生达成如下职业能力目标：

1．掌握基于三层架构实现对数据的增、删、改、查功能

2．灵活使用 Postman 测试后端接口功能

3．掌握数据表的关联分析和修改

4．灵活使用基础类

任务情景

本工作单元的任务就是完成工作单元 2 中设计的用户和角色管理功能的后端接口，具体的后端接口详情见表 4-1。

表 4-1　用户和角色管理功能的后端接口详情

接 口 功 能	HTTP 请求方式	URL 地址	支 持 格 式
用户管理的接口			
获取用户列表	GET	/system/user/list	JSON
新增用户数据	POST	/system/user	JSON
修改用户数据	PUT	/system/user	JSON
删除用户数据	DELETE	/system/user/{userIds}	JSON
根据用户编号获取用户信息与角色列表	GET	/system/user/{userIds}, /system/user/	JSON
角色管理的接口			
获取角色列表	GET	/system/role/list	JSON
新增角色数据	POST	/system/role	JSON
修改角色数据	PUT	/system/role	JSON
删除角色数据	DELETE	/system/role/{roleIds}	JSON
根据角色编号获取详细信息	GET	/system/role/{roleId}	JSON
修改角色状态	PUT	/system/role/changeStatus	JSON

前置知识

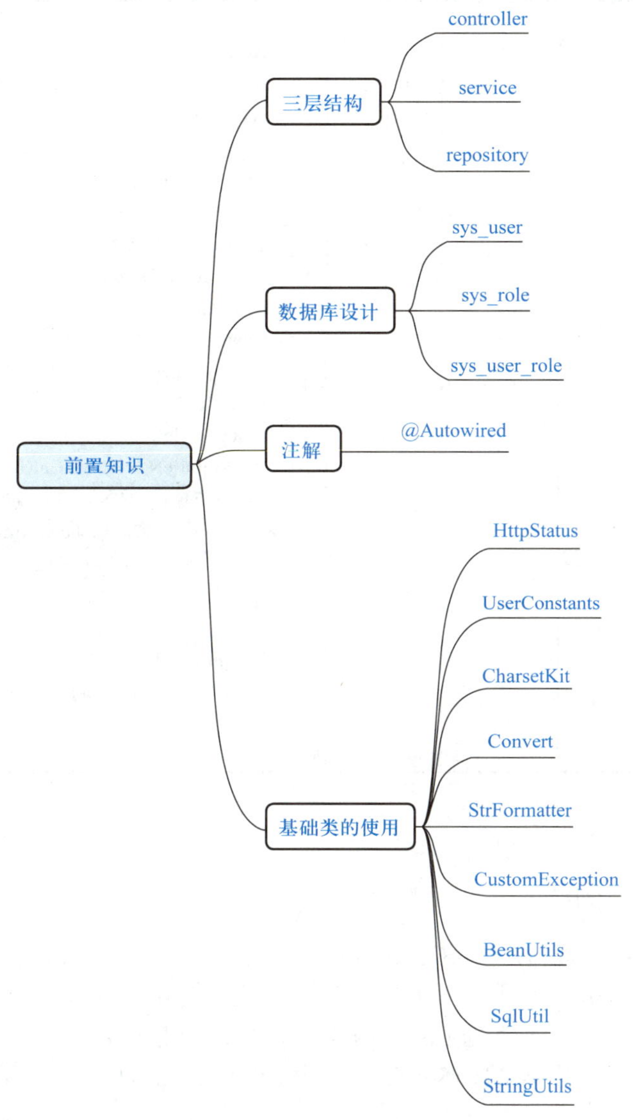

任务1　实现用户列表接口

微课 4-1　实现用户列表接口

任务分析

通过本任务，完成表 4-1 中的获取用户列表的接口功能，开发人员将根据接口描述编码。所有接口功能的设计遵循 IPO（input-process-output）模式：输入参数 + 逻辑处理 + 输出返回值，该功能基于 IPO 模式，用户列表的接口功能见表 4-2。

表 4-2　用户列表的接口功能

URL 地址	HTTP 请求方式	支持格式	输入参数	输出返回值

/system/user/list	GET	JSON	无参数	[{
				"createBy": "admin",
				"createTime": "2018-03-16 03:33:00",
				"updateBy": "ry",
				"updateTime": "2020-03-15 09:33:47",
				"remark": " 管理员 ",
				"userId": 1,
				"userName": "admin",
				"nickName": "alex",
				"userType": "00",
				"email": "alex@163.com",
				"phonenumber": "15888888888",
				"sex": "1",
				"avatar": "https://wx.qlogo.cn/mmopen/vi_32/1yzTKJKIfurhDI29RqibEicNOoH0WiaCuKb6jWppVu4uzWovO0d1ICAwuW4rB4zfUxVvGHfNuxXLHu44t3yBkgbicQ/132",
				"password": "$2a$10$boXFAiZ4OdtZiT2owx.xx.F848I4rh4JCQQDvgAaEwiktcFRh8Ile",
				"status": "0",
				"delFlag": "0",
				"loginIp": "127.0.0.1",
				"loginDate": "2018-03-16T03:33:00.000+00:00"
				}
				]

任务实施

步骤 1　编写控制层代码

在 controller 包下创建 SysUserController 类，通过 @RequestMapping+@GetMapping 映射“/system/user/list”URL 地址以及 GET 请求方式，使用 @RestController 表示返回数据格式为“JSON”，输出返回值为 List<SysUser> 对象，使用 @Autowired 注解加载业务层 SysUserService 对象，调用 SysUserService 的 selectUserList() 方法实现查询用户表全部数据的业务逻辑代码。

```
package edu.friday.controller;
import edu.friday.model.SysUser;
import edu.friday.service.SysUserService;
import org.springframework.beans.factory.annotation.Autowired;
import org.springframework.web.bind.annotation.GetMapping;
import org.springframework.web.bind.annotation.RequestMapping;
import org.springframework.web.bind.annotation.RestController;
import java.util.List;
/**
 * 用户表控制层
 */
```

```
@RestController
@RequestMapping("/system/user")
public class SysUserController {
    @Autowired
SysUserService userService;
/**

 * 显示用户表所有数据

 */

   @GetMapping("/list")
    public List<SysUser> list() {
        return userService.selectUserList();
    }
}
```

步骤 2　编写业务层代码

1）在“/src/main/java/edu/friday”包下创建 service 包，在 service 包下创建步骤 1 中使用的 SysUserService 接口。

```
package edu.friday.service;
import edu.friday.model.SysUser;
import java.util.List;

/**
 * 用户表业务层接口
 */
public interface SysUserService {
    List<SysUser> selectUserList();
}
```

2）在 service 包下创建 impl 子包，编写 SysUserServiceImpl 实现类，调用数据控制层 SysUserRepository 的 findAll() 方法获取用户表数据。

```
package edu.friday.service.impl;
import edu.friday.model.SysUser;
import edu.friday.repository.SysUserRepository;
import edu.friday.service.SysUserService;
import org.springframework.beans.factory.annotation.Autowired;
import org.springframework.stereotype.Service;
import java.util.List;

/**
 * 用户表业务层
 */
@Service
public class SysUserServiceImpl implements SysUserService {
```

```
        @Autowired
        SysUserRepository sysUserRepository;
        /**
         * 显示所有用户
         */
        @Override
        public List<SysUser> selectUserList() {
            return sysUserRepository.findAll();
        }
    }
```

步骤 3 编写数据控制层代码

创建 SysUserRepository 步骤同工作单元 3 的任务 2 的步骤 4。

步骤 4 运行调试

在 IDEA 中运行 FridayApplication，打开浏览器并输入 http://localhost:8080/system/user/list，运行效果如图 4-1 所示。

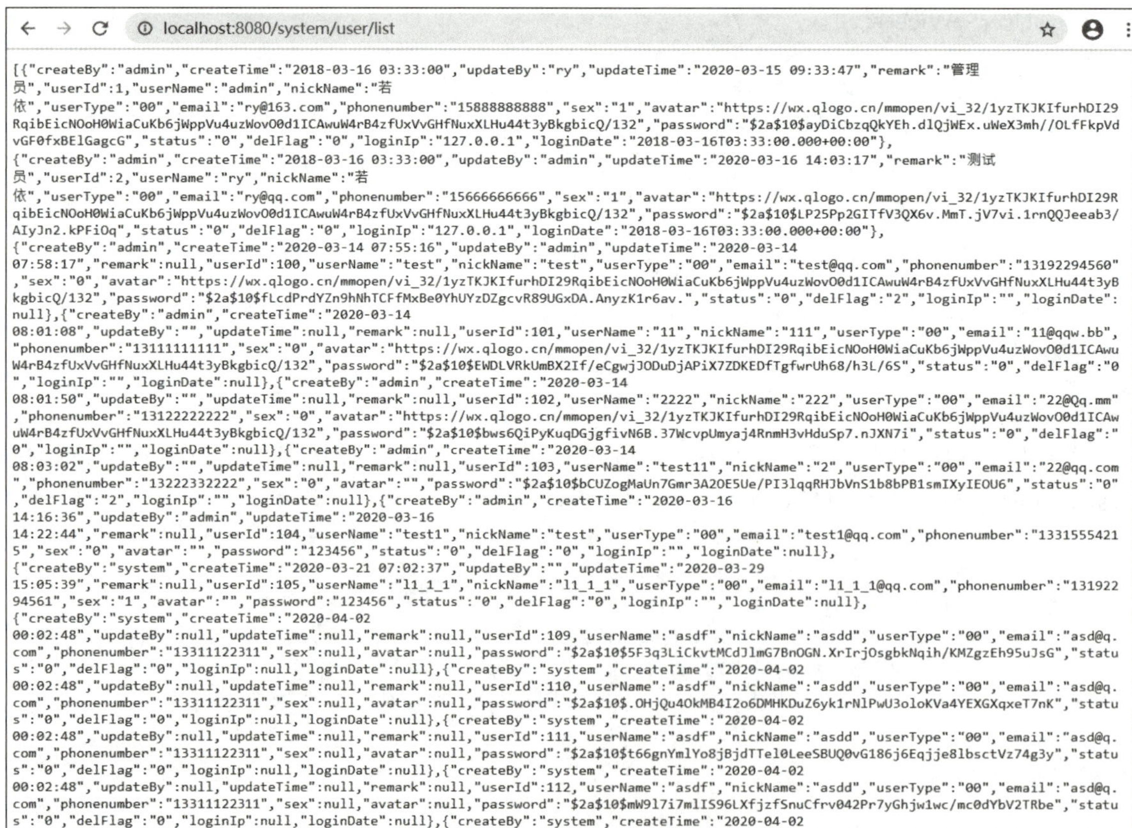

图 4-1　以 JSON 形式展示 sys_user 表中所有数据

在浏览器中查看 JSON 数据时，格式较混乱，不容易看清数据结构。Postman 软件显示

JSON 数据格式的效果会更好，本书均使用 Postman 测试后端接口功能。Postman 的安装见工作单元 1。

打开 Postman 软件，注册并登录，在 GET 请求里输入网址 http://localhost:8080/system/user/list，单击右边的“Send”按钮，JSON 样式的显示如图 4-2 所示。

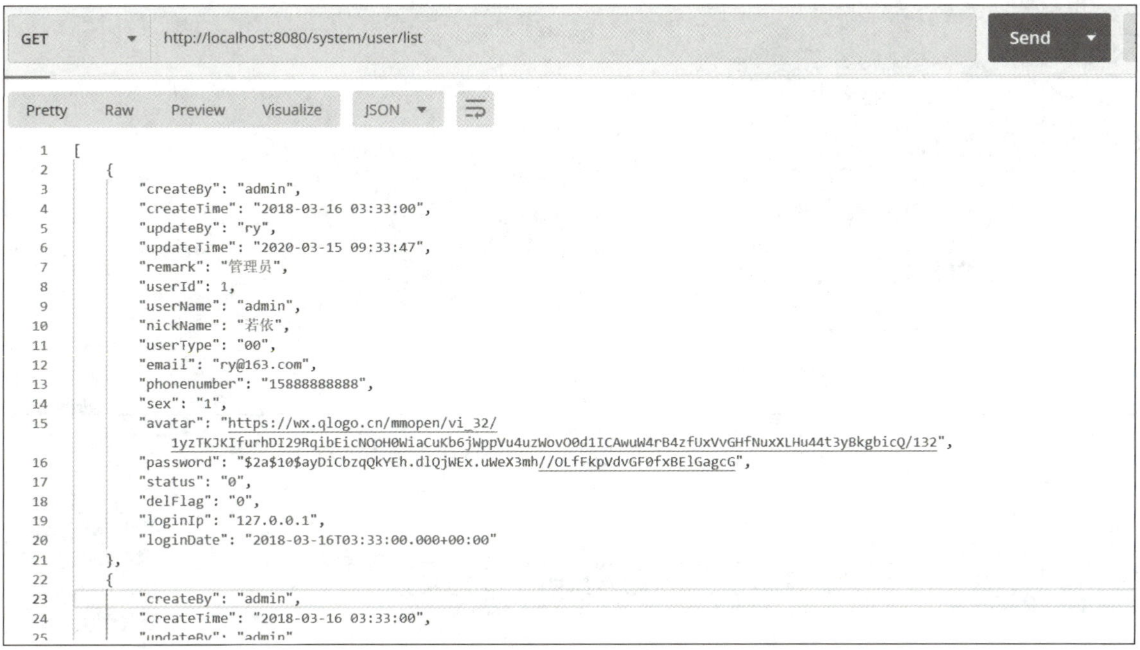

GET　http://localhost:8080/system/user/list　Send

Pretty　Raw　Preview　Visualize　JSON


```
[
    {
        "createBy": "admin",
        "createTime": "2018-03-16 03:33:00",
        "updateBy": "ry",
        "updateTime": "2020-03-15 09:33:47",
        "remark": "管理员",
        "userId": 1,
        "userName": "admin",
        "nickName": "若依",
        "userType": "00",
        "email": "ry@163.com",
        "phonenumber": "15888888888",
        "sex": "1",
        "avatar": "https://wx.qlogo.cn/mmopen/vi_32/
            1yzTKJKIfurhDI29RqibEicNOoH0WiaCuKb6jWppVu4uzWovO0d1ICAwuW4rB4zfUxVvGHfNuxXLHu44t3yBkgbicQ/132",
        "password": "$2a$10$ayDiCbzqQkYEh.dlQjWEx.uWeX3mh//OLfFkpVdvGF0fxBElGagcG",
        "status": "0",
        "delFlag": "0",
        "loginIp": "127.0.0.1",
        "loginDate": "2018-03-16T03:33:00.000+00:00"
    },
    {
        "createBy": "admin",
        "createTime": "2018-03-16 03:33:00",
        "updateBy": "admin"
```


图 4-2　JSON 样式的显示

任务评价

技　能　点	知　识　点	自我评价（不熟悉 / 基本掌握 / 熟练掌握 / 灵活运用）
编写控制层代码	Spring 常用注解 Spring MVC 常用注解	
编写业务层代码		
编写数据访问层代码		

拓展练习

仿照本任务的操作步骤实现获取角色列表的接口功能，见表 4-3。

表 4-3　获取角色列表的接口功能

接 口 功 能	HTTP 请求方式	URL 地址	支 持 格 式
获取角色列表	GET	/system/role/list	JSON

完成后重启项目 FridayApplication，在 Postman 中访问网址 http://localhost:8080/system/role/list，结果如图 4-3 所示。

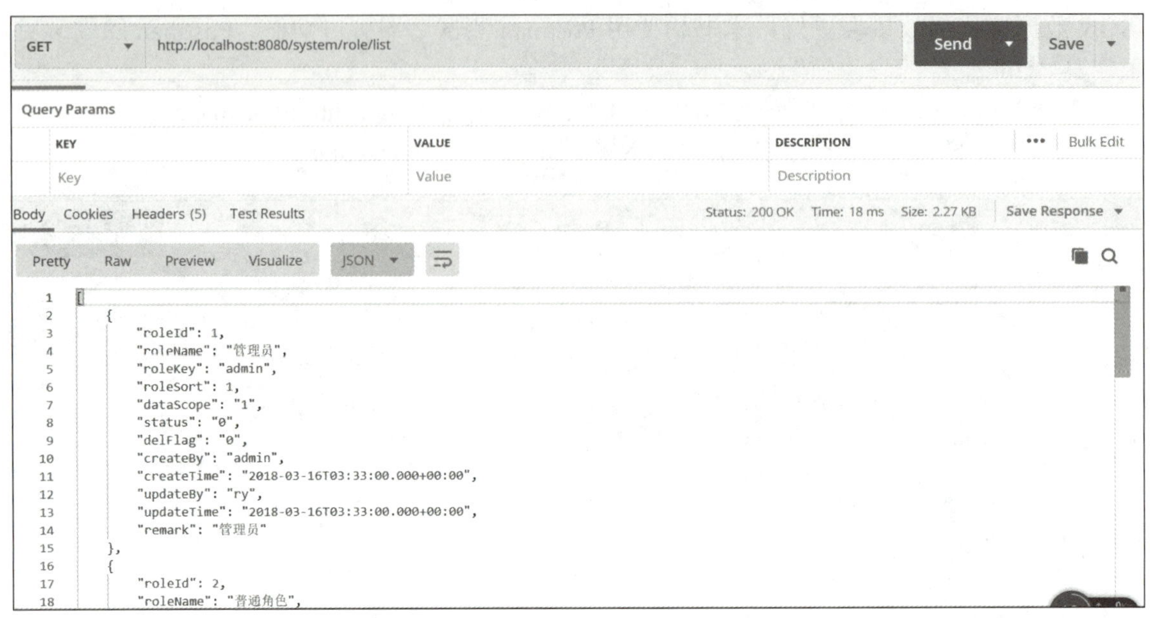

图 4-3　以 JSON 形式展示 sys_role 表中所有数据

任务 2　实现用户列表的查询与分页接口

微课 4-2　实现用户列表的查询与分页接口

任务分析

本任务是基于任务 1 的升级，使用户列表接口具备数据分页查询和模糊查询的功能。

根据接口设计的 IPO 模式，开发人员基于工作单元 2 中的用户管理 UI 图，分析输入和输出值如下：

1）分页查询。对数据进行分页时，需要知道当前页数和每页显示多少条记录来进行数据查询，所以增加 page 和 size 作为输入参数，同时需要输出返回值为用户列表数据以及总的数据条数。

2）模糊查询。用户管理页面对用户名称、手机号和状态进行模糊查询，所以增加 userName、phonenumber 和 status 作为输入参数，同时需要输出返回值为用户列表数据。

通过上面的分析，分页和模糊查询功能的用户列表接口描述见表 4-4。

表 4-4　分页和模糊查询功能的用户列表接口描述

URL 地址	HTTP 请求方式	支持格式	输入参数	输出返回值
/system/user/list	GET	JSON	分页参数：page、size；查询参数：userName、phonenumber、status	{ "total": 2, "rows": [{ "createBy": "admin",

（续）

URL 地址	HTTP 请求方式	支持格式	输入参数	输出返回值
/system/user/list	GET	JSON	分页参数：page、size；查询参数：userName、phonenumber、status	"createTime": "2018-03-16 03:33:00", "updateBy": "ry", "updateTime": "2020-03-15 09:33:47", "remark": " 管理员 ", "userId": 1, "userName": "admin", "nickName": "alex", "userType": "00", "email": "alex@163.com", "phonenumber": "15888888888", "sex": "1", "avatar": "https://wx.qlogo.cn/mmopen/vi_32/1yzTKJKIfurhDI29RqibEicNOoH0WiaCuKb6jWppVu4uzWovO0d1ICAwuW4rB4zfUxVvGHfNuxXLHu44t3yBkgbicQ/132", "password": "$2a$10$boXFAiZ4OdtZiT2owx.xx.F848I4rh4JCQQDvgAaEwiktcFRh8Ile", "status": "0", "delFlag": "0", "loginIp": "127.0.0.1", "loginDate": "2018-03-16T03:33:00.000+00:00" }], "code": 200, "msg": 0 }

任务实施

步骤1　编写视图对象代码

1）在 model 包下创建 vo 包，在 vo 包下增加 SysUserVO 类，将用户管理页面的所有数据封装以及进行数据格式验证。

```
package edu.friday.model.vo;
import com.fasterxml.jackson.annotation.JsonFormat;
import com.fasterxml.jackson.annotation.JsonProperty;
import lombok.AllArgsConstructor;
import lombok.Data;
import lombok.NoArgsConstructor;
import javax.validation.constraints.Email;
import javax.validation.constraints.NotBlank;
import javax.validation.constraints.Size;
import java.io.Serializable;
import java.util.Date;
import java.util.List;

/**
 * 用户对象 sys_user
 */
```

```
@Data
@NoArgsConstructor
@AllArgsConstructor
public class SysUserVO implements Serializable {
    private static final long serialVersionUID = 1L;

    /**
     * 用户 ID
     */
    private Long userId;

    /**
     * 用户账号
     */
    @NotBlank(message = "用户账号不能为空")
    @Size(min = 0, max = 30, message = "用户账号长度不能超过 30 个字符")
    private String userName;

    /**
     * 用户昵称
     */
    @Size(min = 0, max = 30, message = "用户昵称长度不能超过 30 个字符")
    private String nickName;

    /**
     * 用户邮箱
     */
    @Email(message = "邮箱格式不正确")
    @Size(min = 0, max = 50, message = "邮箱长度不能超过 50 个字符")
    private String email;

    /**
     * 手机号码
     */
    @Size(min = 0, max = 11, message = "手机号码长度不能超过 11 个字符")
    private String phonenumber;

    /**
     * 用户性别
     */
    private String sex;
    /**
     * 用户类型
     */
    private String userType;

    /**
     * 用户头像
     */
    private String avatar;

    /**
     * 密码
     */
    @JsonProperty
```

```
private String password;

/**
 * 盐加密
 */
private String salt;

/**
 * 账号状态（0 正常，1 停用）
 */
private String status;

/**
 * 删除标识（0 代表存在，2 代表删除）
 */
private String delFlag;

/**
 * 最后登录 IP
 */
private String loginIp;

/**
 * 最后登录时间
 */
private Date loginDate;

/**
 * 创建者
 */
private String createBy;

/**
 * 创建时间
 */
@JsonFormat(pattern = "yyyy-MM-dd HH:mm:ss")
private Date createTime;

/**
 * 更新者
 */
private String updateBy;

/**
 * 更新时间
 */
@JsonFormat(pattern = "yyyy-MM-dd HH:mm:ss")
private Date updateTime;

/**
 * 备注
 */
private String remark;

/**
```

```
     * 角色对象
     */
    private List<SysRoleVO> roles;

    /**
     * 角色组
     */
    private Long[] roleIds;

    public SysUserVO(Long userId) {
        this.userId = userId;
    }

    public boolean isAdmin() {
        return isAdmin(this.userId);
    }

    public static boolean isAdmin(Long userId) {
        return userId != null && 1L == userId;
    }
}
```

2）在 pom.xml 文件中引入格式验证依赖，用于注解参数的验证。

```
<dependencies>
    ...
    <dependency>
        <groupId>javax.validation</groupId>
        <artifactId>validation-api</artifactId>
        <version>2.0.1.Final</version>
    </dependency>
</dependencies>
```

步骤 2 编写控制层代码

修改 SysUserController 中 list() 方法的输入参数、返回值以及程序逻辑，具体如下。

- 输入参数：增加两个参数 SysUserVO 和 Spring Data JPA 提供的 Pageable，分别用于封装查询参数 userName、phonenumber、status 和分页参数 page、size。
- 返回值：使用 TableDataInfo 作为分页列表数据的返回值，封装了任务分析中的返回值数据列表 rows、数据总条数 total、状态码 code 以及消息 msg。
- 程序逻辑：程序将 page 和 size 参数封装到 Pageable 对象，Pageable 对象作为参数传递给 selectUserList 方法，实现分页功能，代码如下。

```
@GetMapping("/list")
public TableDataInfo list(SysUserVO user, Pageable page) {
    int pageNum = page.getPageNumber() - 1;
    pageNum = pageNum <= 0 ? 0 : pageNum;
    page = PageRequest.of(pageNum, page.getPageSize());
    return userService.selectUserList(user, page);
}
```

步骤 3　编写业务层代码

1）修改 SysUserService 接口的 selectUserList 抽象方法。

```
public TableDataInfo selectUserList(SysUserVO user, Pageable page);
```

2）修改 SysUserServiceImpl 实现类中的 selectUserList 方法，通过使用 Spring Data JPA 的 ExampleMatcher 匹配器构建模糊查询条件，相当于 SQL 的匹配条件，通过由实体 SysUser 对象和 ExampleMatcher 匹配器共同创建的 Example 对象真正实现查询功能，代码如下。

```
public TableDataInfo selectUserList(SysUserVO user, Pageable page) {
    SysUser sysuser = new SysUser();
    BeanUtils.copyPropertiesIgnoreEmpty(user, sysuser);
    sysuser.setDelFlag("0");
    ExampleMatcher exampleMatcher = ExampleMatcher.matching()
            .withMatcher("userName", ExampleMatcher.GenericPropertyMatchers.contains())
            .withMatcher("phonenumber", ExampleMatcher.GenericPropertyMatchers.contains());
    Example<SysUser> example = Example.of(sysuser, exampleMatcher);
    Page<SysUser> rs = sysUserRepository.findAll(example, page);

    return TableDataInfo.success(rs.toList(), rs.getTotalElements());
}
```

步骤 4　运行调试

1）验证分页功能。在 IDEA 中运行 FridayApplication，打开 Postman 并输入网址 http://localhost:8080/system/user/list?page=1&size=10，运行效果如图 4-4 所示。

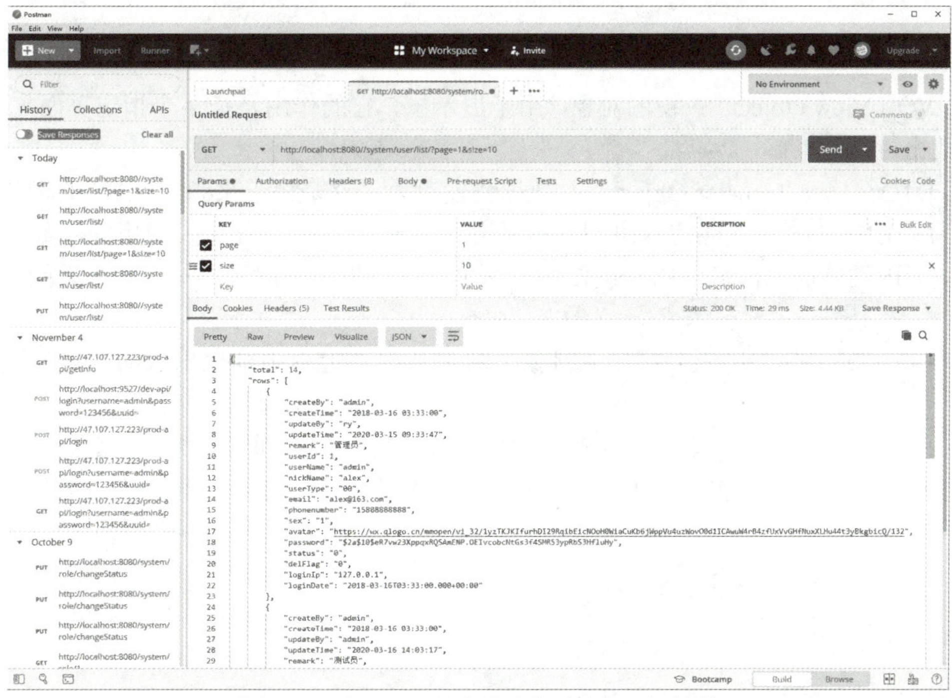

图 4-4　以 JSON 形式展示分页功能运行效果

2）验证查询功能。Postman 中测试效果，输入网址 http://localhost:8080/system/user/list?userName=a，返回 8 个用户名里含有字母 a 的数据，运行效果如图 4-5 所示。

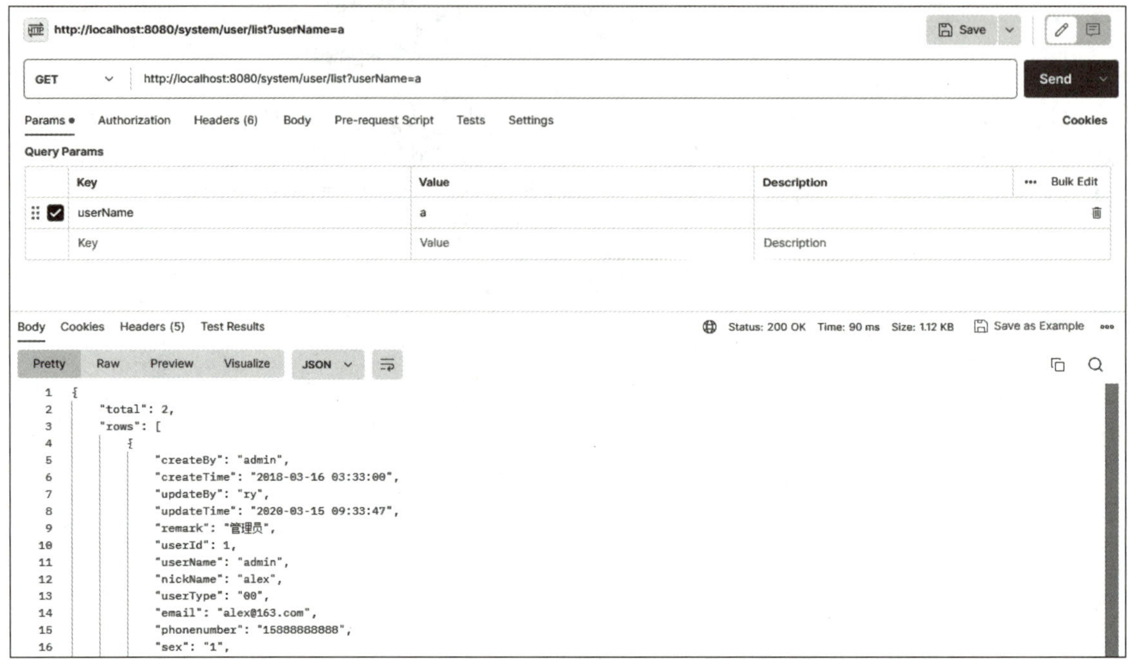

图 4-5　以 JSON 形式展示查询功能运行效果

知识小结

1. 实体对象

➢ VO（View Object）：视图对象，用于展示层，它的作用是把某个指定页面（或组件）的所有数据封装起来。

➢ DTO（Data Transfer Object）：数据传输对象，这个概念来源于 J2EE 的设计模式，原本是为了 EJB 的分布式应用提供粗粒度的数据实体，以减少分布式调用的次数，从而提高分布式调用的性能和降低网络负载，但在这里，其泛指用于展示层与服务层之间的数据传输对象。

➢ DO（Domain Object）：领域对象，就是从现实世界中抽象出来的有形或无形的业务实体。

➢ PO（Persistent Object）：持久化对象，它跟持久层（通常是关系型数据库）的数据结构形成一一对应的映射关系，如果持久层是关系型数据库，那么，数据表中的每个（或若干个）字段就对应 PO 的一个（或若干个）属性。

2. 使用 Spring Data JPA 的 Pageable 实现分页

3. 使用 Spring Data JPA 的 Example 实现动态查询

Example 查询的核心组成部分具体如下。

➢ 实体对象：即本任务中的 SysUser 对象，对应 sys_user 表。在构建查询条件时，一个实体对象代表的是查询条件中的数据表。

➢ 匹配器：即 ExampleMatcher 对象，它是匹配实体对象的，使用实体对象进行查询，本任务中要查询用户名字中有“A”的数据，使用匹配器构造模糊查询条件代码为：withMatcher("userName", ExampleMatcher.GenericPropertyMatchers.contains())。

➢ 实例：即 Example 对象，由实体对象和匹配器共同创建，代表的是完整的查询条件。程序调用 Spring Data JPA 提供的 findAll（Example）即可执行查询功能。

任务评价

技　能　点	知　识　点	自我评价（不熟悉 / 基本掌握 / 熟练掌握 / 灵活运用）
编写视图对象代码	Spring Data JPA 的 Pageable 分页 Spring Data JPA 的 Example 动态查询	
编写分页代码		
编写查询代码		

拓展练习

仿照本任务的操作步骤，实现获取角色列表且具备分页和查询功能的接口，见表 4-5。

表 4-5　获取角色列表且具备分页和查询功能的接口描述

接口功能	HTTP 请求方式	URL 地址	支持格式
获取角色列表、分页和查询	GET	/system/role/list	JSON

完成后重启项目 FridayApplication，在 Postman 中访问网址 http://localhost:8080/system/role/list?roleName= 测试，结果如图 4-6 所示。

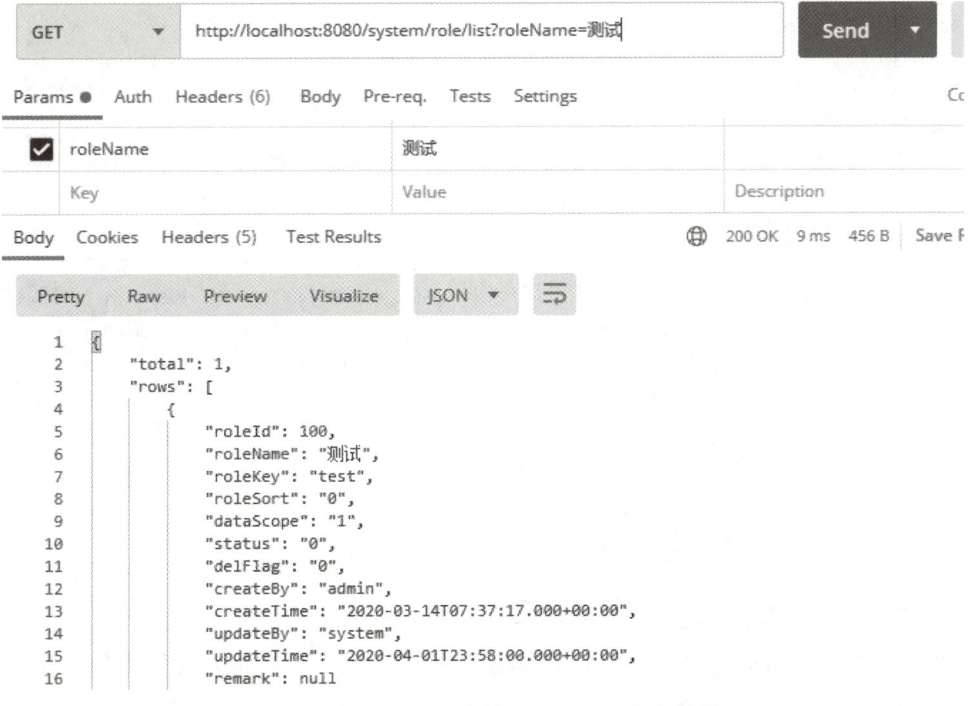

图 4-6　以 JSON 形式展示 sys_role 表中数据

提示： 测试时如果出现 total 值为 0 的情况，是因为没有完成工作单元 3 的任务，请修改 model 包下 SysRole 实体类的 roleId 类型，改“long”为“Long”。

任务 3　实现用户新增接口

微课 4-3　实现用户新增接口

任务分析

通过本任务，完成表 4-1 中的新增用户数据的接口功能，业务逻辑是保存用户数据到 sys_user 表，同时把该用户对应的角色关系保存到 sys_user_role 表中。

开发人员编写的接口输入和返回值描述，见表 4-6。

表 4-6　接口输入和返回值描述

URL 地址	HTTP 请求方式	支持格式	输入参数	输出返回值
/system/user	POST	JSON	SysUserVO	{"msg": " 操作成功 ", "code": 200}

任务实施

步骤 1　编写控制层代码

在 SysUserController 中添加 add 方法，其中输入参数使用 @RequestBody 将前端页面传过来的参数自动封装到 SysUserVO 对象中，返回值使用 RestResult 对象返回 msg 和 code 值，首先进行业务逻辑规则检查，检查用户账号、手机、邮箱的唯一性，通过检查之后调用 insertUser 方法保存数据。

```
@PostMapping
public RestResult add(@RequestBody SysUserVO user) {
        if (UserConstants.NOT_UNIQUE.equals(userService.checkUserNameUnique(user.
getUserName()))) {
            return RestResult.error(" 新增用户 " + user.getUserName() + " 失败，登录账号已存在 ");
        } else if (UserConstants.NOT_UNIQUE.equals(userService.checkPhoneUnique(user))) {
            return RestResult.error(" 新增用户 " + user.getUserName() + " 失败，手机号码已存在 ");
        } else if (UserConstants.NOT_UNIQUE.equals(userService.checkEmailUnique(user))) {
            return RestResult.error(" 新增用户 " + user.getUserName() + " 失败，邮箱账号已存在 ");
        }
        user.setCreateBy("system");
        boolean flag = userService.insertUser(user);
        return toAjax(flag ? 1 : 0);
}
```

步骤 2　编写业务逻辑检查规则代码

1）在 SysUserService 接口中增加如下所示的抽象方法，以实现步骤 1 中检查用户账号、手机、邮箱的唯一性。

```
/**
  * 校验用户名称是否唯一
  * @param userName 用户名称
  * @return 结果
  */
 String checkUserNameUnique(String userName);
/**
  * 校验用户手机是否唯一
  * @param userInfo 用户信息
  * @return
  */
 String checkPhoneUnique(SysUserVO userInfo);

 /**
  * 校验 Email 是否唯一
  * @param userInfo 用户信息
  * @return
  */
 String checkEmailUnique(SysUserVO userInfo);
```

2）在 SysUserServiceImpl 中增加对应的实现方法，代码如下。

```
@Override
public String checkUserNameUnique(String userName) {
    SysUser sysuser = new SysUser();
    sysuser.setUserName(userName);
    return count(sysuser);
}

@Override
public String checkPhoneUnique(SysUserVO userInfo) {
    SysUser user = new SysUser();
    BeanUtils.copyProperties(userInfo, user);
    return checkUnique(user);
}

@Override
public String checkEmailUnique(SysUserVO userInfo) {
    SysUser user = new SysUser();
    BeanUtils.copyProperties(userInfo, user);
    return checkUnique(user);
}

@Override
public String count(SysUser sysuser) {
    Example<SysUser> example = Example.of(sysuser);
    long count = sysUserRepository.count(example);
    if (count > 0) {
        return UserConstants.NOT_UNIQUE;
    }
    return UserConstants.UNIQUE;
}
```

```
@Override
public String checkUnique(SysUser user) {
    Long userId = StringUtils.isNull(user.getUserId()) ? -1L : user.getUserId();
    Example<SysUser> example = Example.of(user);
    SysUser info = findOne(example);
    if (StringUtils.isNotNull(info) && info.getUserId().longValue() != userId.longValue()) {
        return UserConstants.NOT_UNIQUE;
    }
    return UserConstants.UNIQUE;
}

@Override
public SysUser findOne(Example<SysUser> example) {
    List<SysUser> list = sysUserRepository.findAll(example, PageRequest.of(0, 1)).toList();
    if (list.isEmpty()) {
        return null;
    }
    return list.get(0);
}
```

步骤 3 编写业务层代码

1）修改 SysUserService 接口，实现步骤 1 中的 insertUser 方法。

```
/**
* 新增并保存用户信息
* @param user 用户信息
* @return 结果
*/
boolean insertUser(SysUserVO user);
```

2）在 SysUserServiceImpl 中增加对应的实现方法。

```
@Override
@Transactional
public boolean insertUser(SysUserVO user) {
    SysUser sysUser = new SysUser();
    BeanUtils.copyProperties(user, sysUser);
    sysUser.setDelFlag("0");
    // 新增用户信息
    sysUserRepository.save(sysUser);
    user.setUserId(sysUser.getUserId());
    // 新增用户与角色管理
    insertUserRole(user);
    return null != sysUser.getUserId();
}
/**
 * 新增用户角色信息
 * @param user 用户对象
 */
@Transactional
public int insertUserRole(SysUserVO user) {
    Long[] roles = user.getRoleIds();
```

```
    if (StringUtils.isNull(roles) || roles.length == 0) {
        return 0;
    }
    Long[] userIds = new Long[roles.length];
    Arrays.fill(userIds, user.getUserId());
    return sysUserRepository.batchInsertUserRole(userIds, roles);
}
```

步骤 4　编写数据控制层代码

1）repository 包下新增 custom 子包，新增 SysUserCustomRepository 接口，定义批处理添加用户角色映射的方法。

```
package edu.friday.repository.custom;
import org.springframework.stereotype.Repository;
/**
 * 用户表数据层
 */
@Repository
public interface SysUserCustomRepository {
    int batchInsertUserRole(Long[] userIds, Long[] roles);
}
```

2）在 repository\custom 下新增 impl 子包，新建 SysUserCustomRepositoryImpl 实现类，使用 EntityManager.createNativeQuery() 构建原生 SQL 的查询语句，然后使用 query.executeUpdate() 执行查询语句。

```
package edu.friday.repository.custom.impl;
import edu.friday.repository.custom.SysUserCustomRepository;
import edu.friday.utils.SqlUtil;
import javax.persistence.EntityManager;
import javax.persistence.PersistenceContext;
import javax.persistence.Query;

public class SysUserCustomRepositoryImpl implements SysUserCustomRepository {

    @PersistenceContext
    private EntityManager entityManager;

    @Override
    public int batchInsertUserRole(Long[] userIds, Long[] roles) {
        int length = userIds.length > roles.length ? roles.length : userIds.length;
        StringBuffer sql = new StringBuffer();
        sql.append(" insert into sys_user_role(user_id, role_id) values ");
        sql.append(SqlUtil.getBatchInsertSqlStr(length,2));
        Query query = entityManager.createNativeQuery(sql.toString());
        int paramIndex = 1;
        for (int i = 0; i < length; i++) {
            query.setParameter(paramIndex++, userIds[i]);
            query.setParameter(paramIndex++, roles[i]);
```

```
            }
            return query.executeUpdate();
        }
    }
```

3）修改 SysUserRepository 接口，多继承 SysUserCustomRepository 接口。

```
    @Repository
    public interface SysUserRepository extends JpaRepository<SysUser, Long> ,
SysUserCustomRepository {
    }
```

步骤 5　运行调试

Postman 下测试新增数据，输入网址 http://localhost:8080/system/user，选择“POST”方式，在 body 里选择“raw”→“JSON”，提交如图 4-7 所示的 JSON 内容，单击“Send”按钮，然后查看 sys_user 和 sys_user_role 表是否有新增数据。

POST　http://localhost:8080/system/user　Send

Params　Authorization　Headers (8)　Body　Pre-request Script　Tests　Settings

none　form-data　x-www-form-urlencoded　raw　binary　GraphQL　JSON


```
{
        "createBy": "admin",
        "createTime": "2018-03-16 03:33:00",
        "updateBy": "ry",
        "updateTime": "2020-03-15 09:33:47",
        "remark": "管理员",
        "userId": 120,
        "userName": "julia",
        "nickName": "julia",
        "userType": "00",
        "email": "ry@163.com",
        "phonenumber": "15888888888",
        "sex": "1",
        "avatar": "https://wx.qlogo.cn/mmopen/vi_32/1yzTKJKIfurhOI29RqibEicNOoH0WiaCuKb6jWppVu4uzWovO0d1ICAwuW4rB4zfUxVvGHfNuxXLHu44t3yBkgbicQ/132",
        "password": "$2a$10$ayDiCbzqQkYEh.d1QjWEx.uWeX3mh//OLfFkpVdvGF0fxBElGagcG",
        "status": "0",
        "delFlag": "0",
        "loginIp": "127.0.0.1",
        "loginDate": "2018-03-16T03:33:00.000+00:00",
        "roleIds":[1,2]
}
```


图 4-7　Postman 测试新增数据

知识小结

Spring Data JPA 的常用 API 如下。

- EntityManager.createNativeQuery() 构建原生 SQL 的查询语句。
- Query.executeQuery 方法：这个方法被用来执行 SELECT 语句，它是使用最多的 SQL 语句。
- Query.executeUpdate 方法：用于执行 INSERT、UPDATE 或 DELETE 语句以及 SQL DDL（数据定义语言）语句，如 CREATE TABLE 和 DROP TABLE。

任务评价

技　能　点	知　识　点	自我评价（不熟悉 / 基本掌握 / 熟练掌握 / 灵活运用）
编写原生 SQL 代码	Spring Data JPA 的查询 API	
运用 Spring Data JPA 的 API 编写查询代码		

拓展练习

仿照本任务的操作步骤，实现新增角色数据的接口，见表 4-7。

表 4-7　新增角色数据的接口

接口功能	HTTP 请求方式	URL 地址	支持格式
新增角色数据	POST	/system/role	JSON

完成后重启项目 FridayApplication，在 Postman 中访问网址 http://localhost:8080/system/role，请求方式为 POST，过程如图 4-8 所示，然后在 sys_role 表中检查是否有新增数据。

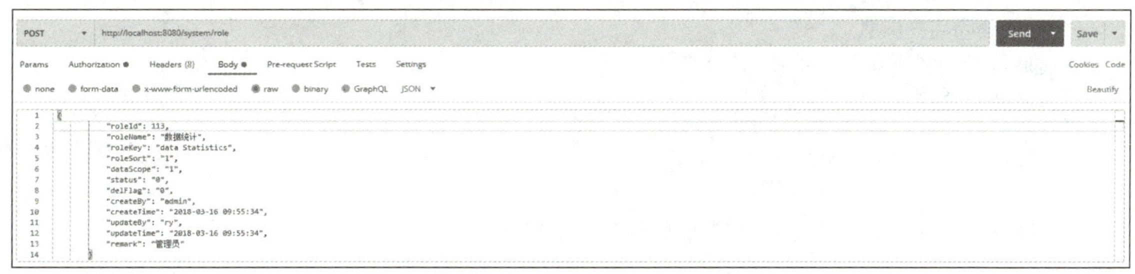

图 4-8　sys_role 表中新增数据

任务 4　实现批处理删除用户接口

微课 4-4　实现批处理删除用户接口

任务分析

通过本任务，完成表 4-8 中删除用户数据的接口功能。

开发人员编写的接口输入参数和输出返回值描述见表 4-8，业务逻辑采用逻辑删除用户，即不真正删除用户数据，而是通过修改用户信息 del_flag=2 标识用户为删除状态，其与角色的关联不删除，当角色查询时会判断用户状态。

表 4-8　删除用户数据的接口功能

URL 地址	HTTP 请求方式	支持格式	输入参数	输出返回值
/system/user/{userIds}	DELETE	JSON	userIds	{"msg": " 操作成功 ", "code": 200}

任务实施

步骤 1　编写控制层代码

在 SysUserController 中添加 remove 方法，其中使用 @PathVariable 将输入参数绑定到请求路径中占位符的值，返回值使用 RestResult 对象返回 msg 和 code 值，业务逻辑调用 deleteUserByIds 方法逻辑删除用户数据。

```
@DeleteMapping("/{userIds}")
public RestResult remove(@PathVariable Long[] userIds) {
    return toAjax(userService.deleteUserByIds(userIds));
}
```

步骤 2 编写业务层代码

1）修改 SysUserService 接口，实现步骤 1 中的 deleteUserByIds 方法。

```
/**
 * 批量删除用户信息
 * @param userIds 需要删除的用户 ID
 * @return 结果
 */
int deleteUserByIds(Long[] userIds);
```

2）在 SysUserServiceImpl 中增加对应的实现方法。

```
@Override
@Transactional
public int deleteUserByIds(Long[] userIds) {
    return sysUserRepository.deleteUserByIds(userIds);
}
```

步骤 3 编写数据控制层代码

修改 SysUserRepository 接口，增加步骤 2 中的 deleteUserByIds 方法，使用 @Modifying 和 @Query 的组合完成数据表的更新功能。

```
@Repository
public interface SysUserRepository extends JpaRepository<SysUser, Long> ,
SysUserCustomRepository {
        @Modifying
        @Query(value = " update sys_user set del_flag = '2' where user_id in :userIds ",
nativeQuery = true)
        int deleteUserByIds(@Param("userIds") Long[] userIds);
}
```

步骤 4 运行调试

Postman 下测试批量删除数据，在删除接口前查询 sys_user 表，如图 4-9 所示，user_id 为 125 和 128 的用户数据的 del_flag 值为 0。

Postman 中输入网址 http://localhost:8080/system/user/125,128，选择“DELETE”方式，调用删除接口返回成功信息如图 4-10 所示。

对象　sys_user @friday (local) - 表

开始事务　文本　筛选　排序　导入　导出

user_id	user_name	nick_name	del_flag	user_type	email	phonenumber	sex	avatar	password	status
1	admin	alex	0	00	alex@163.com	15888888888	1	https://wx.qlogo.cn/mmo	$2a$10$boXFAiZ4OdtZiT2	0
2	元春	元春	0	00	yr@qq.com	15666666666	1	https://wx.qlogo.cn/mmo	$2a$10$LP25Pp2GITfV3Q	0
100	迎春	迎春	2	00	test@qq.com	13192294560	0	https://wx.qlogo.cn/mmo	$2a$10$fLcdPrdYZn9hNh	0
101	探春	探春	0	00	11@qqw.bb	13111111111	0	https://wx.qlogo.cn/mmo	$2a$10$EWDLVRkUmBX2	0
102	惜春	惜春	0	00	22@Qq.mm	13122222222	0	https://wx.qlogo.cn/mmo	$2a$10$bws6QiPyKuqDGj	0
103	秦可卿	秦可卿	2	00	22@qq.com	13222332222	0		$2a$10$bCUZogMaUn7G	0
104	晴雯	晴雯	0	00	test1@qq.com	13315554215	0		123456	0
105	贾政	贾政	0	00	l1_1_1@qq.com	13192294561	1		123456	0
109	贾链	贾链	0	00	asd@q.com	13311122311	(Null)	(Null)	$2a$10$5F3q3LiCkvtMCd.	0
111	冷子兴	冷子兴	0	00	asd@q.com	13311122311	(Null)	(Null)	$2a$10$t66gnYmlYo8jBjd	0
112	贾雨村	贾雨村	0	00	asd@q.com	13311122311	(Null)	(Null)	$2a$10$mW9l7i7mlIS96L	0
113	贾宝玉	贾宝玉	0	00	asdasd@qq.com	13122321123	(Null)	(Null)	$2a$10$/47VtkXAXxw15J	0
114	王熙凤	王熙凤	0	00	asdasda@qq.com	13412311222	(Null)	https://wx.qlogo.cn/mmo	$2a$10$wXRCV6ite.IjeRuh	0
115	妙玉	妙玉	0	00	del_flag@qq.com	15312112311	(Null)	https://wx.qlogo.cn/mmo	$2a$10$5KlITzUxdpmeST	0
124	234	2413213	2	00	2@qq.com	13523423523	(Null)	(Null)	$2a$10$2b23r.tLlZyQP02	0
125	甄士隐	甄士隐	0	00	a@qq.com	13512312312	(Null)	(Null)	$2a$10$AtE16A4t8gbVfB1	0
126	123	asd	2	00	a@qq.com	13512312312	(Null)	(Null)	$2a$10$3gKAwPznJyyMD	0
127	123123	11	2	00	11@qq.com	13512312312	(Null)	(Null)	$2a$10$UAMCS4ZFBq3u1	0
128	alex123123	222	0	00	222@qq.com	13512321312	(Null)	(Null)	$2a$10$boXFAiZ4OdtZiT2	0
134	10112	10113	2	(Null)	11@qq.com	13545679122	0	(Null)		0

图 4-9　sys_user 表中数据

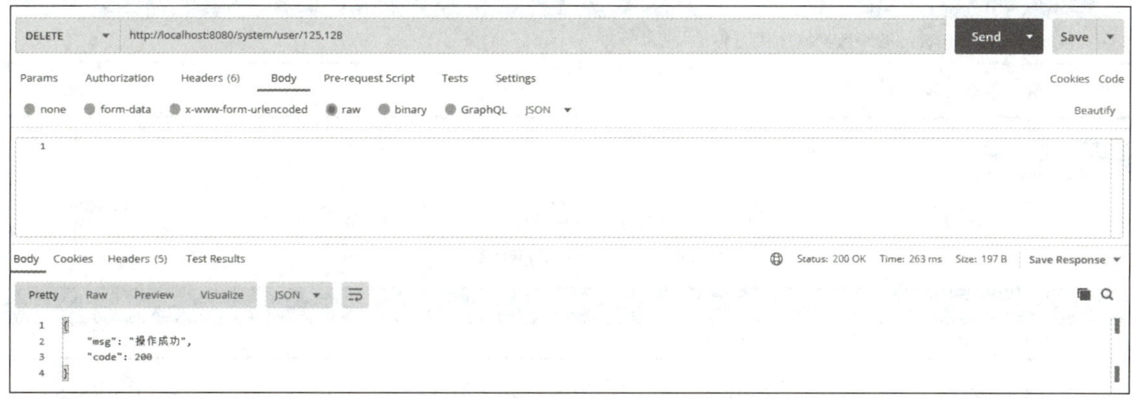

图 4-10　调用删除接口

查看数据表 sys_user 里 user_id 为 125 和 128 的用户数据的 del_flag 值被修改为 2，表示逻辑删除成功，如图 4-11 所示。

对象　sys_user @friday (local) - 表

开始事务　文本　筛选　排序　导入　导出

user_id	user_name	nick_name	del_flag	user_type	email	phonenumber	sex	avatar	password	status
1	admin	alex	0	00	alex@163.com	15888888888	1	https://wx.qlogo.cn/mmo	$2a$10$boXFAiZ4OdtZiT2	0
2	元春	元春	0	00	yr@qq.com	15666666666	1	https://wx.qlogo.cn/mmo	$2a$10$LP25Pp2GITfV3Q	0
100	迎春	迎春	2	00	test@qq.com	13192294560	0	https://wx.qlogo.cn/mmo	$2a$10$fLcdPrdYZn9hNh	0
101	探春	探春	0	00	11@qqw.bb	13111111111	0	https://wx.qlogo.cn/mmo	$2a$10$EWDLVRkUmBX2	0
102	惜春	惜春	0	00	22@Qq.mm	13122222222	0	https://wx.qlogo.cn/mmo	$2a$10$bws6QiPyKuqDGj	0
103	秦可卿	秦可卿	2	00	22@qq.com	13222332222	0		$2a$10$bCUZogMaUn7G	0
104	晴雯	晴雯	0	00	test1@qq.com	13315554215	0		123456	0
105	贾政	贾政	0	00	l1_1_1@qq.com	13192294561	1		123456	0
109	贾链	贾链	0	00	asd@q.com	13311122311	(Null)	(Null)	$2a$10$5F3q3LiCkvtMCd.	0
111	冷子兴	冷子兴	0	00	asd@q.com	13311122311	(Null)	(Null)	$2a$10$t66gnYmlYo8jBjd	0
112	贾雨村	贾雨村	0	00	asd@q.com	13311122311	(Null)	(Null)	$2a$10$mW9l7i7mlIS96L	0
113	贾宝玉	贾宝玉	0	00	asdasd@qq.com	13122321123	(Null)	(Null)	$2a$10$/47VtkXAXxw15J	0
114	王熙凤	王熙凤	0	00	asdasda@qq.com	13412311222	(Null)	https://wx.qlogo.cn/mmo	$2a$10$wXRCV6ite.IjeRuh	0
115	妙玉	妙玉	0	00	del_flag@qq.com	15312112311	(Null)	https://wx.qlogo.cn/mmo	$2a$10$5KlITzUxdpmeST	0
124	234	2413213	2	00	2@qq.com	13523423523	(Null)	(Null)	$2a$10$2b23r.tLlZyQP02	0
125	甄士隐	甄士隐	2	00	a@qq.com	13512312312	(Null)	(Null)	$2a$10$AtE16A4t8gbVfB1	0
126	123	asd	2	00	a@qq.com	13512312312	(Null)	(Null)	$2a$10$3gKAwPznJyyMD	0
127	123123	11	2	00	11@qq.com	13512312312	(Null)	(Null)	$2a$10$UAMCS4ZFBq3u1	0
128	alex123123	222	2	00	222@qq.com	13512321312	(Null)	(Null)	$2a$10$boXFAiZ4OdtZiT2	0
134	10112	10113	2	(Null)	11@qq.com	13545679122	0	(Null)		0

图 4-11　del_flag 值为 2

知识小结

1. 物理删除和逻辑删除的区别

➢ 物理删除：直接删除数据，执行 DELETE 操作。

➢ 逻辑删除：通过更新数据标识位改变数据的状态，执行 UPDATE 操作，update del_flag = 2 标识为删除状态。

2. Spring Data JPA 的查询注解

➢ @Query：步骤 3 中使用编码 Query 对象的方式进行增删改查的操作，Spring Data JPA 也提供了注解的方式完成增删改查的操作。

➢ @Modifying：如果是删除或修改操作，需要加入 @Modifying 注解进行修饰，表示 UPDATE 或 DELETE 操作。

任务评价

技能点	知识点	自我评价（不熟悉 / 基本掌握 / 熟练掌握 / 灵活运用）
运用 @Query+@Modifying 编写数据操作代码	逻辑删除原理 Spring Data JPA 的数据操作注解	
完成数据的逻辑删除功能		

拓展练习

仿照本任务的操作步骤实现角色数据接口的删除，删除角色数据接口描述见表 4-9。

表 4-9　删除角色数据接口描述

接口功能	HTTP 请求方式	URL 地址	支持格式
删除角色数据	DELETE	/system/role/{roleIds}	JSON

完成后重启项目 FridayApplication，在 Postman 中访问网址 http://localhost:8080/system/role/110，结果如图 4-12 所示。

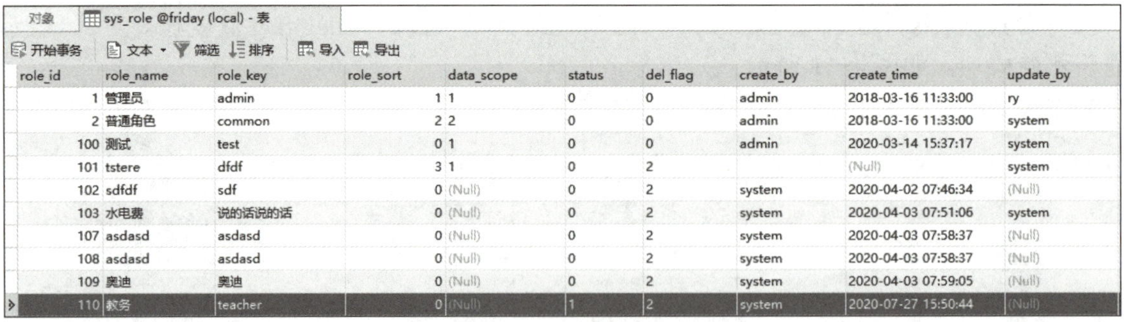

对象　sys_role @friday (local) - 表

开始事务　文本　筛选　排序　导入　导出

role_id	role_name	role_key	role_sort	data_scope	status	del_flag	create_by	create_time	update_by
1	管理员	admin	1	1	0	0	admin	2018-03-16 11:33:00	ry
2	普通角色	common	2	2	0	0	admin	2018-03-16 11:33:00	system
100	测试	test	0	1	0	0	admin	2020-03-14 15:37:17	system
101	tstere	dfdf	3	1	0	2		(Null)	system
102	sdfdf	sdf	0	(Null)	0	2	system	2020-04-02 07:46:34	(Null)
103	水电费	说的话说的话	0	(Null)	0	2	system	2020-04-03 07:51:06	system
107	asdasd	asdasd	0	(Null)	0	2	system	2020-04-03 07:58:37	(Null)
108	asdasd	asdasd	0	(Null)	0	2	system	2020-04-03 07:58:37	(Null)
109	奥迪	奥迪	0	(Null)	0	2	system	2020-04-03 07:59:05	(Null)
110	教务	teacher	0	(Null)	1	2	system	2020-07-27 15:50:44	(Null)

图 4-12　sys_role 表中删除数据

任务 5　实现修改用户接口

微课 4-5　实现修改用户接口

任务分析

通过本任务，完成表 4-1 中修改用户数据的接口功能。

开发人员编写的接口输入参数和输出返回值描述如下，业务逻辑由于 sys_user 表和 sys_user_role 表的级联关系，采用先删除 sys_user_role 表的关联数据，然后更新 sys_user 表数据，最后保存最新的用户角色映射关系到 sys_user_role 表，实现逻辑，见表 4-10。

表 4-10　修改用户数据接口功能描述

URL 地址	HTTP 请求方式	支 持 格 式	输 入 参 数	输出返回值
/system/user	PUT	JSON	SysUserVO	{"msg": " 操作成功 ", "code": 200}

任务实施

步骤 1　编写控制层代码

在 SysUserController 中添加 edit 方法，其中使用 @Validated 配合 SysUserVO 的校验注解验证数据格式，返回值使用 RestResult 对象返回 msg 和 code 值，业务逻辑调用 updateUser 方法更新用户数据和用户角色映射数据。

```
/**
 * 修改用户
 */
@PutMapping
public RestResult edit(@Validated @RequestBody SysUserVO user) {
    if (UserConstants.NOT_UNIQUE.equals(userService.checkPhoneUnique(user))) {
        return RestResult.error(" 修改用户 " + user.getUserName() + " 失败，手机号码已存在 ");
    } else if (UserConstants.NOT_UNIQUE.equals(userService.checkEmailUnique(user))) {
        return RestResult.error(" 修改用户 " + user.getUserName() + " 失败，邮箱账号已存在 ");
    }
    user.setUpdateBy("system");
    boolean flag = userService.updateUser(user);
    return toAjax(flag ? 1 : 0);
}
```

步骤 2　编写业务层代码

1）修改 SysUserService 接口，实现步骤 1 中的 updateUser 方法。

```
/**
 * 修改并保存用户信息
 * @param user 用户信息
 * @return 结果
 */
boolean updateUser(SysUserVO user);
```

2）在 SysUserServiceImpl 中增加对应的实现方法。

```
@Override
@Transactional
public boolean updateUser(SysUserVO user) {
```

```
        Long userId = user.getUserId();
        Optional<SysUser> op = sysUserRepository.findById(userId);
        if (!op.isPresent()) {
            return false;
        }
        // 删除用户与角色关联
        sysUserRepository.deleteUserRoleByUserId(userId);
        SysUser sysUser = op.get();
        BeanUtils.copyPropertiesIgnoreNull(user, sysUser);
        // 更新用户信息
        sysUserRepository.save(sysUser);
        // 新增用户与角色关联
        insertUserRole(user);
        return null != sysUser.getUserId();
    }
```

步骤 3 编写数据控制层代码

修改 SysUserRepository 接口，增加步骤 2 中的 deleteUserRoleByUserId 方法。

```
@Modifying
@Query(value = " delete from sys_user_role where user_id=:userId ", nativeQuery = true)
int deleteUserRoleByUserId(@Param("userId")Long userId);
```

步骤 4 运行调试

Postman 下测试更新用户数据，在更新接口前查询 sys_user 表，如图 4-13 所示，user_id 为 128 的用户邮箱为 222@qq.com。

对象 | sys_user @friday (local) - 表

开始事务 文本 筛选 排序 导入 导出

user_id	user_name	nick_name	user_type	email	phonenumber	sex	avatar	password	status	del_flag
1	admin	alex	00	alex@163.com	15888888888	1	https://wx.qlogo.cn/mmo	$2a$10$boXFAiZ4OdtZiT2	0	0
2	元春	元春	00	yr@qq.com	15666666666	1	https://wx.qlogo.cn/mmo	$2a$10$LP25Pp2GITfV3Q	0	0
100	迎春	迎春	00	test@qq.com	13192294560	0	https://wx.qlogo.cn/mmo	$2a$10$fLcdPrdYZn9hNh	0	2
101	探春	探春	00	11@qqw.bb	13111111111	0	https://wx.qlogo.cn/mmo	$2a$10$EWDLVRkUmBX2	0	0
102	惜春	惜春	00	22@Qq.mm	13122222222	0	https://wx.qlogo.cn/mmo	$2a$10$bws6QiPyKuqDGj	0	0
103	秦可卿	秦可卿	00	22@qq.com	13222332222	0		$2a$10$bCUZogMaUn7G	0	2
104	晴雯	晴雯	00	test1@qq.com	13315554215	0		123456	0	0
105	贾政	贾政	00	l1_1_1@qq.com	13192294561	1		123456	0	0
109	贾链	贾链	00	asd@q.com	13311122311	(Null)	(Null)	$2a$10$5F3q3LiCkvtMCd.	0	0
111	冷子兴	冷子兴	00	asd@q.com	13311122311	(Null)	(Null)	$2a$10$t66gnYmlYo8jBjd	0	0
112	贾雨村	贾雨村	00	asd@q.com	13311122311	(Null)	(Null)	$2a$10$mW9l7i7mlIS96L	0	0
113	贾宝玉	贾宝玉	00	asdasd@qq.com	13122321123	(Null)	(Null)	$2a$10$/47VtkXAXxw15J	0	0
114	王熙凤	王熙凤	00	asdasda@qq.com	13412311222	(Null)	https://wx.qlogo.cn/mmo	$2a$10$wXRCV6ite.ljeRu	0	0
115	妙玉	妙玉	00	del_flag@qq.com	15312112311	(Null)	https://wx.qlogo.cn/mmo	$2a$10$5KlITzUxdpmeST	0	0
124	234	2413213	00	2@qq.com	13523423523	(Null)	(Null)	$2a$10$2b23r.tLlZyQP02	0	2
125	甄士隐	甄士隐	00	a@qq.com	13512312312	(Null)	(Null)	$2a$10$AtE16A4t8gbVfB1	0	2
126	123	asd	00	a@qq.com	13512312312	(Null)	(Null)	$2a$10$3gKAwPznJyyMD.	0	2
127	123123	11	00	11@qq.com	13512312312	(Null)	(Null)	$2a$10$UAMCS4ZFBq3u1	0	2
128	alex123123	222	00	222@qq.com	13512321312	(Null)	(Null)	$2a$10$boXFAiZ4OdtZiT2	0	2
134	10112	10113	(Null)	11@qq.com	13545679122	0	(Null)		0	2

图 4-13 sys_user 表的数据

Postman 中输入网址 http://localhost:8080/system/user，选择“PUT”方式，下面选择“Body”→“raw”→“JSON”，放入 JSON 数据，然后单击“Send”按钮，返回成功信息，如图 4-14 所示。

图 4-14　调用更新接口

查看数据表 sys_user 里 user_id 为 128 的用户的 email 值被修改为 alex@qq.com，如图 4-15 所示。

对象 | sys_user @friday (local) - 表

开始事务　文本　筛选　排序　导入　导出

user_id	user_name	nick_name	user_type	email	phonenumber	sex	avatar	password	status	del_flag
1	admin	alex	00	alex@163.com	15888888888	1	https://wx.qlogo.cn/mmo	$2a$10$boXFAiZ4OdtZiT2	0	0
2	元春	元春	00	yr@qq.com	15666666666	1	https://wx.qlogo.cn/mmo	$2a$10$LP25Pp2GITfV3Q	0	0
100	迎春	迎春	00	test@qq.com	13192294560	0	https://wx.qlogo.cn/mmo	$2a$10$fLcdPrdYZn9hNh	0	2
101	探春	探春	00	11@qqw.bb	13111111111	0	https://wx.qlogo.cn/mmo	$2a$10$EWDLVRkUmBX2	0	0
102	惜春	惜春	00	22@Qq.mm	13122222222	0	https://wx.qlogo.cn/mmo	$2a$10$bws6QiPyKuqDGj	0	0
103	秦可卿	秦可卿	00	22@qq.com	13222332222	0		$2a$10$bCUZogMaUn7G	0	2
104	晴雯	晴雯	00	test1@qq.com	13315554215	0		123456	0	0
105	贾政	贾政	00	l1_1_1@qq.com	13192294561	1		123456	0	0
109	贾链	贾链	00	asd@q.com	13311122311	(Null)	(Null)	$2a$10$5F3q3LiCkvtMCd.	0	0
111	冷子兴	冷子兴	00	asd@q.com	13311122311	(Null)	(Null)	$2a$10$t66gnYmlYo8jBjd	0	0
112	贾雨村	贾雨村	00	asd@q.com	13311122311	(Null)	(Null)	$2a$10$mW9l7i7mlIS96L	0	0
113	贾宝玉	贾宝玉	00	asdasd@qq.com	13122321123	(Null)	(Null)	$2a$10$/47VtkXAXxw15J	0	0
114	王熙风	王熙风	00	asdasda@qq.com	13412311222	(Null)	https://wx.qlogo.cn/mmo	$2a$10$wXRCV6ite.ljeRu	0	0
115	妙玉	妙玉	00	del_flag@qq.com	15312112311	(Null)	https://wx.qlogo.cn/mmo	$2a$10$5KlITzUxdpmeST	0	0
124	234	2413213	00	2@qq.com	13523423523	(Null)	(Null)	$2a$10$2b23r.tLlZyQP02	0	2
125	甄士隐	甄士隐	00	a@qq.com	13512312312	(Null)	(Null)	$2a$10$AtE16A4t8gbVfB1	0	2
126	123	asd	00	a@qq.com	13512312312	(Null)	(Null)	$2a$10$3gKAwPznJyyMD	0	2
127	123123	11	00	11@qq.com	13512312312	(Null)	(Null)	$2a$10$UAMCS4ZFBq3u1	0	2
128	Alex666	Alex666	00	alex@qq.com	15888888888	(Null)	(Null)	$2a$10$boXFAiZ4OdtZiT2	0	2
134	10112	10113	(Null)	11@qq.com	13545679122	0	(Null)		0	2

图 4-15　email 值为 alex@qq.com

任务评价

技　能　点	知　识　点	自我评价（不熟悉 / 基本掌握 / 熟练掌握 / 灵活运用）
完成数据的修改功能	MySQL 外键与级联操作	

拓展练习

仿照本任务的操作步骤，实现角色数据接口的修改，业务逻辑由于 sys_role 表和 sys_role_menu 表的级联关系，采用先删除 sys_role_menu 表的关联数据，然后更新 sys_role 表数据，最后保存最新的角色菜单映射关系到 sys_role_menu 表，实现逻辑，见表 4-11。

表 4-11　修改角色数据的接口描述

接 口 功 能	HTTP 请求方式	URL 地址	支 持 格 式
修改角色数据	PUT	/system/role	JSON

完成后重启项目 FridayApplication，在 Postman 中访问网址 http://localhost:8080/system/role，请求方式为 PUT，运行结果如图 4-16 所示。

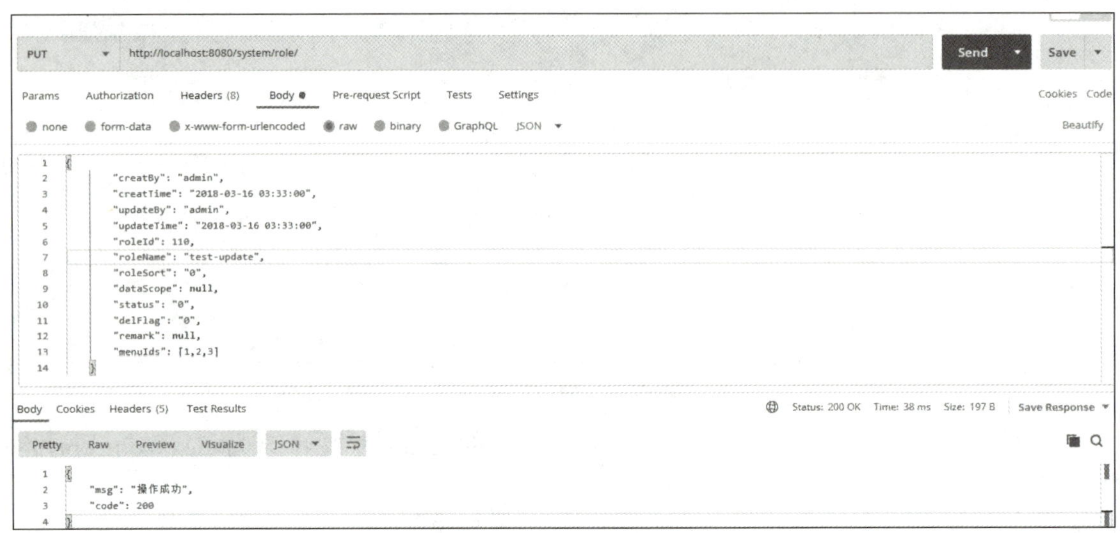

图 4-16　调用修改角色数据的接口

查看数据表 sys_role 里 role_id 为 110 的 role_name 值被修改为 test-update，如图 4-17 所示。

role_id	role_name	role_key	role_sort	data_scope	status	del_flag	create_by	create_time	update_by	update_time
1	管理员	admin	1	1	0	0	admin	2018-03-16 11:33:00	ry	2018-03-16 11:33:00
2	普通角色	common	2	2	0	0	admin	2018-03-16 11:33:00	system	2020-04-02 07:58:06
100	测试	test	0	1	0	0	admin	2020-03-14 15:37:17	system	2020-04-02 07:58:00
101	tstere	dfdf	3	1	0	2		(Null)	system	2020-04-07 00:44:49
102	sdfdf	sdf	0	(Null)	0	2	system	2020-04-02 07:46:34	(Null)	(Null)
103	水电费	说的话说的话	0	(Null)	0	2	system	2020-04-03 07:51:06	system	2020-04-03 07:54:39
107	asdasd	asdasd	0	(Null)	0	2	system	2020-04-03 07:58:37	(Null)	(Null)
108	asdasd	asdasd	0	(Null)	0	2	system	2020-04-03 07:58:37	(Null)	(Null)
109	奥迪	奥迪	0	(Null)	0	2	system	2020-04-03 07:59:05	(Null)	(Null)
110	test-update	teacher	0	(Null)	0	0	system	2020-07-27 15:50:44	system	(Null)

图 4-17　role_name 值被修改为 test-update

查看数据表 sys_role_menu 里新增了 3 条 role_id 为 110 的数据，如图 4-18 所示。

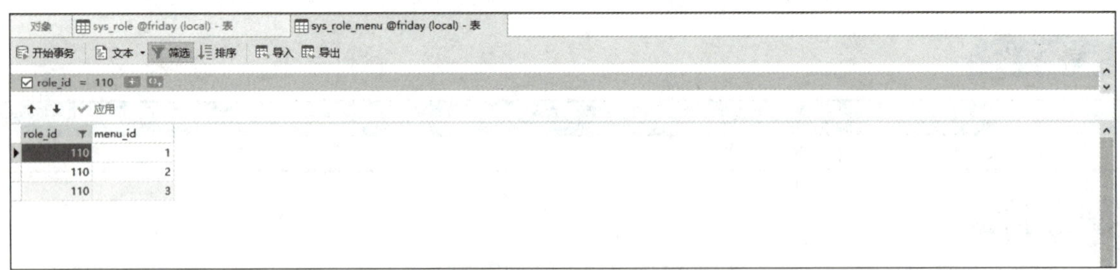

role_id	menu_id
110	1
110	2
110	3

图 4-18　sys_role_menu 表数据

任务 6　实现获取用户信息与角色列表接口

微课 4-6　实现获取用户信息与角色列表接口

任务分析

通过本任务，完成实现表 4-12 中的获取用户信息与角色列表接口功能。

开发人员编写的接口输入参数和输出返回值描述如下，此接口主要用于在新增用户时返回角色列表数据 roles，在修改用户时返回当前用户数据 data 和用户角色对应关系列表 roleIds，输出返回值见表 4-12。

表 4-12　获取用户信息与角色列表接口功能详情描述

URL 地址	HTTP 请求方式	支持格式	输入参数	输出返回值

/system/user/{userId}, /system/user/	GET	JSON	userId	{
				"msg": " 操作成功 ",
				"code": 200,
				"roleIds": [
				1
				],
				"data": {
				"createBy": "admin",
				"createTime": "2018-03-16 03:33:00",
				"updateBy": "ry",
				"updateTime": "2020-03-15 09:33:47",
				"remark": " 管理员 ",
				"userId": 1,
				"userName": "admin",
				"nickName": "alex",
				"userType": "00",
				"email": "alex@163.com",
				"phonenumber": "15888888888",
				"sex": "1",
				"avatar": "https://wx.qlogo.cn/mmopen/vi_32/1yzTKJKIfurhDI29RqibEicNOoH0WiaCuKb6jWppVu4uzWovO0d1ICAwuW4rB4zfUxVvGHfNuxXLHu44t3yBkgbicQ/132",
				"password": "$2a$10$boXFAiZ4OdtZiT2owx.xx.F848I4rh4JCQQDvgAaEwiktcFRh8Ile",
				"status": "0",
				"delFlag": "0",
				"loginIp": "127.0.0.1",
				"loginDate": "2018-03-16T03:33:00.000+00:00"
				},
				"roles": [
				{
				"roleId": 1,
				"roleName": " 管理员 ",
				"roleKey": "admin",
				"roleSort": 1,
				"dataScope": "1",
				"status": "0",
				"delFlag": "0",
				"createBy": "admin",
				"createTime": "2018-03-16T03:33:00.000+00:00",
				"updateBy": "ry",
				"updateTime": "2018-03-16T03:33:00.000+00:00",
				"remark": " 管理员 "
				}
				]
				}

任务实施

步骤 1 编写控制层代码

在 SysUserController 类中增加 getInfo 方法，业务逻辑调用 roleService.selectRoleAll 方法返回角色列表数据，调用 userService.selectUserById 方法返回用户信息数据，调用 roleService.selectRoleListByUserId 方法返回用户角色映射关系列表。

```
@Autowired
private SysRoleService roleService;

/**
 * 根据用户编号获取详细信息
 */
@GetMapping(value = {"/", "/{userId}"})
public RestResult getInfo(@PathVariable(value = "userId", required = false) Long userId) {
    RestResult ajax = RestResult.success();
    ajax.put("roles", roleService.selectRoleAll());
    if (StringUtils.isNotNull(userId)) {
        ajax.put(RestResult.DATA_TAG, userService.selectUserById(userId));
        ajax.put("roleIds", roleService.selectRoleListByUserId(userId));
    }
    return ajax;
}
```

步骤 2 编写用户业务层代码

1）修改 SysUserService 接口，实现步骤 1 中的 selectUserById 方法。

```
/**
 * 通过用户 ID 查询用户
 * @param userId 用户 ID
 * @return 用户对象信息
 */
SysUser selectUserById(Long userId);
```

2）在 SysUserServiceImpl 中增加对应的实现方法。

```
@Override
public SysUser selectUserById(Long userId) {
    SysUser sysuser = new SysUser();
    sysuser.setUserId(userId);
    sysuser.setDelFlag("0");
    return sysUserRepository.findOne(Example.of(sysuser)).get();
}
```

步骤 3 编写角色业务层代码

1）在 service 包下新增 SysRoleService 接口，实现步骤 1 中的 selectRoleAll 和 selectRoleListByUserId 方法。

```
package edu.friday.service;
import edu.friday.model.SysRole;
import java.util.List;
/**
 * 角色业务层接口
 */
public interface SysRoleService {
    List<SysRole> selectRoleAll();
    List<Long> selectRoleListByUserId(Long userId);
}
```

2）在 impl 子包下新增 SysRoleServiceImpl 实现类，SysRoleServiceImpl 中增加对应的实现方法。

```
package edu.friday.service.impl;
import edu.friday.model.SysRole;
import edu.friday.repository.SysRoleRepository;
import edu.friday.service.SysRoleService;
import org.springframework.beans.factory.annotation.Autowired;
import org.springframework.stereotype.Service;
import java.util.List;

/**
 * 角色表业务层
 */
@Service
public class SysRoleServiceImpl implements SysRoleService {
    @Autowired
    SysRoleRepository sysRoleRepository;

    /**
     * 查询所有角色
     * @return 角色列表
     */
    @Override
    public List<SysRole> selectRoleAll() {
        return sysRoleRepository.findAll();
    }
    /**
     * 根据用户 ID 获取角色选择框列表
     * @param userId 用户 ID
     * @return 选中角色 ID 列表
     */
    @Override
    public List<Long> selectRoleListByUserId(Long userId) {
        return sysRoleRepository.selectRoleIdsByUserId(userId);
    }
}
```

步骤 4　编写角色数据控制层代码

在 repository 包下新增 SysRoleRepository 接口，增加步骤 3 中的 selectRoleIdsByUserId 方法。

```
package edu.friday.repository;
import edu.friday.model.SysRole;
import org.springframework.data.jpa.repository.JpaRepository;
import org.springframework.data.jpa.repository.Query;
import org.springframework.data.repository.query.Param;
import org.springframework.stereotype.Repository;
import java.util.List;
/**
 * 角色表数据层
 */
@Repository
public interface SysRoleRepository extends JpaRepository<SysRole, Long> {
    final String JOIN_USER_ROLE = " left join sys_user_role ur on ur.role_id = r.role_id ";
    final String JOIN_USER = "left join sys_user u on u.user_id = ur.user_id ";

    @Query(value = " select r.role_id from sys_role r " + JOIN_USER_ROLE + JOIN_USER +
            "WHERE r.del_flag = '0' and u.user_id = :userId ", nativeQuery = true)
    List<Long> selectRoleIdsByUserId(@Param("userId") Long userId);
}
```

步骤 5 运行调试

Postman 中测试获取角色列表的 API 接口，选择“GET”模式，在浏览器 URL 地址栏中输入 http://localhost:8080/system/user/，单击“Send”按钮，如图 4-19 所示，返回角色列表数据。

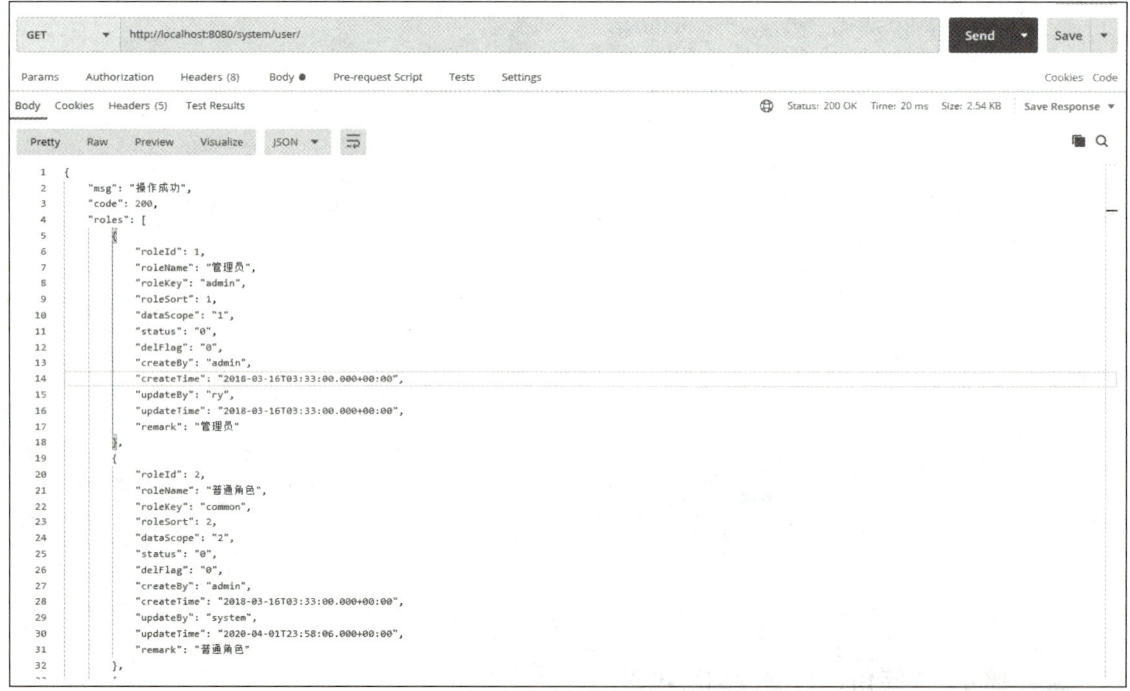

图 4-19 角色列表数据

Postman 中测试获取用户信息和用户角色对应管理列表的 API 接口，选择“GET”模式，在浏览器 URL 地址栏中输入 http://localhost:8080/system/user/1，单击“Send”按钮，如图 4-20 所示，返回当前用户信息和角色列表数据。

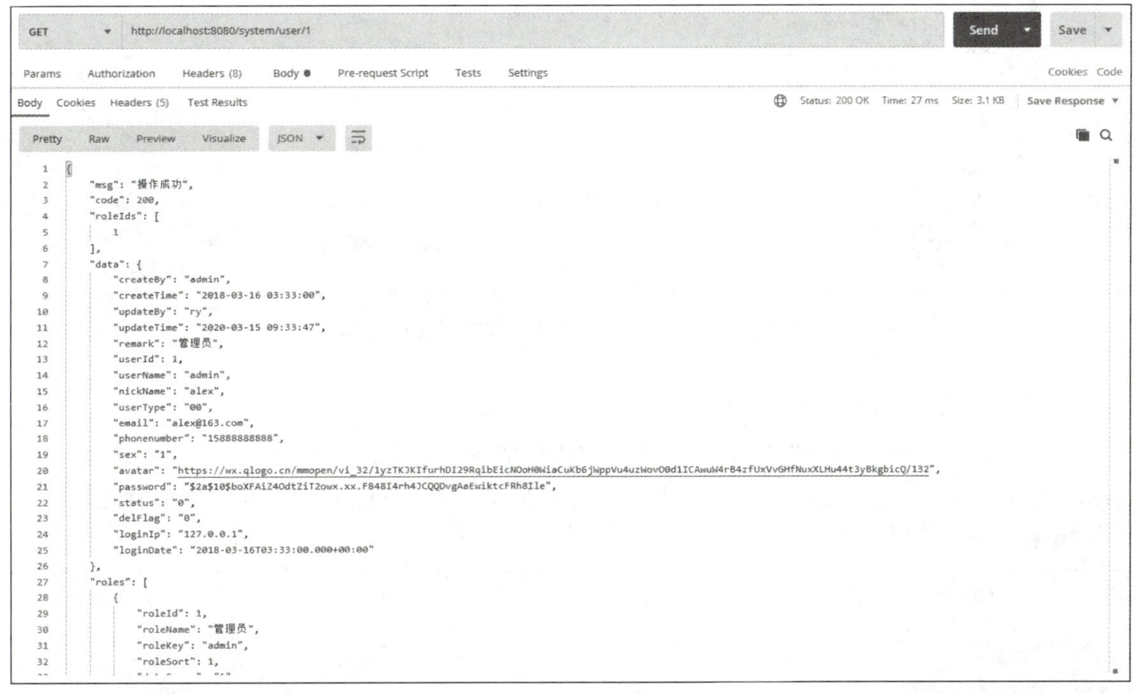

图 4-20　用户信息以及角色列表数据

任务评价

技　能　点	知　识　点	自我评价（不熟悉 / 基本掌握 / 熟练掌握 / 灵活运用）
完成数据的多表关联查询功能	数据库左外连接	

拓展练习

仿照本任务的操作步骤，实现表 4-13 中的接口功能。

表 4-13　获取角色编号和修改角色状态接口描述

接 口 功 能	HTTP 请求方式	URL 地址	支 持 格 式
根据角色编号获取详细信息	GET	/system/role/{roleId}	JSON
修改角色状态	PUT	/system/role/changeStatus	JSON

1. 根据角色编号获取角色详细信息

Postman 中选择“GET”模式，在浏览器 URL 地址栏中输入 http://localhost:8080/system/role/1，单击“Send”按钮，运行结果如图 4-21 所示。

```
GET  http://localhost:8080/system/role/1
{
    "msg": "操作成功",
    "code": 200,
    "data": {
        "roleId": 1,
        "roleName": "管理员",
        "roleKey": "admin",
        "roleSort": "1",
        "dataScope": "1",
        "status": "0",
        "delFlag": "0",
        "createBy": "admin",
        "createTime": "2018-03-16T03:33:00.000+00:00",
        "updateBy": "ry",
        "updateTime": "2018-03-16T03:33:00.000+00:00",
        "remark": "管理员"
    }
}
```

图 4-21　调用角色编号获取详细信息的接口

2. 修改角色状态

Postman 中选择“PUT”模式，浏览器 URL 地址栏中输入 http://localhost:8080/system/role/changeStatus，单击“Send”按钮，更新 sys_role 数据表的 status 字段值，值由 0 更新为 1，运行结果如图 4-22 所示。

```
PUT  http://localhost:8080/system/role/changeStatus
{
    "roleId": 110,
    "roleName": "管理员",
    "roleKey": "admin",
    "roleSort": "0",
    "dataScope": "1",
    "status": "1",
    "delFlag": "0",
    "remark": "管理员"
}

{
    "msg": "操作成功",
    "code": 200
}
```

图 4-22　修改角色状态

单元小结

在后端接口开发中，对数据表的增、删、改、查操作是后端开发的基础能力。能够基于三层架构使用 Spring 生态框架完成对数据的操作，以及使用 Postman 工具验证接口功能是 1+X 职业技能等级证书（Java 应用开发）的职业技能要求。通过完成本工作单元的任务和后面的实战强化，熟练掌握开发增、删、改、查功能的后端接口的职业技能。

实战强化

按照本工作单元步骤，编写实训项目“诚品书城”的产品展示、购物车后端接口。

工作单元 5

实现登录认证和鉴权

职业能力

本工作单元主要实现用户的登录认证和鉴权功能，最终希望学生达成如下职业能力目标：

1. 掌握使用 Spring Security 实现登录认证的流程
2. 掌握使用 Spring Security 鉴别用户是否具有控制访问权限
3. 掌握使用 JWT 实现无状态、分布式的 Web 应用授权

任务情景

企业应用系统对安全性有较高的要求，因此业务系统必备的安全性功能往往就是对用户的认证和鉴权。

- 认证：指验证某个用户是否为系统中的合法主体，也就是说用户能否访问该系统。
- 鉴权：指验证某个用户是否有权限执行某个操作，如果未授权则不能访问该系统。

为解决对用户的认证和鉴权问题，可引用 Spring Security 框架作为权限管理系统的安全访问控制解决方案。Spring Security 框架为基于 Spring 的企业应用系统提供声明式的安全访问控制功能，其在进行用户认证以及授予权限的时候，通过各种各样的拦截器来控制权限的访问，从而实现安全控制，同时减少了为系统安全而编写大量重复代码的工作，图 5-1 为登录认证和访问授权的流程。

在图 5-1 中，登录认证与鉴权的过程中将面临下面 3 个问题：

1）如何配置管理系统的权限？

2）如何实现用户的登录认证以及授权功能？

3）如何实现未授权的访问控制，即系统的鉴权功能？

基于上述 3 个问题，本工作单元的具体任务如下：

1）实现基于 Spring Security 框架的权限控制功能。

2）实现基于 JWT 的登录认证功能。

3）实现基于 JWT 的访问鉴权功能。

4）用户登录认证通过后获取登录用户授权信息的接口功能。

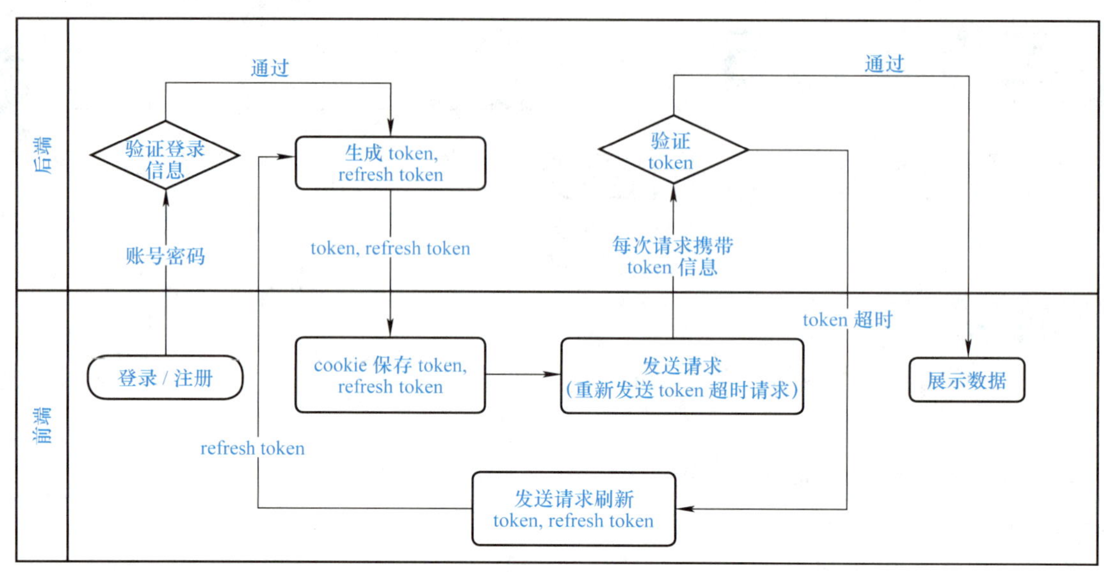

图 5-1　登录认证和访问授权的流程

登录功能的后端接口详情见表 5-1。

表 5-1　登录功能的后端接口详情

接 口 功 能	HTTP 请求方式	URL 地址	支持格式
登录验证	POST	/login	JSON
获取登录用户的授权信息	GET	/getInfo	JSON

前置知识

微课 5-1　实现基于 Spring Security 的权限控制功能

任务 1　实现基于 Spring Security 的权限控制功能

任务分析

在权限管理系统中，应用系统需要对资源进行配置，决定了哪些资源无须授权可以直接访问，哪些资源需要授权才可以访问。本任务结合 Spring Security 框架的 WebSecurityConfigurerAdapter 配置类，来配置资源访问权限和验证用户权限信息，实现系统的权限控制功能。

任务实施

步骤 1　导入 Spring Security 依赖

在 pom.xml 中引入 Spring Security 和 fastjson 的依赖。

```
<dependencies>
    ...
    <dependency>
        <groupId>org.springframework.boot</groupId>
        <artifactId>spring-boot-starter-security</artifactId>
    </dependency>
    <dependency>
        <groupId>com.alibaba</groupId>
        <artifactId>fastjson</artifactId>
        <version>1.2.56</version>
    </dependency>
</dependencies>
```

步骤 2　使用 Spring Security 配置访问权限

在包“edu.friday.config”下创建一个继承 WebSecurityConfigurerAdapter 抽象类的 SecurityConfig 类，重写 configure（HttpSecurity http）方法，用来配置资源访问权限和验证用户权限信息，本任务中关于 Spring Security 框架的相关知识请参考知识小结。

```
/**
 * 配置 Spring Security
 */
@EnableGlobalMethodSecurity(prePostEnabled = true, securedEnabled = true)
public class SecurityConfig extends WebSecurityConfigurerAdapter {
    /**
     * 自定义用户认证逻辑
     */
    @Autowired
    private UserDetailsService userDetailsService;
    /**
     * 认证失败处理类
     */
    @Autowired
    private AuthenticationEntryPointImpl unauthorizedHandler;
    /**
     * 解决 无法直接注入 AuthenticationManager
     * @return
     * @throws Exception
     */
    @Bean
    @Override
    public AuthenticationManager authenticationManagerBean() throws Exception {
        return super.authenticationManagerBean();
    }
```

```
    @Override
    protected void configure(HttpSecurity httpSecurity) throws Exception {
        httpSecurity
                // CSRF 禁用，因为不使用 session
                .csrf().disable()
                // 认证失败处理类
                .exceptionHandling().authenticationEntryPoint(unauthorizedHandler).and()
                // 基于 token，所以不需要 session
                .sessionManagement().sessionCreationPolicy(SessionCreationPolicy.STATELESS).and()
                // 过滤请求
                .authorizeRequests()
                // 对于登录，允许匿名访问
                .antMatchers("/login" , "/profile/avatar/**").anonymous()
                .antMatchers(
                        HttpMethod.GET,
                        "/*.html" ,
                        "/**/*.html" ,
                        "/**/*.css" ,
                        "/**/*.js"
                ).permitAll()
                .antMatchers("/profile/**").anonymous()
                // 除了上面之外的所有请求都需要鉴权认证
                .anyRequest().authenticated()
                .and()
                .headers().frameOptions().disable();
    }
    /**
     * 强散列哈希加密实现
     */
    @Bean
    public BCryptPasswordEncoder bCryptPasswordEncoder() {
        return new BCryptPasswordEncoder();
    }
    /**
     * 身份认证接口
     */
    @Override
    protected void configure(AuthenticationManagerBuilder auth) throws Exception {
        auth.userDetailsService(userDetailsService).passwordEncoder(bCryptPasswordEncoder());
    }
}
```

步骤 3 创建认证失败处理类

在包“edu.friday.common.security.handle”下创建 AuthenticationEntryPointImpl 认证失败处理类，用来解决匿名用户访问无权限资源时的异常，该类必须要实现 AuthenticationEntryPoint 接口，该接口的作用是当用户请求了一个受保护的系统资源时，用户没有通过认证则抛出异常，AuthenticationEntryPoint.Commence() 方法就会被调用返回认证失败的消息。

```
/**
 * 认证失败处理类
 */
@Component
public class AuthenticationEntryPointImpl implements AuthenticationEntryPoint, Serializable {
    private static final long serialVersionUID = -8970718410437077606L;

    @Override
    public void commence(HttpServletRequest request, HttpServletResponse response,
AuthenticationException e)
            throws IOException {
        int code = HttpStatus.UNAUTHORIZED;
        String msg = StringUtils.format(" 请求访问：{}，认证失败，无法访问系统资源 ", request.
getRequestURI());
        ServletUtils.renderString(response, JSON.toJSONString(RestResult.error(code, msg)));
    }
}
```

步骤 4　实现获取用户和权限信息的业务层代码

1）在包“edu.friday.common.security.service”下创建 UserDetailsServiceImpl 类，并实现 UserDetailsService 接口，该接口的 loadUserByUsername() 方法用来将从数据库中获取到的用户、权限和菜单等信息封装到 Spring Security 框架提供的 UserDetails 接口中，这里使用 LoginUser 继承 UserDetails 接口来保存获取的用户、权限等信息。

```
/**
 * 获取用户信息和权限信息
 *
 */
@Service
public class UserDetailsServiceImpl implements UserDetailsService {
    private static final Logger log = LoggerFactory.getLogger(UserDetailsServiceImpl.class);

    @Autowired
    private SysUserService userService;

    @Autowired
    private SysPermissionService permissionService;

    @Override
    public UserDetails loadUserByUsername(String username) throws UsernameNotFoundException {
        SysUser user = userService.selectUserByUserName(username);
        if (StringUtils.isNull(user)) {
            log.info(" 登录用户：{} 不存在 .", username);
            throw new UsernameNotFoundException(" 登录用户： " + username + " 不存在 ");
        } else if (UserStatus.DELETED.getCode().equals(user.getDelFlag())) {
            log.info(" 登录用户：{} 已被删除 .", username);
            throw new BaseException(" 对不起，您的账号： " + username + " 已被删除 ");
        } else if (UserStatus.DISABLE.getCode().equals(user.getStatus())) {
            log.info(" 登录用户：{} 已被停用 .", username);
```

```
            throw new BaseException(" 对不起，您的账号：" + username + " 已停用 ");
        }
        return createLoginUser(user);
    }

    public UserDetails createLoginUser(SysUser user) {
        SysUserVO sysUserVO = new SysUserVO();
        BeanUtils.copyProperties(user, sysUserVO);
        return new LoginUser(sysUserVO, permissionService.getMenuPermission(sysUserVO));
    }
}
```

2）在包“edu.friday.common.security”下增加 UserDetailsServiceImpl 中调用的 LoginUser 类，并实现 UserDetails 接口。

```
/**
 * 登录用户身份权限
 */
@Data
@NoArgsConstructor
@AllArgsConstructor
public class LoginUser implements UserDetails {
    private static final long serialVersionUID = 1L;
    // 用户唯一标识
    private String token;
    // 登录时间
    private Long loginTime;
    // 过期时间
    private Long expireTime;
    // 登录 IP 地址
    private String ipaddr;
    // 登录地点
    private String loginLocation;
    // 浏览器类型
    private String browser;
    // 操作系统
    private String os;
    // 权限列表
    private Set<String> permissions;
    // 用户信息
    private SysUserVO user;

    public LoginUser(SysUserVO user, Set<String> permissions) {
        this.user = user;
        this.permissions = permissions;
    }

    @JsonIgnore
    @Override
    public String getPassword() {
        return user.getPassword();
```

```
    }

    @Override
    public String getUsername() {
        return user.getUserName();
    }

    /**
     * 账户是否未过期，过期无法验证
     */
    @JsonIgnore
    @Override
    public boolean isAccountNonExpired() {
        return true;
    }

    /**
     * 指定用户是否解锁，锁定的用户无法进行身份验证
     * @return
     */
    @JsonIgnore
    @Override
    public boolean isAccountNonLocked() {
        return true;
    }

    /**
     * 指示是否已过期的用户的凭据（密码），过期的凭据防止认证
     * @return
     */
    @JsonIgnore
    @Override
    public boolean isCredentialsNonExpired() {
        return true;
    }

    /**
     * 是否可用，禁用的用户不能身份验证
     * @return
     */
    @JsonIgnore
    @Override
    public boolean isEnabled() {
        return true;
    }

    @Override
    public Collection<? extends GrantedAuthority> getAuthorities() {
        return null;
    }
}
```

步骤 5 实现获取用户信息的业务层代码

1）在 SysUserService 接口增加步骤 4 中调用的 selectUserByUserName 抽象方法。

```
SysUser selectUserByUserName(String userName);
```

2）在 SysUserServiceImpl 中增加对应的实现方法。

```
@Override
public SysUser selectUserByUserName(String userName) {
    SysUser sysuser = new SysUser();
    sysuser.setUserName(userName);
    sysuser.setDelFlag("0");
    Example<SysUser> example = Example.of(sysuser);
    return findOne(example);
}
```

步骤 6 实现获取权限信息的业务层代码

1）在包“edu.friday.common.security.service”下增加 UserDetailsServiceImpl 中调用的 SysPermissionService 类，用于获取角色和菜单权限信息。

```
/**
 * 用户权限处理
 */
@Component
public class SysPermissionService {
    @Autowired
    private SysRoleService roleService;
    @Autowired
    private SysMenuService menuService;
    /**
     * 获取角色数据权限
     * @param user 用户信息
     * @return 角色权限信息
     */
    public Set<String> getRolePermission(SysUserVO user) {
        Set<String> roles = new HashSet<String>();
        // 管理员拥有所有权限
        if (user.isAdmin()) {
            roles.add( “admin” );
        } else {
            roles.addAll(roleService.selectRolePermissionByUserId(user.getUserId()));
        }
        return roles;
    }
    /**
     * 获取菜单数据权限
     * @param user 用户信息
     * @return 菜单权限信息
```

```
     */
    public Set<String> getMenuPermission(SysUserVO user) {
        Set<String> roles = new HashSet<String>();
        // 管理员拥有所有权限
        if (user.isAdmin()) {
            roles.add( "*:*:*" );
        } else {
            roles.addAll(menuService.selectMenuPermsByUserId(user.getUserId()));
        }
        return roles;
    }
}
```

2）在 SysRoleService 中增加 SysPermissionService 类中调用的 selectRolePermissionByUserId (Long userId) 方法，用于根据用户 ID 查询权限信息。

```
// SysRoleService 接口中增加如下方法
Set<String> selectRolePermissionByUserId(Long userId);

// SysRoleServiceImpl 类中实现新增的方法
/**
 * 根据用户 ID 查询权限
 * @param userId 用户 ID
 * @return 权限列表
 */
@Override
public Set<String> selectRolePermissionByUserId(Long userId) {
    List<SysRole> perms = sysRoleRepository.selectRoleByUserId(userId);
    Set<String> permsSet = new HashSet<>();
    for (SysRole perm : perms) {
        if (StringUtils.isNotNull(perm)) {
            permsSet.addAll(Arrays.asList(perm.getRoleKey().trim().split(",")));
        }
    }
    return permsSet;
}
```

3）在包“edu.friday.service”下增加 SysPermissionService 类中调用的 SysMenuService 接口的 selectMenuPermsByUserId 方法。

```
/**
 * 菜单业务层
 */
public interface SysMenuService {
    /**
     * 根据用户 ID 查询权限
     * @param userId 用户 ID
     * @return 权限列表
     */
    Set<String> selectMenuPermsByUserId(Long userId);
}
```

4）在包“edu.friday.service.impl”下增加 SysMenuServiceImpl 类，编写 selectMenuPermsByUserId 方法用于根据用户 ID 查询菜单权限。

```
/**
 * 菜单 业务层处理
 */
@Service
public class SysMenuServiceImpl implements SysMenuService {
    @Autowired
    SysMenuRepository sysMenuRepository;
    /**
     * 根据用户 ID 查询权限
     * @param userId 用户 ID
     * @return 权限列表
     */
    @Override
    public Set<String> selectMenuPermsByUserId(Long userId) {
        List<String> perms = sysMenuRepository.selectMenuPermsByUserId(userId);
        Set<String> permsSet = new HashSet<>();
        for (String perm : perms) {
            if (StringUtils.isNotEmpty(perm)) {
                permsSet.addAll(Arrays.asList(perm.trim().split(",")));
            }
        }
        return permsSet;
    }
}
```

5）在包“edu.friday.repository”下增加 SysMenuRepository 接口的 selectMenuPermsByUserId 方法，用于根据用户 ID 查询菜单权限数据。

```
/**
 * 菜单表 数据层
 */
@Repository
public interface SysMenuRepository extends JpaRepository<SysMenu, Long> {
    String JOIN_ROLE_MENU = " left join sys_role_menu rm on m.menu_id = rm.menu_id ";
    String JOIN_USER_ROLE = " left join sys_user_role ur on rm.role_id = ur.role_id ";

    @Query(value = " select distinct m.perms from sys_menu m "
            + JOIN_ROLE_MENU + JOIN_USER_ROLE + " where ur.user_id = :userId ", nativeQuery = true)
    List<String> selectMenuPermsByUserId(@Param("userId") Long userId);
}
```

步骤 7 运行调试

使用 Postman 访问下面 URL 地址 http://localhost:8080/system/user/list，如图 5-2 所示。因为引入了 Spring Security 进行权限管理，所以无法在未登录授权的情况下访问资源。在下面两个任务分别实现登录认证和鉴权功能后，在授权的情况下访问资源才能获取资源数据。

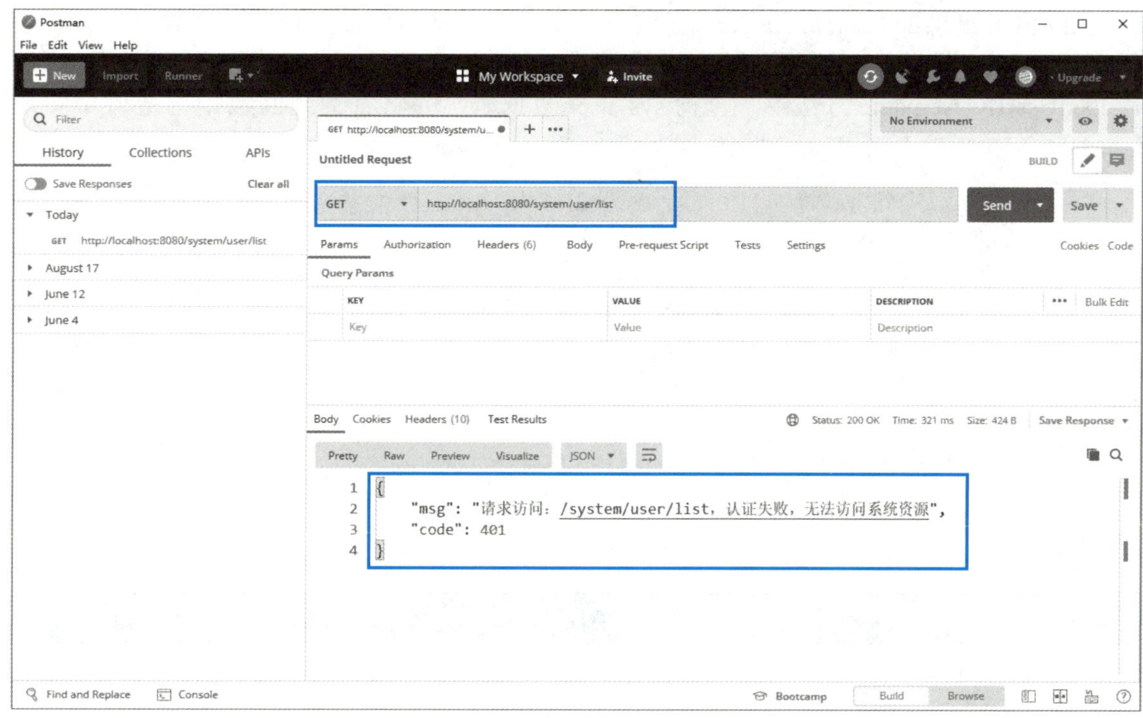

图 5-2　未授权无法访问资源

知识小结

1. 使用 Spring Security 的 configure(HttpSecurity httpSecurity) 配置类进行资源访问控制配置

此为权限管理的核心配置，Spring Security 配置项作用见表 5-2。

表 5-2　Spring Security 配置项作用

配置项名称	配置项作用
anyRequest	匹配所有请求路径
antMatchers	根据通配符匹配 request 路径
anonymous	匿名可以访问
denyAll	用户不能访问
permitAll	用户可以任意访问
authenticated	用户登录后可访问
hasAnyAuthority	如果有参数，参数表示权限，则其中任何一个权限可以访问
hasAnyRole	如果有参数，参数表示角色，则其中任何一个角色可以访问
hasAuthority	如果有参数，参数表示权限，则其权限可以访问
hasRole	如果有参数，参数表示角色，则其角色可以访问

2. 使用 UserDetailsService 接口获取用户和权限信息

当不做任何配置的时候，用户名和密码是按照 Spring Security 规则自动生成的，而在真实业务系统项目中，用户名和密码都是从数据库中查询出来的，因此要通过实现 Spring

Security 框架提供的 UserDetailsService 接口，实现自定义的认证逻辑，来获取用户和权限信息。例如，本任务中使用 UserDetailsServiceImpl 类实现 UserDetailsService 接口，然后在 loadUserByUsername 方法中，查询 sys_user 数据表，并将查询到的数据构造一个实现 UserDetails 接口的 LoginUser 对象返回即可。

3. 使用 Spring Security 的 configure(AuthenticationManagerBuilder auth) 配置身份验证的实现对象

此处的实现逻辑是绑定 UserDetailsService 接口的实现类，来获取用户信息和 BCryptPasswordEncoder 类进行密码加密处理。当后续进行表单登录认证的时候，系统自动初始化这两个对象进行认证，配置的代码如下：

```
auth.userDetailsService(userDetailsService).passwordEncoder(bCryptPasswordEncoder());
```

任务评价

技　能　点	知　识　点	自我评价（不熟悉 / 基本掌握 / 熟练掌握 / 灵活运用）
使用 Spring Security 配置资源的访问控制逻辑	Spring Security 的工作原理 Spring Security UserDetailsService 接口实现	
使用 Spring Security 实现自定义的用户验证逻辑		

任务 2　实现基于 JWT 的登录认证功能

微课 5-2　实现基于 JWT 的登录认证功能

任务分析

通过本任务，完成登录验证的接口功能。登录验证的接口描述见表 5-3。

对于开发人员编写的接口输入和返回值，业务逻辑分为 3 个核心步骤：首先是通过 AuthenticationManager 的 authenticate 方法进行用户验证并返回用户信息；其次使用 Redis 工具类将用户信息存储到 Redis 数据库中；最后将保存用户的键值以 JSON Web Token（缩写 JWT）的数据格式返回前端。

表 5-3　登录验证的接口描述

URL 地址	HTTP 请求方式	支持格式	输入参数	输出返回值
/login	POST	JSON	username、password	{"msg": " 操作成功 ", "code": 200}

任务实施

步骤 1　导入 JWT、Redis 和连接池的依赖

在 pom.xml 中导入 JWT、Redis 和连接池的依赖。

```
<properties>
    ...
    <jwt.version>0.9.0</jwt.version>
</properties>
<dependencies>
    ...
    <!--Token 生成与解析 -->
    <dependency>
        <groupId>io.jsonwebtoken</groupId>
        <artifactId>jjwt</artifactId>
        <version>${jwt.version}</version>
    </dependency>
    <!-- redis 缓存操作 -->
    <dependency>
        <groupId>org.springframework.boot</groupId>
        <artifactId>spring-boot-starter-data-redis</artifactId>
    </dependency>
    <!-- pool 对象池 -->
    <dependency>
        <groupId>org.apache.commons</groupId>
        <artifactId>commons-pool2</artifactId>
    </dependency>
</dependencies>
```

步骤 2　配置 Redis 连接和 Token 信息

修改 resources 下的 application.yml 文件，添加如下配置参数，用于配置 Redis 连接和配置 Token 令牌的密钥和有效期信息。

```
spring:
  ...
  # redis 配置
  redis:
    # 地址
    host: localhost
    # 端口，默认为 6379
    port: 6379
    # 密码
    password:
    # 连接超时时间
    timeout: 10s
    lettuce:
      pool:
        # 连接池中的最小空闲连接
        min-idle: 0
        # 连接池中的最大空闲连接
        max-idle: 8
        # 连接池的最大数据库连接数
        max-active: 8
        # # 连接池最大阻塞等待时间（使用负值表示没有限制）
```

```
            max-wait: -1ms
# token 配置
token:
    # 令牌自定义标识
    header: Authorization
    # 令牌秘钥
    secret: (FRIDAY:)_$^11244^%$_(IS:)_@@++--(COOL:)_++++_.sds_(GUY:)
    # 令牌有效期（默认 30 分钟）
    expireTime: 60
```

步骤 3 编写登录接口的控制层代码

在包“edu.friday.controller”下新增 SysLoginController 类用于实现登录功能的 API 接口。

```
/**
 * 登录验证
 */
@RestController
public class SysLoginController {
    @Autowired
    private SysLoginService loginService;
    /**
     * 登录方法
     * @param username 用户名
     * @param password 密码
     * @return 结果
     */
    @PostMapping({"/login", "/"})
    public RestResult login(String username, String password, String uuid) {
        RestResult ajax = RestResult.success();
        // 生成令牌
        String token = loginService.login(username, password, "123", uuid);
        ajax.put(Constants.TOKEN, token);
        return ajax;
    }
}
```

步骤 4 编写登录接口的业务层代码

在包“edu.friday.common.security.service”下新增步骤 3 中使用的 SysLoginService 类，用于对登录用户的验证，将 username 和 password 封装到 UsernamePasswordAuthenticationToken 对象中，然后将该实例给 AuthenticationManager 进行验证，只要返回 Authentication 对象，就代表用户通过验证了。

```
/**
 * 登录校验方法
 */
@Component
public class SysLoginService {
```

```
    @Autowired
    private TokenService tokenService;
    @Resource
    private AuthenticationManager authenticationManager;
    /**
     * 登录验证
     * @param username 用户名
     * @param password 密码
     * @param code     验证码
     * @param uuid     唯一标识
     * @return 结果
     */
    public String login(String username, String password, String code, String uuid) {
        Authentication authentication = null;
        try {
            // 该方法会去调用 UserDetailsServiceImpl.loadUserByUsername 方法
            authentication = authenticationManager.authenticate(new UsernamePasswordAuthentica-
tionToken(username, password));
        } catch (Exception e) {
            if (e instance of BadCredentialsException) {
                throw new UserPasswordNotMatchException();
            } else {
                throw new CustomException(e.getMessage());
            }
        }
        LoginUser loginUser = (LoginUser) authentication.getPrincipal();
        // 生成 Token
        return tokenService.createToken(loginUser);
    }
}
```

步骤 5　编写验证异常代码

在包“edu.friday.common.exception.user”下新增步骤 4 中使用的 UserPasswordNotMatch_Exception 类和它的 UserException 父类，用于用户密码不正确或不符合规范异常的情况。

```
// UserPasswordNotMatchException 类
package edu.friday.common.exception.user;
/**
 * 用户密码不正确或不符合规范异常类
 */
public class UserPasswordNotMatchException extends UserException {
    private static final long serialVersionUID = 1L;
    public UserPasswordNotMatchException() {
        super("user.password.not.match" , null);
    }
}

// UserException 类
package edu.friday.common.exception.user;
```

```
import edu.friday.common.exception.BaseException;
/**
 * 用户信息异常类
 */
public class UserException extends BaseException {
    private static final long serialVersionUID = 1L;
    public UserException(String code, Object[] args) {
        super("user", code, args, null);
    }
}
```

步骤 6 编写网络令牌 Token 的业务层代码

在包“edu.friday.common.security.service”下新增步骤 4 中使用的 TokenService 类以及相关的方法，用于创建 JSON 网络令牌并将其存储在 Redis 数据库中。

```
/**
 * Token 验证处理
 */
@Component
public class TokenService {
    // 令牌自定义标识
    @Value("${token.header}")
    private String header;

    // 令牌秘钥
    @Value("${token.secret}")
    private String secret;

    // 令牌有效期（默认 30 分钟）
    @Value("${token.expireTime}")
    private int expireTime;

    protected static final long MILLIS_SECOND = 1000;
    protected static final long MILLIS_MINUTE = 60 * MILLIS_SECOND;
    private static final Long MILLIS_MINUTE_TEN = 20 * 60 * 1000L;

    @Autowired
    private RedisCache redisCache;
    /**
     * 创建令牌
     * @param loginUser 用户信息
     * @return 令牌
     */
    public String createToken(LoginUser loginUser) {
        String token = IdUtils.fastUUID();
        loginUser.setToken(token);
        refreshToken(loginUser);
        Map<String, Object> claims = new HashMap<>();
        claims.put(Constants.LOGIN_USER_KEY, token);
```

```
        return createToken(claims);
    }

    /**
     * 刷新令牌有效期
     * @param loginUser 登录信息
     */
    public void refreshToken(LoginUser loginUser) {
        loginUser.setLoginTime(System.currentTimeMillis());
        loginUser.setExpireTime(loginUser.getLoginTime() + expireTime * MILLIS_MINUTE);
        // 根据 uuid 将 loginUser 缓存
        String userKey = getTokenKey(loginUser.getToken());
        redisCache.setCacheObject(userKey, loginUser, expireTime, TimeUnit.MINUTES);
    }

    /**
     * 从数据声明生成令牌
     * @param claims 数据声明
     * @return 令牌
     */
    private String createToken(Map<String, Object> claims) {
        String token = Jwts.builder()
                .setClaims(claims)
                .signWith(SignatureAlgorithm.HS512, secret).compact();
        return token;
    }

    private String getTokenKey(String uuid) {
        return Constants.LOGIN_TOKEN_KEY + uuid;
    }
}
```

步骤 7　编写 RedisCache 工具类代码

1）在包“edu.friday.utils”下增加步骤 6 中调用的 RedisCache 工具类，用于在 Redis 数据库将网络令牌 Token 进行存储和读取操作。

```
/**
 * spring redis 工具类
 **/
@Component
public class RedisCache {
    @Autowired
    public RedisTemplate redisTemplate;

    /**
     * 缓存基本的对象，Integer、String、实体类等
     * @param key   缓存的键值
     * @param value 缓存的值
     * @return 缓存的对象
     */
```

```
    public <T> ValueOperations<String, T> setCacheObject(String key, T value) {
        ValueOperations<String, T> operation = redisTemplate.opsForValue();
        operation.set(key, value);
        return operation;
    }

    /**
     * 缓存基本的对象，Integer、String、实体类等
     * @param key      缓存的键值
     * @param value    缓存的值
     * @param timeout  时间
     * @param timeUnit 时间颗粒度
     * @return 缓存的对象
     */
    public <T> ValueOperations<String, T> setCacheObject(String key, T value, Integer timeout, TimeUnit
timeUnit) {
        ValueOperations<String, T> operation = redisTemplate.opsForValue();
        operation.set(key, value, timeout, timeUnit);
        return operation;
    }

    /**
     * 获得缓存的基本对象
     * @param key 缓存键值
     * @return 缓存键值对应的数据
     */
    public <T> T getCacheObject(String key) {
        ValueOperations<String, T> operation = redisTemplate.opsForValue();
        return operation.get(key);
    }
}
```

2）在包“edu.friday.config”下增加RedisConfig类，用于配置Redis的序列化和反序列化方式，当存储数据到Redis的时候，键值（key）和值（value）都是通过Spring框架的Serializer序列化到数据库的，根据业务需要，这里配置用String来序列化和反序列化key的值，用FastJson来序列化和反序列化value的值。

```
/**
 * redis 配置
 */
@Configuration
@EnableCaching
public class RedisConfig extends CachingConfigurerSupport {
    @Bean
    @SuppressWarnings(value = {"unchecked" , "rawtypes"})
    public RedisTemplate<Object, Object> redisTemplate(RedisConnectionFactory connectionFactory) {
        RedisTemplate<Object, Object> template = new RedisTemplate<>();
        template.setConnectionFactory(connectionFactory);
        FastJson2JsonRedisSerializer serializer = new FastJson2JsonRedisSerializer(Object.class);
```

```
            ObjectMapper mapper = new ObjectMapper();
            mapper.setVisibility(PropertyAccessor.ALL, JsonAutoDetect.Visibility.ANY);
            mapper.enableDefaultTyping(ObjectMapper.DefaultTyping.NON_FINAL);
            serializer.setObjectMapper(mapper);
            template.setValueSerializer(serializer);
            // 使用 StringRedisSerializer 来序列化和反序列化 Redis 的 key 值
            template.setKeySerializer(new StringRedisSerializer());
            template.afterPropertiesSet();
            return template;
        }
    }
```

3）在包“edu.friday.config”下增加 FastJson2JsonRedisSerializer 类，用于使用 FastJson 来序列化和反序列化 Redis 的值。

```
auth.userDetailsService(userDetailsService).passwordEncoder(bCryptPasswordEncoder());

/**
 * Redis 使用 FastJson 序列化
 */
public class FastJson2JsonRedisSerializer<T> implements RedisSerializer<T> {
    @SuppressWarnings("unused")
    private ObjectMapper objectMapper = new ObjectMapper();
    public static final Charset DEFAULT_CHARSET = Charset.forName("UTF-8");
    private Class<T> clazz;
    static {
        ParserConfig.getGlobalInstance().setAutoTypeSupport(true);
    }

    public FastJson2JsonRedisSerializer(Class<T> clazz) {
        super();
        this.clazz = clazz;
    }

    public byte[] serialize(T t) throws SerializationException {
        if (t == null) {
          return new byte[0];
        }
        return JSON.toJSONString(t, SerializerFeature.WriteClassName).getBytes(DEFAULT_CHARSET);
    }

    public T deserialize(byte[] bytes) throws SerializationException {
        if (bytes == null || bytes.length <= 0) {
          return null;
        }
        String str = new String(bytes, DEFAULT_CHARSET);
        return JSON.parseObject(str, clazz);
    }

    public void setObjectMapper(ObjectMapper objectMapper) {
        Assert.notNull(objectMapper, " 'objectMapper' must not be null");
```

```
        this.objectMapper = objectMapper;
    }

    protected JavaType getJavaType(Class<?> clazz) {
        return TypeFactory.defaultInstance().constructType(clazz);
    }
}
```

步骤 8 运行调试

Postman 下测试登录接口，选择“POST”模式，输入 URL 地址 http://localhost:8080/login，Params 输入传输参数 username=“admin”和 password=“123456”，单击“Send”按钮，返回值中包含认证成功后的 Token 值，如图 5-3 所示。

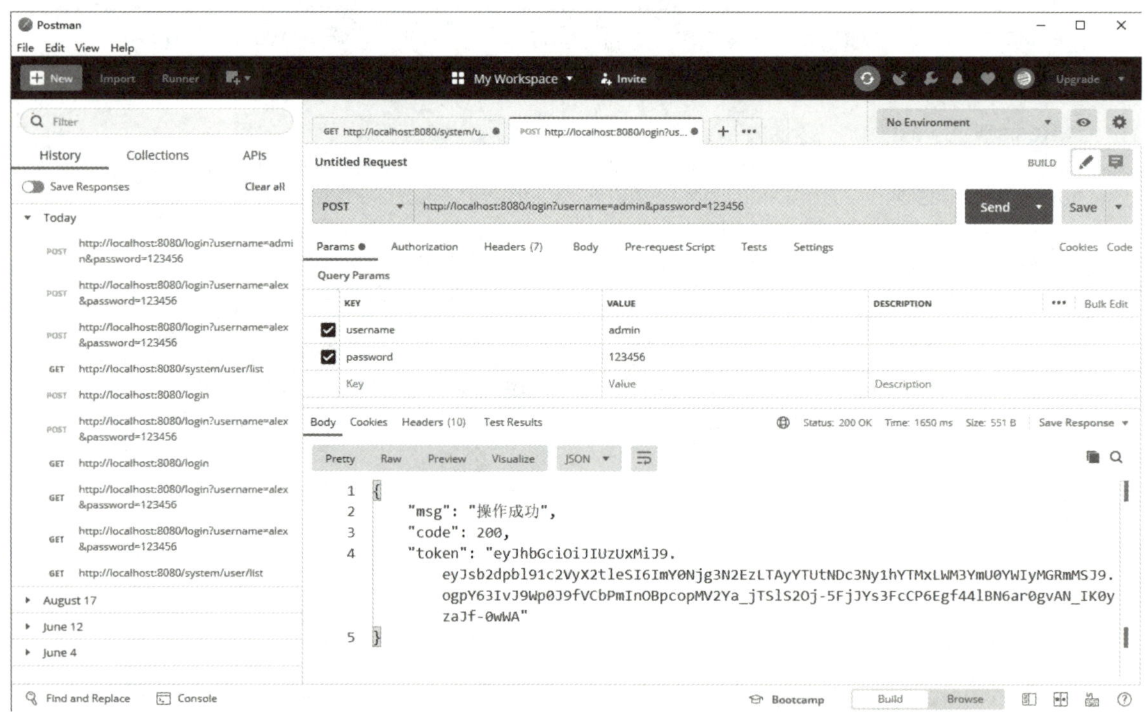

图 5-3 认证成功返回 Token 值

知识小结

1. Spring Security 框架的核心认证组件

➢ AuthenticationManager 认证管理器：该接口的作用是对用户的未授权凭据进行认证，认证通过则返回授权状态的凭据 Authentication，否则将抛出认证异常的 AuthenticationException。

➢ Authentication 认证凭证：Authentication 对象是 Spring Security 框架提供的用户信息认证对象。其有未认证和已认证两种状态。在本任务中，封装从客户端获取的用户名和密码并作为参数传入认证管理器（AuthenticationManager）的 authenticate 方法时，Authentication 对象是未认证的状态。Spring Security 框架将未认证的 Authentication 对象和 UserDetailsService.loadUserByUsername 方法获取的 UserDetails 进行匹配，成功后将 UserDetails 中的用户权限信息复制到 Authentication 对象中，此时 Authentication 对象是已认证的状态。

2. JWT 核心知识

JWT，全称是 JSON Web Token，是一种 JSON 风格的轻量级的授权和身份认证规范，可实现无状态、分布式的 Web 应用授权。JWT 是一个加密后的接口访问密码，并且该密码里面包含用户名等信息，通过用户名信息既可以知道用户的身份，又可以知道该用户是否具有访问当前应用系统的权限。

JWT 的原理是，使用 Spring Security 框架对用户认证成功以后，生成一个 JSON 对象，同时为了防止用户篡改数据，会加上签名，发回给客户端，之后，用户与服务端通信的时候，都要发回这个 JSON 对象，后端服务完全靠这个对象来认定用户身份。

JWT 的数据结构包含三个部分，它们之间用“.”连接：Header.Payload.Signature：

➢ Header：Header 用于描述 JWT 的最基本信息，由声明类型和声明加密的算法两部分组成，如 {"typ":"JWT","alg":"HS256"} 表明类型为“JWT”，加密算法是 HS256 算法。

➢ Payload：该部分是一个 JSON 对象，用来存放实际需要传递的数据，如本任务中登录验证成功后的用户信息的键值。

➢ Signature：该部分是对 Header 和 Payload 的签名，防止用户篡改数据。在后端服务器指定一个密钥，然后使用 Header 里面指定的签名算法产生签名。

把 Header、Payload 和 Signature 三个部分拼成一个字符串，每个部分之间用“.”分隔，就可以返回给用户。

任务评价

技　能　点	知　识　点	自我评价（不熟悉 / 基本掌握 / 熟练掌握 / 灵活运用）
使用 Spring Security 框架完成用户登录验证	JWT 的核心知识 Redis 的数据操作 Spring Security 的安全认证原理	
使用 JWT 生成 Token 认证令牌		
使用 RedisTemplate 存储和读取 Redis 数据		

任务 3　实现基于 JWT 的访问鉴权功能

微课 5-3　实现基于 JWT 的访问鉴权功能

任务分析

用户登录成功后，系统返回生成的 Token 值，以后用户每次请求都会在 header 中加入“Authorization:token 值”，后台系统会使用过滤器拦截请求，验证 Token 值，如果通过验证则返回请求的资源，否则就返回无权访问的错误页，完整的基于 JWT 的用户认证流程如图 5-4 所示。

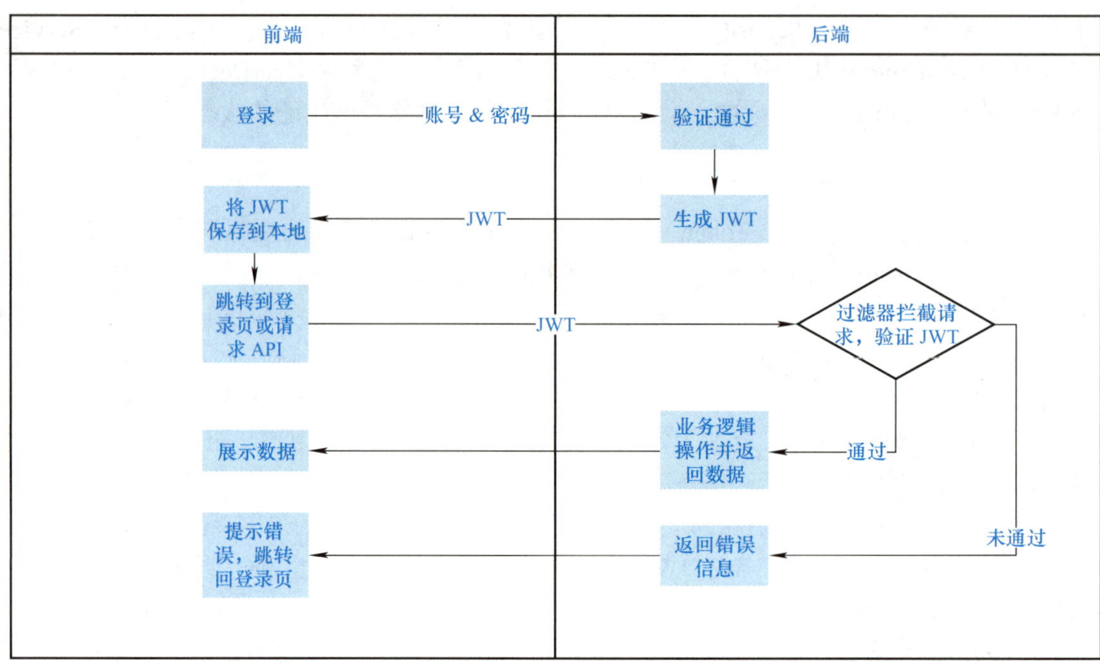

图 5-4　基于 JWT 的用户认证流程

任务实施

步骤 1　创建验证 Token 的过滤器

在包“edu.friday.common.security.filter”下新增 JwtAuthenticationTokenFilter 过滤器，用于验证客户端携带 Token 的有效性。JwtAuthenticationTokenFilter 需继承 OncePerRequestFilter，确保在一次请求中只通过一次 filter，而不需要重复执行，通过调用 SecurityContextHolder.getContext().setAuthentication 方法将 AuthenticationManager 返回的 Authentication 对象赋予当前的 SecurityContext，建立安全上下文。

```
/**
 * Token 过滤器 验证 Token 有效性
 */
@Component
public class JwtAuthenticationTokenFilter extends OncePerRequestFilter {
    @Autowired
    private TokenService tokenService;

    @Override
    protected void doFilterInternal(HttpServletRequest request, HttpServletResponse response, FilterChain
chain)
            throws ServletException, IOException {
        LoginUser loginUser = tokenService.getLoginUser(request);
        if (StringUtils.isNotNull(loginUser) && StringUtils.isNull(SecurityUtils.getAuthentication())) {
            tokenService.verifyToken(loginUser);
```

```
                UsernamePasswordAuthenticationToken authenticationToken = new UsernamePassw-
ordAuthenticationToken(loginUser, null, loginUser.getAuthorities());
                authenticationToken.setDetails(new WebAuthenticationDetailsSource().buildDetails(request));
                SecurityContextHolder.getContext().setAuthentication(authenticationToken);
            }
            chain.doFilter(request, response);
        }
    }
```

步骤 2　编写网络令牌 Token 的业务层代码

在 TokenService 类下新增步骤 1 中使用的 getLoginUser、verifyToken 方法，分别用于从客户端的 Token 中获取用户信息和验证令牌的有效性。

```
/**
 * 获取用户身份信息
 * @return 用户信息
 */
public LoginUser getLoginUser(HttpServletRequest request) {
    // 获取请求携带的令牌
    String token = getToken(request);
    if (StringUtils.isNotEmpty(token)) {
        Claims claims = parseToken(token);
        // 解析对应的权限以及用户信息
        String uuid = (String) claims.get(Constants.LOGIN_USER_KEY);
        String userKey = getTokenKey(uuid);
        LoginUser user = redisCache.getCacheObject(userKey);
        return user;
    }
    return null;
}

/**
 * 获取请求 Token
 * @param request
 * @return token
 */
private String getToken(HttpServletRequest request) {
    String token = request.getHeader(header);
    if (StringUtils.isNotEmpty(token) && token.startsWith(Constants.TOKEN_PREFIX)) {
        token = token.replace(Constants.TOKEN_PREFIX, "");
    }
    return token;
}

/**
 * 从令牌中获取数据声明
 * @param token 令牌
```

```
 * @return 数据声明
 */
private Claims parseToken(String token) {
    return Jwts.parser()
            .setSigningKey(secret)
            .parseClaimsJws(token)
            .getBody();
}

/**
 * 验证令牌有效期，相差不足 20 分钟，自动刷新缓存
 * @param loginUser 令牌
 * @return 令牌
 */
public void verifyToken(LoginUser loginUser) {
    long expireTime = loginUser.getExpireTime();
    long currentTime = System.currentTimeMillis();
    if (expireTime - currentTime <= MILLIS_MINUTE_TEN) {
        refreshToken(loginUser);
    }
}
```

步骤 3 编写安全服务工具类

在包“edu.friday.utils.security”下新增步骤 1 中使用的 SecurityUtils 类，用于安全服务工具类。

```
/**
 * 安全服务工具类
 */
public class SecurityUtils {
    /**
     * 获取用户账户
     **/
    public static String getUsername() {
        try {
            return getLoginUser().getUsername();
        } catch (Exception e) {
            throw new CustomException(" 获取用户账户异常 ", HttpStatus.UNAUTHORIZED);
        }
    }

    /**
     * 获取用户
     **/
    public static LoginUser getLoginUser() {
        try {
            return (LoginUser) getAuthentication().getPrincipal();
        } catch (Exception e) {
```

```
            throw new CustomException(" 获取用户信息异常 ", HttpStatus.UNAUTHORIZED);
        }
    }

    /**
     * 获取 Authentication
     */
    public static Authentication getAuthentication() {
        return SecurityContextHolder.getContext().getAuthentication();
    }

    /**
     * 生成 BCryptPasswordEncoder 密码
     * @param password 密码
     * @return 加密字符串
     */
    public static String encryptPassword(String password) {
        BCryptPasswordEncoder passwordEncoder = new BCryptPasswordEncoder();
        return passwordEncoder.encode(password);
    }

    /**
     * 判断密码是否相同
     * @param rawPassword    真实密码
     * @param encodedPassword 加密后字符
     * @return 结果
     */
    public static boolean matchesPassword(String rawPassword, String encodedPassword) {
        BCryptPasswordEncoder passwordEncoder = new BCryptPasswordEncoder();
        return passwordEncoder.matches(rawPassword, encodedPassword);
    }

    /**
     * 是否为管理员
     * @param userId 用户 ID
     * @return 结果
     */
    public static boolean isAdmin(Long userId) {
        return userId != null && 1L == userId;
    }
}
```

步骤 4　SecurityConfig 类中配置过滤器

在 SecurityConfig 类中配置新增的 JwtAuthenticationTokenFilter 类，使用 httpSecurity.addFilterBefore 方法将这个新增的过滤器增加到链路中。

```
/**
 * token 认证过滤器
```

```
 */
@Autowired
private JwtAuthenticationTokenFilter authenticationTokenFilter;

@Override
protected void configure(HttpSecurity httpSecurity) throws Exception {
    ...
    // 添加 JWT filter
    httpSecurity.addFilterBefore(authenticationTokenFilter, UsernamePasswordAuthenticationFilter.class);
}
```

步骤 5 运行调试

1）执行任务 2 中的运行测试，获取认证成功后的 Token 值。Postman 下选择“POST”模式，输入 URL 地址 http://localhost:8080/login，Params 输入传输参数 username=“admin”和 password=“123456”，单击“Send”按钮，获取认证成功后的 Token 值，如图 5-5 所示。

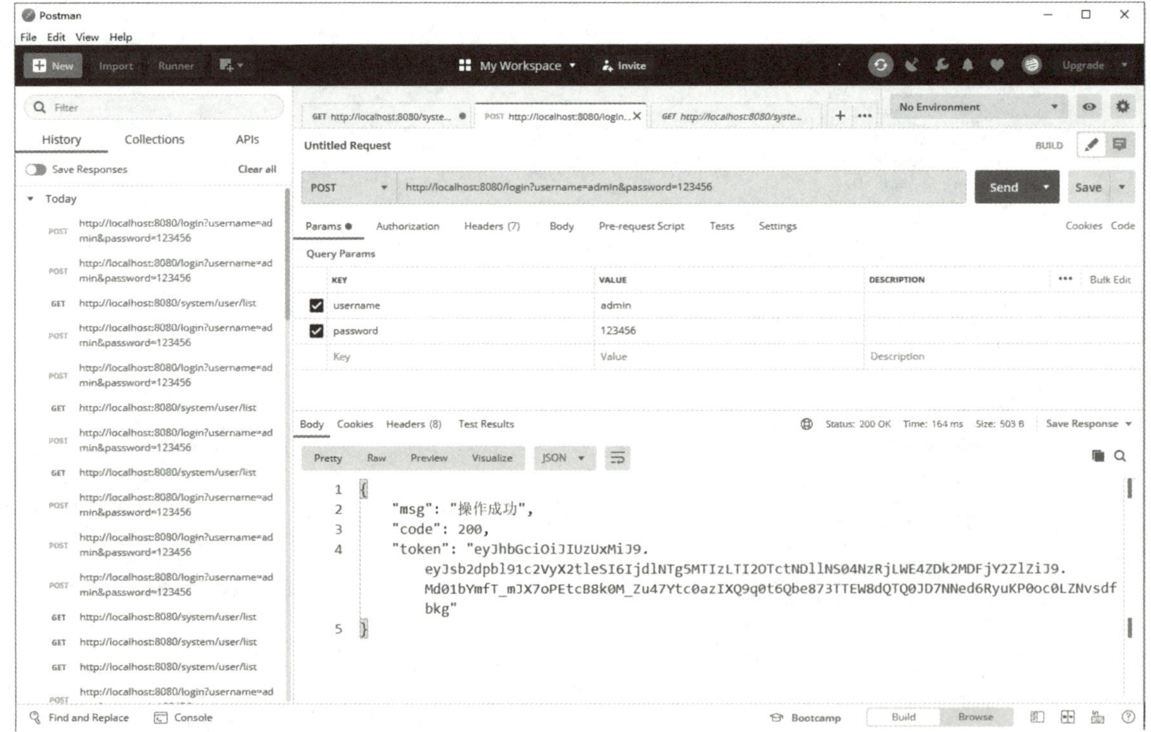

图 5-5 获取认证成功后的 Token 值

2）Postman 下选择“GET”模式，输入 URL：http://localhost:8080/system/user/list，在 Headers 下设置 Authorization 的值为上一步返回的 Token 的值，单击“Send”按钮，登录认证之后，可以访问授权的资源，如图 5-6 所示。

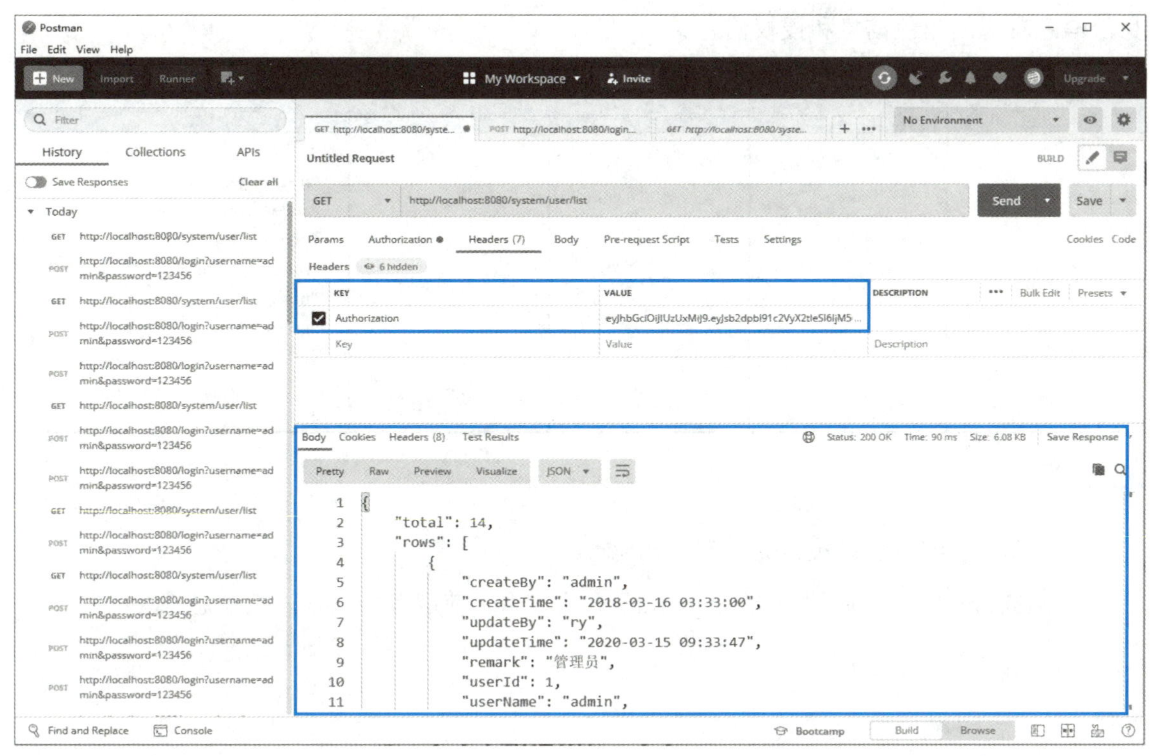

图 5-6　访问授权的资源

知识小结

OncePerRequestFilter 能够确保在一次请求中只通过一次过滤器拦截，而不需要重复地执行。此方法是为了兼容不同的 Web 容器，并不是所有的 Web 容器都只过滤一次，servlet 版本不同，执行过程也不同，因此，为了兼容各种运行环境和版本，Spring 框架中过滤器大多都继承 OncePerRequestFilter 以确保一次请求一个过滤拦截。

任务评价

技　能　点	知　识　点	自我评价（不熟悉 / 基本掌握 / 熟练掌握 / 灵活运用）
使用过滤器完成系统的鉴权功能	SpringBoot 框架的 OncePerRequestFilter 工作原理	

任务 4　实现获取登录用户授权信息接口功能

微课 5-4　实现获取登录用户授权信息接口功能

任务分析

通过本任务，完成表 5-4 中的获取登录用户授权信息和获取菜单路由信息的接口功能。开发人员编写的接口输入和返回值描述见表 5-4，此接口主要用于获取当前登录用户的身份信息对应的角色信息和授权信息。

表 5-4　获取用户授权信息的接口描述

URL 地址	HTTP 请求方式	支持格式	输入参数	输出返回值

/getInfo	GET	JSON	认证之后的 token 值	{
				"msg": " 操作成功 ",
				"code": 200,
				"permissions": ["*:*:*"],
				"roles": ["admin"],
				"user": {
				"userId":1,
				...
				}
				}

任务实施

步骤 1　编写登录接口的控制层代码

在 SysLoginController 类中增加 getInfo 方法，用于获取登录用户授权信息。

```
public class SysLoginController {
    ...
    /**
     * 获取登录用户授权信息
     * @return 用户信息
     */
    @GetMapping("getInfo")
    public RestResult getInfo() {
        LoginUser loginUser = tokenService.getLoginUser(ServletUtils.getRequest());
        SysUserVO user = loginUser.getUser();
        // 角色集合
        Set<String> roles = permissionService.getRolePermission(user);
        // 权限集合
        Set<String> permissions = permissionService.getMenuPermission(user);
        RestResult ajax = RestResult.success();
        ajax.put("user", user);
        ajax.put("roles", roles);
        ajax.put("permissions", permissions);
        return ajax;
    }
}
```

步骤 2　运行调试

1）执行任务 2 中的运行测试，获取认证成功后的 Token 值。Postman 下选择“POST”模式，输入 URL 地址 http://localhost:8080/login，Params 输入传输参数 username=“admin”和 password=“123456”，单击“Send”按钮获取认证成功后的 Token 值，如图 5-7 所示。

2）Postman 下选择“GET”模式，输入 URL 地址 http://localhost:8080/getInfo，在 Headers 下设置 Authorization 的值为上一步返回的 Token 的值，单击“Send”按钮，登录认

证之后，可以获取登录用户授权信息，如图 5-8 所示。

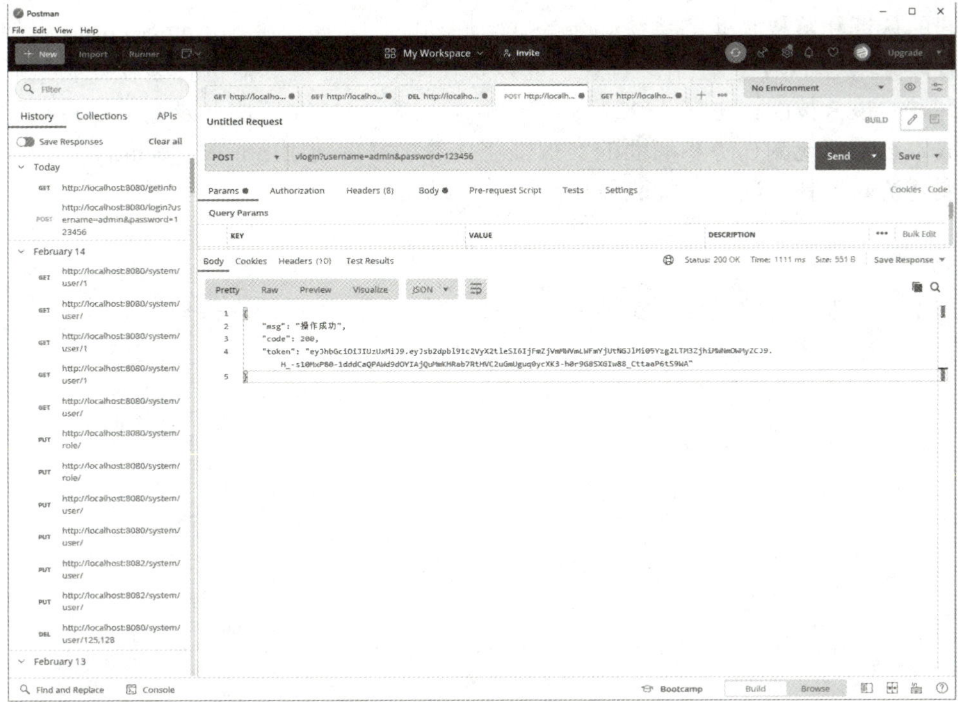

图 5-7　认证成功后的 Token 值

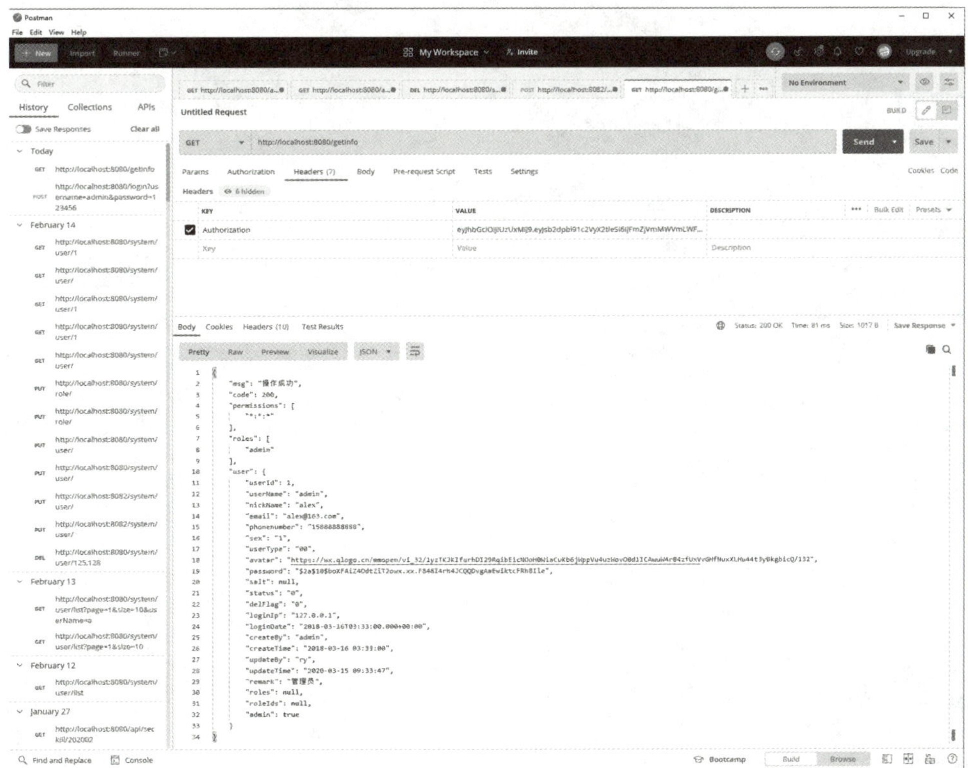

图 5-8　登录用户授权信息

单元小结

用户的认证和鉴权是业务系统必备的安全性功能，能够使用 Spring Security 框架配置资源访问权限、结合 JWT 进行用户登录认证以及使用 Spring 过滤器进行鉴权控制是后端安全性功能开发的核心技能，也是 1+X 职业技能等级证书（JAVA 应用开发）的职业技能要求。通过完成本工作单元的任务和后面的实战强化，熟练掌握使用 Spring Security 框架构建企业系统的安全访问控制解决方案的职业技能。

实战强化

按照本工作单元步骤，编写实训项目“诚品书城”的登录后端接口。

工作单元 6

实现菜单管理接口

职业能力

本工作单元主要完成项目的菜单管理功能接口，最终希望学生达成的职业能力目标是：

1. 熟练掌握基于三层架构的分层开发方法
2. 熟练掌握控制层、业务层和数据控制层的功能实现

任务情景

本工作单元共分为 4 个任务，使用开发人员常用的分层开发方法，完成本工作单元中设计的菜单管理和菜单相关功能的接口，具体的接口功能详情见表 6-1。

表 6-1　菜单管理和菜单相关功能的接口

序　号	接口功能	HTTP 请求方式	URL 地址	支持格式
菜单管理功能接口				
1	获取菜单列表	GET	/system/menu/list	JSON
2	新增菜单数据	POST	/system/menu	JSON
3	修改菜单数据	PUT	/system/menu	JSON
4	删除菜单数据	DELETE	/system/menu/{menuIds}	JSON
5	根据菜单编号获取详细信息	GET	/system/menu/{menuId}	JSON
角色管理功能接口				
6	获取树状菜单列表	GET	/system/menu/treeselect	JSON
7	根据角色编号获取树状菜单列表	GET	/system/menu/roleMenuTreeselect/{roleIds}	JSON
登录授权功能接口				
8	获取菜单路由信息	GET	/getRouters	JSON

本工作单元采用“分而治之”的分层开发方法，即三层架构中，每层只关注自己的职责，上一层通过调用下一层的接口来使用其功能，而不用关心下一层的代码实现细节，这样每层只关注自己功能的实现。完成上一层的全部功能编码后，再编写下一层代码。在企业项目开发中，这种基于三层架构的分层开发方法有利于提高开发效率和代码质量。本工作单元的具体任务如下：

1. 实现菜单后端接口层
2. 实现菜单后端业务层
3. 实现菜单后端数据控制层
4. 测试验证菜单后端接口

前置知识

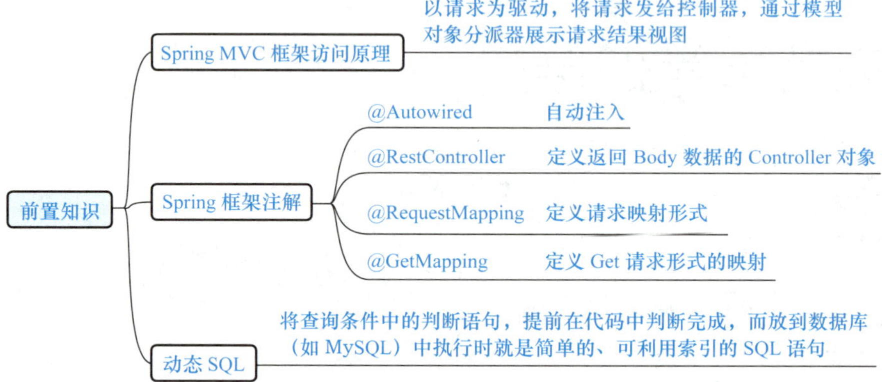

任务 1 实现菜单后端接口层

微课 6-1　实现菜单后端接口层

任务分析

本任务将完成表 6-1 中全部接口功能的控制层代码。控制层的编码主要关注接口输入参数、返回值以及定义调用业务层接口这 3 个方面。开发人员编写的接口输入参数和返回值描述见表 6-2。

表 6-2　接口输入参数和返回值描述

接口功能	URL 地址	HTTP 请求方式	支持格式	输入参数	输出返回值
获取菜单列表	/system/menu/list	GET	JSON	SysMenuVO	RestResult
新增菜单数据	/system/menu	POST	JSON	SysMenu	RestResult
修改菜单数据	/system/menu	PUT	JSON	SysMenu	RestResult
删除菜单数据	/system/menu/{menuIds}	DELETE	JSON	menuIds	RestResult
根据菜单编号获取详细信息	/system/menu/{menuId}	GET	JSON	menuId	RestResult
获取树状菜单列表	/system/menu/treeselect	GET	JSON	SysMenuVO	RestResult
根据角色编号获取树状菜单列表	/system/menu/roleMenuTreeselect/{roleIds}	GET	JSON	roleIds	RestResult
获取菜单路由信息	/getRouters	GET	JSON	无	RestResult

任务实施

步骤 1　编写菜单控制层代码

在包“edu.friday.controller”下新增 SysMenuController 类，根据表 6-2 中描述的输入参数和返回值实现菜单接口。

```
/**
 * 菜单信息
 */
@RestController
@RequestMapping("/system/menu")
public class SysMenuController extends BaseController {
    @Autowired
    private SysMenuService menuService;
    @Autowired
    private TokenService tokenService;
    /**
     * 获取菜单列表
     */
    @GetMapping("/list")
    public RestResult list(SysMenuVO menu) {
        LoginUser loginUser = tokenService.getLoginUser(ServletUtils.getRequest());
        Long userId = loginUser.getUser().getUserId();
        List<SysMenu> menus = menuService.selectMenuList(menu, userId);
        return RestResult.success(menus);
    }
    /**
     * 根据菜单编号获取详细信息
     */
    @GetMapping(value = "/{menuId}")
    public RestResult getInfo(@PathVariable Long menuId) {
        return RestResult.success(menuService.selectMenuById(menuId));
    }
    /**
     * 获取菜单下拉树列表
     */
    @GetMapping("/treeselect")
    public RestResult treeselect(SysMenuVO menu) {
        LoginUser loginUser = tokenService.getLoginUser(ServletUtils.getRequest());
        Long userId = loginUser.getUser().getUserId();
        List<SysMenu> menus = menuService.selectMenuList(menu, userId);
        List<SysMenuVO> menuVOS = BeanUtils.copyProperties(menus, SysMenuVO.class);
        return RestResult.success(menuService.buildMenuTreeSelect(menuVOS));
    }
    /**
     * 加载对应角色菜单列表树
     */
    @GetMapping(value = "/roleMenuTreeselect/{roleId}")
    public RestResult roleMenuTreeselect(@PathVariable("roleId") Long roleId) {
        LoginUser loginUser = tokenService.getLoginUser(ServletUtils.getRequest());
        Long userId = loginUser.getUser().getUserId();
        SecurityUtils.getAuthentication();
        List<SysMenu> menus = menuService.selectMenuList(userId);
        RestResult ajax = RestResult.success();
        ajax.put("checkedKeys", menuService.selectMenuListByRoleId(roleId));
```

```
        List<SysMenuVO> menuVOS = BeanUtils.copyProperties(menus, SysMenuVO.class);
        ajax.put("menus", menuService.buildMenuTreeSelect(menuVOS));
        return ajax;
    }
    /**
     * 新增菜单
     */
    @PostMapping
    public RestResult add(@Validated @RequestBody SysMenu menu) {
        if (UserConstants.NOT_UNIQUE.equals(menuService.checkMenuNameUnique(menu))) {
            return RestResult.error(" 新增菜单 " + menu.getMenuName() + " 失败，名称已存在 ");
        }
        menu.setCreateBy("system");
        return toAjax(menuService.insertMenu(menu));
    }
    /**
     * 修改菜单
     */
    @PutMapping
    public RestResult edit(@Validated @RequestBody SysMenu menu) {
        if (UserConstants.NOT_UNIQUE.equals(menuService.checkMenuNameUnique(menu))) {
            return RestResult.error(" 修改菜单 " + menu.getMenuName() + " 失败，名称已存在 ");
        }
        menu.setUpdateBy("system");
        return toAjax(menuService.updateMenu(menu));
    }
    /**
     * 删除菜单
     */
    @DeleteMapping("/{menuId}")
    public RestResult remove(@PathVariable("menuId") Long menuId) {
        if (menuService.hasChildByMenuId(menuId)) {
            return RestResult.error(" 存在子菜单，不允许删除 ");
        }
        if (menuService.checkMenuExistRole(menuId)) {
            return RestResult.error(" 菜单已分配，不允许删除 ");
        }
        return toAjax(menuService.deleteMenuById(menuId));
    }
}
```

步骤 2 更新登录接口的控制层代码

在 SysLoginController 类中增加 getRouters 方法，实现表 6-2 中的用于实现获取菜单路由信息功能的接口。

```
public class SysLoginController {
    ...
    /**
     * 获取路由信息
```

```
     * @return 路由信息
     */
    @GetMapping("getRouters")
    public RestResult getRouters() {
        LoginUser loginUser = tokenService.getLoginUser(ServletUtils.getRequest());
        // 用户信息
        SysUserVO user = loginUser.getUser();
        List<SysMenuVO> menus = menuService.selectMenuTreeByUserId(user.getUserId());
        return RestResult.success(menuService.buildMenus(menus));
    }
}
```

步骤 3　编写菜单的表现层对象

在包“edu.friday.model.vo”下新增步骤 1 和 2 中使用的表现层对象 SysMenuVO 类。

```
/**
 * 菜单权限表 sys_menu
 */
@Data
@NoArgsConstructor
@AllArgsConstructor
public class SysMenuVO implements Serializable
{
    private static final long serialVersionUID = 1L;
    /** 菜单 ID */
    @Id
    @GeneratedValue(strategy = GenerationType.IDENTITY)
    @Column(name = "menuId")
    private Long menuId;

    /** 菜单名称 */
    @NotBlank(message = " 菜单名称不能为空 ")
    @Size(min = 0, max = 50, message = " 菜单名称长度不能超过 50 个字符 ")
    private String menuName;

    /** 父菜单名称 */
    private String parentName;

    /** 父菜单 ID */
    private Long parentId;

    /** 显示顺序 */
    @NotBlank(message = " 显示顺序不能为空 ")
    private String orderNum;

    /** 路由地址 */
    @Size(min = 0, max = 200, message = " 路由地址不能超过 200 个字符 ")
    private String path;
```

```
    /** 组件路径 */
    @Size(min = 0, max = 200, message = " 组件路径不能超过 255 个字符 ")
    private String component;

    /** 是否为外链（0 是 1 否） */
    private String isFrame;

    /** 类型（M 目录 C 菜单 F 按钮） */
    @NotBlank(message = " 菜单类型不能为空 ")
    private String menuType;

    /** 菜单状态 :0 显示 ,1 隐藏 */
    private String visible;

    /** 权限字符串 */
    @Size(min = 0, max = 100, message = " 权限标识长度不能超过 100 个字符 ")
    private String perms;

    /** 菜单图标 */
    private String icon;

    /** 创建者 */
    private String createBy;

    /** 创建时间 */
    @JsonFormat(pattern = "yyyy-MM-dd HH:mm:ss")
    private Date createTime;

    /** 更新者 */
    private String updateBy;

    /** 更新时间 */
    @JsonFormat(pattern = "yyyy-MM-dd HH:mm:ss")
    private Date updateTime;

    /** 备注 */
    private String remark;

    /** 子菜单 */
    private List<SysMenuVO> children = new ArrayList<SysMenuVO>();
}
```

步骤 4　更新菜单实体映射对象

在包“edu.friday.model”下更新实体对象 SysMenu。

```
/**
 * 菜单权限表 sys_menu
 */
@Entity
```

```
@Table(name = "sys_menu")
@Data
@NoArgsConstructor
@AllArgsConstructor
public class SysMenu extends BaseModel {
    /** 菜单 ID */
    @Id
    @GeneratedValue(strategy = GenerationType.IDENTITY)
    private Long menuId;
    /** 菜单名称 */
    private String menuName;
    /** 父菜单 ID */
    private Long parentId;
    /** 显示顺序 */
    private String orderNum;
    /** 路由地址 */
    private String path;
    /** 组件路径 */
    private String component;
    /** 是否为外链（0 是 1 否） */
    private String isFrame;
    /** 类型（M 目录 C 菜单 F 按钮） */
    private String menuType;
    /** 菜单状态 :0 显示 ,1 隐藏 */
    private String visible;
    /** 权限字符串 */
    private String perms;
    /** 菜单图标 */
    private String icon;
}
```

任务评价

技　能　点	知　识　点	自我评价（不熟悉 / 基本掌握 / 熟练掌握 / 灵活运用）
根据接口描述实现控制层代码功能	Spring MVC 工作原理	

任务 2　实现菜单后端业务层

微课 6-2　实现菜单后端业务层

任务分析

编写控制层定义的业务层接口代码，根据控制层的输入参数和返回值实现业务层的逻辑。

任务实施

步骤 1　编写业务层接口代码

根据任务 1 控制层中调用的业务层接口方法，编写 SysMenuService 接口的抽象方法。

```
/**
 * 菜单业务层
 */
public interface SysMenuService {
    /**
     * 根据用户查询系统菜单列表
     * @param userId 用户 ID
     * @return 菜单列表
     */
    List<SysMenu> selectMenuList(Long userId);
    /**
     * 根据用户查询系统菜单列表
     * @param menu 菜单信息
     * @param userId 用户 ID
     * @return 菜单列表
     */
    List<SysMenu> selectMenuList(SysMenuVO menu, Long userId);
    /**
     * 根据用户 ID 查询权限
     * @param userId 用户 ID
     * @return 权限列表
     */
    Set<String> selectMenuPermsByUserId(Long userId);
    /**
     * 根据用户 ID 查询菜单树信息
     * @param userId 用户 ID
     * @return 菜单列表
     */
    List<SysMenuVO> selectMenuTreeByUserId(Long userId);
    /**
     * 根据角色 ID 查询菜单树信息
     * @param roleId 角色 ID
     * @return 选中菜单列表
     */
    List<Long> selectMenuListByRoleId(Long roleId);
    /**
     * 根据菜单 ID 查询信息
     * @param menuId 菜单 ID
     * @return 菜单信息
     */
    SysMenu selectMenuById(Long menuId);
    /**
     * 是否存在菜单子节点
     * @param menuId 菜单 ID
     * @return 结果 true 存在 false 不存在
     */
    boolean hasChildByMenuId(Long menuId);
    /**
     * 查询菜单是否存在角色
     * @param menuId 菜单 ID
     * @return 结果 true 存在 false 不存在
```

```
     */
    boolean checkMenuExistRole(Long menuId);
    /**
     * 校验菜单名称是否唯一
     * @param menu 菜单信息
     * @return 结果
     */
    String checkMenuNameUnique(SysMenu menu);
    /**
     * 新增保存菜单信息
     * @param menu 菜单信息
     * @return 结果
     */
    int insertMenu(SysMenu menu);
    /**
     * 修改保存菜单信息
     * @param menu 菜单信息
     * @return 结果
     */
    int updateMenu(SysMenu menu);
    /**
     * 删除菜单管理信息
     * @param menuId 菜单 ID
     * @return 结果
     */
    int deleteMenuById(Long menuId);
    /**
     * 构建前端路由所需要的菜单
     * @param menus 菜单列表
     * @return 路由列表
     */
    List<RouterVo> buildMenus(List<SysMenuVO> menus);
    /**
     * 构建前端所需要的树结构
     * @param menus 菜单列表
     * @return 树结构列表
     */
    List<SysMenuVO> buildMenuTree(List<SysMenuVO> menus);
    /**
     * 构建前端所需要的下拉树结构
     * @param menus 菜单列表
     * @return 下拉树结构列表
     */
    List<TreeSelect> buildMenuTreeSelect(List<SysMenuVO> menus);
}
```

步骤 2　编写表现层对象

1）在包“edu.friday.common.result”下新增本任务步骤 1 中使用的树型结构实体类 TreeSelect。

```
/**
 * 树型结构实体类
 */
@Data
@NoArgsConstructor
@AllArgsConstructor
public class TreeSelect implements Serializable {
    private static final long serialVersionUID = 1L;
    /** 节点 ID */
    private Long id;
    /** 节点名称 */
    private String label;
    /** 子节点 */
    @JsonInclude(JsonInclude.Include.NON_EMPTY)
    private List<TreeSelect> children;

    public TreeSelect(SysMenuVO menu) {
        this.id = menu.getMenuId();
        this.label = menu.getMenuName();
        this.children = menu.getChildren().stream().map(TreeSelect::new).collect(Collectors.toList());
    }
}
```

2）在包“edu.friday.model.vo”下新增步骤 1 中使用的路由配置信息实体类 RouterVo。

```
/**
 * 路由配置信息
 */
@Data
@NoArgsConstructor
@AllArgsConstructor
@JsonInclude(JsonInclude.Include.NON_EMPTY)
public class RouterVo
{
    /** 路由名字 */
    private String name;
    /** 路由地址 */
    private String path;
    /** 是否隐藏路由，当设置 true 的时候该路由不会在侧边栏出现 */
    private String hidden;
    /** 重定向地址，当设置 noRedirect 时，该路由在面包屑导航中不可被单击 */
    private String redirect;
    /** 组件地址 */
    private String component;
    /** 当该路由下的子路由大于 1 个时，变为嵌套的模式 */
    private Boolean alwaysShow;
    /** 其他元素 */
    private MetaVo meta;
    /** 子路由 */
```

```
    private List<RouterVo> children;
}
```

3）在包“edu.friday.model.vo”下新增步骤 1 中使用的路由显示信息实体类 MetaVo。

```
/**
 * 路由显示信息
 */
@Data
@NoArgsConstructor
@AllArgsConstructor
public class MetaVo {
    /** 设置该路由的名字 */
    private String title;
    /** 设置该路由的图标 */
    private String icon;
}
```

步骤 3 编写业务层实现代码

1）编写 SysMenuServiceImpl 类实现 SysMenuService 接口，用于处理菜单管理业务层逻辑，实现菜单数据列表新增、修改、删除以及根据菜单编号获取详细信息等功能，该业务逻辑与用户管理接口的业务逻辑相似，代码如下。

```
/**
 * 菜单业务层处理
 */
@Service
public class SysMenuServiceImpl implements SysMenuService {
    @Autowired
    SysMenuRepository sysMenuRepository;
    /**
     * 查询系统菜单列表
     * @param menu 菜单信息
     * @return 菜单列表
     */
    @Override
    public List<SysMenu> selectMenuList(SysMenuVO menu, Long userId) {
        List<SysMenu> menuList = null;
        SysMenu sysMenu = new SysMenu();
        BeanUtils.copyPropertiesIgnoreEmpty(menu, sysMenu);
        Sort sort = Sort.by("parentId" , "orderNum");
        ExampleMatcher exampleMatcher = ExampleMatcher.matching()
        .withMatcher("menuName" , ExampleMatcher.GenericPropertyMatchers.contains());
        Example example = Example.of(sysMenu, exampleMatcher);
        // 管理员显示所有菜单信息
        if (SysUserVO.isAdmin(userId)) {
          menuList = sysMenuRepository.findAll(example, sort);
        } else {
          menuList = sysMenuRepository.selectMenuListByUserId(sysMenu, userId);
```

```
        }
        return menuList;
    }
    /**
     * 根据菜单 ID 查询信息
     * @param menuId 菜单 ID
     * @return 菜单信息
     */
    @Override
    public SysMenu selectMenuById(Long menuId) {
        Optional<SysMenu> op = sysMenuRepository.findById(menuId);
        return op.isPresent() ? op.get() : null;
    }
    /**
     * 新增保存菜单信息
     * @param menu 菜单信息
     * @return 结果
     */
    @Override
    public int insertMenu(SysMenu menu) {
        sysMenuRepository.save(menu);
        return (menu != null && null != menu.getMenuId()) ? 1 : 0;
    }
    /**
     * 修改保存菜单信息
     * @param menu 菜单信息
     * @return 结果
     */
    @Override
    public int updateMenu(SysMenu menu) {
        SysMenu sysMenu = new SysMenu();
        Optional<SysMenu> op = sysMenuRepository.findById(menu.getMenuId());
        if (!op.isPresent()) {
            return 0;
        }
        sysMenu = op.get();
        BeanUtils.copyPropertiesIgnoreNull(menu, sysMenu);
        sysMenuRepository.save(sysMenu);
        return 1;
    }
    /**
     * 删除菜单管理信息
     * @param menuId 菜单 ID
     * @return 结果
     */
    @Override
    public int deleteMenuById(Long menuId) {
        sysMenuRepository.deleteById(menuId);
        return 1;
    }
}
```

2）编写业务逻辑检查规则代码，在 SysMenuServiceImpl 类中增加 checkMenuNameUnique、hasChildByMenuId 和 checkMenuExistRole 方法，分别用于校验菜单名称是否唯一、菜单是否存在子节点和菜单是否已映射角色信息，代码如下。

```
/**
 * 校验菜单名称是否唯一
 * @param menu 菜单信息
 * @return 结果
 */
@Override
public String checkMenuNameUnique(SysMenu menu) {
    Long menuId = StringUtils.isNull(menu.getMenuId()) ? -1L : menu.getMenuId();
    SysMenu sysMenu = new SysMenu();
    sysMenu.setParentId(menuId);
    sysMenu.setMenuName(menu.getMenuName());
    Page<SysMenu> list = sysMenuRepository.findAll(Example.of(sysMenu), PageRequest.of(0, 1));
    SysMenu info = null;
    if (list.getTotalElements() > 0) {
        info = list.toList().get(0);
    }
    if (StringUtils.isNotNull(info) && info.getMenuId().longValue() != menuId.longValue()) {
        return UserConstants.NOT_UNIQUE;
    }
    return UserConstants.UNIQUE;
}
/**
 * 是否存在菜单子节点
 * @param menuId 菜单 ID
 * @return 结果
 */
@Override
public boolean hasChildByMenuId(Long menuId) {
    SysMenu sysMenu = new SysMenu();
    sysMenu.setParentId(menuId);
    long result = sysMenuRepository.count(Example.of(sysMenu));
    return result > 0 ? true : false;
}
/**
 * 查询菜单使用数量
 * @param menuId 菜单 ID
 * @return 结果
 */
@Override
public boolean checkMenuExistRole(Long menuId) {
    long result = sysMenuRepository.checkMenuExistRole(menuId);
    return result > 0 ? true : false;
}
```

3）该业务逻辑较为复杂的是构建树型菜单数据，在 SysMenuServiceImpl 类中首先构建 recursionFn 方法递归查询菜单子节点，然后使用 buildMenuTree 方法调用 recursionFn 递归方法构建前端所需要的菜单树型结构，代码如下。

```
/**
 * 构建前端所需要的下拉树结构
 * @param menus 菜单列表
 * @return 下拉树结构列表
 */
@Override
public List<TreeSelect> buildMenuTreeSelect(List<SysMenuVO> menus) {
    List<SysMenuVO> menuTrees = buildMenuTree(menus);
    List<TreeSelect> rs = menuTrees.stream().map(TreeSelect::new).collect(Collectors.toList());
    return rs;
}
/**
 * 构建前端所需要的树结构
 * @param menus 菜单列表
 * @return 树结构列表
 */
@Override
public List<SysMenuVO> buildMenuTree(List<SysMenuVO> menus) {
    List<SysMenuVO> returnList = new ArrayList<SysMenuVO>();
    for (Iterator<SysMenuVO> iterator = menus.iterator(); iterator.hasNext(); ) {
        SysMenuVO t = (SysMenuVO) iterator.next();
        // 根据传入的某个父节点 ID，遍历该父节点的所有子节点
        if (t.getParentId() == 0) {
            recursionFn(menus, t);
            returnList.add(t);
        }
    }
    if  (returnList.isEmpty()) {
        returnList = menus;
    }
    return returnList;
}
/**
 * 递归列表
 * @param list
 * @param t
 */
private void recursionFn(List<SysMenuVO> list, SysMenuVO t) {
    // 得到子节点列表
    List<SysMenuVO> childList = getChildList(list, t);
    t.setChildren(childList);
    for (SysMenuVO tChild : childList) {
        if (hasChild(list, tChild)) {
            // 判断是否有子节点
            Iterator<SysMenuVO> it = childList.iterator();
            while (it.hasNext()) {
                SysMenuVO n = (SysMenuVO) it.next();
                recursionFn(list, n);
            }
```

```
            }
        }
    }
    /**
     * 得到子节点列表
     */
    private List<SysMenuVO> getChildList(List<SysMenuVO> list, SysMenuVO t) {
        List<SysMenuVO> tlist = new ArrayList<SysMenuVO>();
        Iterator<SysMenuVO> it = list.iterator();
        while (it.hasNext()) {
            SysMenuVO n = (SysMenuVO) it.next();
            if (n.getParentId().longValue() == t.getMenuId().longValue()) {
                tlist.add(n);
            }
        }
        return tlist;
    }

    /**
     * 判断是否有子节点
     */
    private boolean hasChild(List<SysMenuVO> list, SysMenuVO t) {
        return getChildList(list, t).size() > 0 ? true : false;
    }
```

4）在 SysMenuServiceImpl 类中增加 selectMenuPermsByUserId 和 buildMenus 等方法，用于登录成功后，获取前端菜单路由信息的树型列表，代码如下。

```
    /**
     * 根据用户 ID 查询菜单
     * @param userId 用户名称
     * @return 菜单列表
     */
    @Override

    public List<SysMenuVO> selectMenuTreeByUserId(Long userId) {
        List<SysMenu> menus = null;
        if (SysUserVO.isAdmin(userId)) {
            menus = sysMenuRepository.selectMenuTreeAll();
        } else {
            menus = sysMenuRepository.selectMenuTreeByUserId(userId);
        }
        List<SysMenuVO> menuVOS = BeanUtils.copyProperties(menus, SysMenuVO.class);
        return getChildPerms(menuVOS, 0);
    }
    /**
     * 根据父节点的 ID 获取所有子节点
     * @param list 分类表
     * @param parentId 传入的父节点 ID
     * @return String
     */
```

```
public List<SysMenuVO> getChildPerms(List<SysMenuVO> list, int parentId) {
    List<SysMenuVO> returnList = new ArrayList<SysMenuVO>();
    for (Iterator<SysMenuVO> iterator = list.iterator(); iterator.hasNext(); ) {
        SysMenuVO t = (SysMenuVO) iterator.next();
        // 根据传入的某个父节点 ID, 遍历该父节点的所有子节点
        if (t.getParentId() == parentId) {
            recursionFn(list, t);
            returnList.add(t);
        }
    }
    return returnList;
}
/**
 * 构建前端路由所需要的菜单
 * @param menus 菜单列表
 * @return 路由列表
 */
@Override
public List<RouterVo> buildMenus(List<SysMenuVO> menus) {
    List<RouterVo> routers = new LinkedList<RouterVo>();
    for (SysMenuVO menu : menus) {
        RouterVo router = new RouterVo();
        router.setName(StringUtils.capitalize(menu.getPath()));
        router.setPath(getRouterPath(menu));
        router.setComponent(StringUtils.isEmpty(menu.getComponent()) ? "Layout" : menu.getComponent());
        router.setMeta(new MetaVo(menu.getMenuName(), menu.getIcon()));
        List<SysMenuVO> cMenus = menu.getChildren();
        if (!cMenus.isEmpty() && cMenus.size() > 0 && "M".equals(menu.getMenuType())) {
            router.setAlwaysShow(true);
            router.setRedirect("noRedirect");
            router.setChildren(buildMenus(cMenus));
        }
        routers.add(router);
    }
    return routers;
}
/**
 * 获取路由地址
 * @param menu 菜单信息
 * @return 路由地址
 */
public String getRouterPath(SysMenuVO menu) {
    String routerPath = menu.getPath();
    // 非外链并且是一级目录
    if (0 == menu.getParentId() && "1".equals(menu.getIsFrame())) {
        routerPath = "/" + menu.getPath();
    }
    return routerPath;
}
```

知识小结

1. 菜单的树型结构

用树型结构表示实体类型及实体间联系的数据模型称为层次模型。层次模型的限制条件为有且仅有一个节点，无父节点，此节点为树的根；其他节点有且仅有一个父节点。层次模型的特点为通过指针实现记录之间的联系，查询效率高。其缺点是只能表示 1:N 的联系。

设计树型结构的数据表要关注下面两点内容：

1）通过某个节点找到所有的父节点。

2）获取某个节点的所有的后继节点，包括子节点的子节点。

2. 递归调用

基于菜单的树型结构，程序要实现层次模型的树型结构，即通过某个节点遍历所有子节点或者子节点的子节点，递归调用是有效的方法，就是在当前的函数中调用当前函数本身并传给相应的参数，这一动作是层层进行的，直到满足一般情况的时候，才停止递归调用，开始从最后一个递归调用返回的过程，本任务通过编写 recursionFn 递归调用方法来构建菜单的树型结构。

任务评价

技　能　点	知　识　点	自我评价（不熟悉 / 基本掌握 / 熟练掌握 / 灵活运用）
完成树型结构设计与代码实现	树型结构设计	

任务 3　实现菜单后端数据控制层

微课 6-3　实现菜单后端数据控制层

任务分析

为业务层定义的数据控制层接口编写代码，实现对 MySQL 数据库的数据操作。

任务实施

步骤 1　编写数据控制层代码

在包“edu.friday.repository”下编写 SysMenuRepository 接口，通过 @Query 注解实现本地 SQL 查询，返回业务层需要的菜单数据。

```
/**
 * 菜单表数据层
 */
@Repository
public interface SysMenuRepository extends JpaRepository<SysMenu, Long>, SysMenuCustomRepository {
    String SELECT_DISTINCT = " select distinct m.menu_id, m.parent_id, m.menu_name, m.path,
m.component, m.visible, ifnull(m.perms,'') as perms, m.is_frame, m.menu_type, m.icon, m.order_num,
```

```
m.create_by, m.create_time, m.update_by, m.update_time, m.remark from sys_menu m ";
        String JOIN_ROLE_MENU = " left join sys_role_menu rm on m.menu_id = rm.menu_id ";
        String JOIN_USER_ROLE = " left join sys_user_role ur on rm.role_id = ur.role_id ";
        String JOIN_ROLE = " left join sys_role ro on ur.role_id = ro.role_id ";
        String JOIN_USER = " left join sys_user u on ur.user_id = u.user_id ";
        String ORDER_BY = " order by m.parent_id, m.order_num ";

        @Query(value = " select distinct m.perms from sys_menu m " + JOIN_ROLE_MENU + JOIN_
USER_ROLE + " where ur.user_id = :userId ", nativeQuery = true)
        List<String> selectMenuPermsByUserId(@Param("userId") Long userId);

        @Query(value = " select count(*) from sys_role_menu where menu_id = :menuId ", nativeQuery
= true)
        long checkMenuExistRole(@Param("menuId") Long menuId);

        @Query(value = SELECT_DISTINCT + "  where m.menu_type in ('M', 'C') and m.visible = 0 " +
ORDER_BY, nativeQuery = true)
        List<SysMenu> selectMenuTreeAll();

        @Query(value = SELECT_DISTINCT + JOIN_ROLE_MENU + JOIN_USER_ROLE + JOIN_ROLE
+ JOIN_USER +
        " where u.user_id = :userId and m.menu_type in ('M', 'C') and m.visible = 0  AND ro.status = 0 "
+ ORDER_BY, nativeQuery = true)
        List<SysMenu> selectMenuTreeByUserId(@Param("userId") Long userId);
    }
```

步骤 2　编写自定义数据控制层代码

1）在包“edu.friday.repository.custom”下编写 SysMenuCustomRepository 接口，定义通过用户 ID 和角色 ID 获取菜单列表数据的方法。

```
/**
 * 菜单表数据层
 */
@Repository
public interface SysMenuCustomRepository {
    List<SysMenu> selectMenuListByUserId(SysMenu sysMenu, Long userId);
    List<SysMenu> selectMenuListByRoleId(Long roleId);
}
```

2）在包“edu.friday.repository.custom.impl”下编写 SysMenuCustomRepository 接口的实现类 SysMenuCustomRepositoryImpl，使用 StringBuffer 实例根据业务逻辑拼接动态 SQL，然后调用 EntityManager.createNativeQuery() 构建原生 SQL 的查询语句，最后通过执行 query.getResultList() 方法执行构建的 SQL 查询语句，获取菜单数据。

```
public class SysMenuCustomRepositoryImpl implements SysMenuCustomRepository {
    String SELECT = " select distinct m.* from sys_menu m ";
    String JOIN_ROLE_MENU = " left join sys_role_menu rm on m.menu_id = rm.menu_id ";
    String JOIN_USER_ROLE = " left join sys_user_role ur on rm.role_id = ur.role_id ";
```

```
    String JOIN_ROLE = " left join sys_role ro on ur.role_id = ro.role_id ";
    String ORDER_BY = " order by parent_id, order_num ";

    @PersistenceContext
    private EntityManager entityManager;

    @Override
    public List<SysMenu> selectMenuListByUserId(SysMenu sysMenu, Long userId) {
        boolean queryMenuName = null != sysMenu && !StringUtils.isBlank(sysMenu.getMenuName());
        boolean queryVisible = null != sysMenu && !StringUtils.isBlank(sysMenu.getVisible());
        StringBuffer sql = new StringBuffer();
        sql.append(SELECT);
        sql.append(JOIN_ROLE_MENU);
        sql.append(JOIN_USER_ROLE);
        sql.append(" where ur.user_id = :userId ");
        if (queryMenuName) {
            sql.append(" AND menu_name like concat('%', :menuName, '%') ");
        }
        if (queryVisible) {
            sql.append(" AND visible = :visible ");
        }
        sql.append(ORDER_BY);
        Query query = entityManager.createNativeQuery(sql.toString(), SysMenu.class).setParameter
("userId", userId);
        if (queryMenuName) {
            query.setParameter("menuName", sysMenu.getMenuName());
        }
        if (queryVisible) {
            query.setParameter("visible", sysMenu.getVisible());
        }
        return query.getResultList();
    }

    @Override
    public List<SysMenu> selectMenuListByRoleId(Long roleId) {
        StringBuffer sql = new StringBuffer();
        sql.append(SELECT);
        sql.append(JOIN_ROLE_MENU);
        sql.append(" where rm.role_id = :roleId ");
        sql.append(" and m.menu_id not in ( ");
        sql.append(" select DISTINCT m.parent_id from sys_menu m ");
        sql.append(" inner join sys_role_menu rm on m.menu_id = rm.menu_id ");
        sql.append(" and rm.role_id = :roleId ");
        sql.append(" ) ");
        sql.append(ORDER_BY);
        Query query = entityManager.createNativeQuery(sql.toString(), SysMenu.class).setParameter
("roleId", roleId);
        return query.getResultList();
    }
}
```

任务评价

技能点	知识点	自我评价（不熟悉 / 基本掌握 / 熟练掌握 / 灵活运用）
编写动态 SQL 实现复杂业务逻辑的数据查询	Spring Data JPA 的本地 SQL 查询	

任务 4 测试验证菜单后端接口

微课 6-4　测试验证菜单后端接口

任务分析

使用 Postman 工具测试表 6-1 中的全部功能接口。

任务实施

执行本任务的步骤前，需先登录接口获取 Token 数据，然后在 Headers 下设置 Authorization 的值为 Token 数据，参见工作单元 5。

步骤 1 验证获取树状菜单列表接口

Postman 下验证获取树状菜单列表接口，选择“GET”模式，输入 URL 地址 http://localhost:8080/system/menu/list，单击“Send”按钮，如图 6-1 所示，返回菜单列表数据。

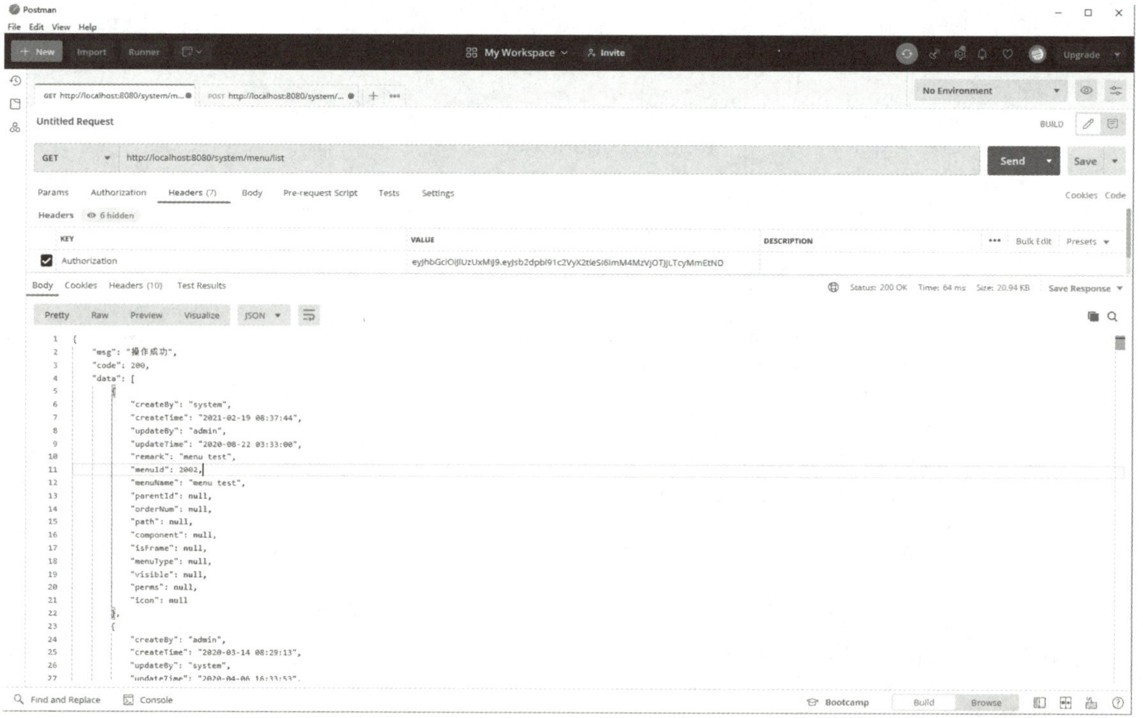

图 6-1　菜单列表数据

步骤 2 验证新增菜单数据接口

Postman 下验证新增菜单数据接口，选择“POST”模式，输入 URL 地址 http://

localhost:8080/system/menu，在“Body”里选择“raw”→“JSON”，提交如图 6-2 中所示 JSON 内容，单击“Send”按钮，然后查看 sys_menu 表是否有新增数据。

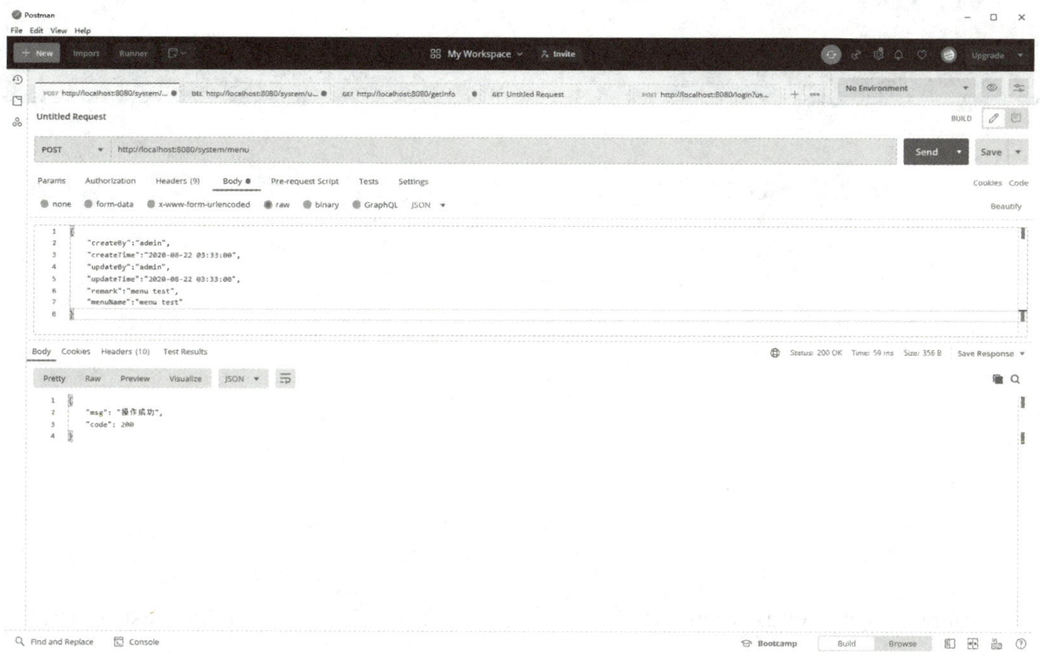

图 6-2　Postman 测试新增数据

步骤 3　验证修改菜单数据接口

Postman 下测试修改菜单数据接口，调用更新接口前的 sys_menu 表，如图 6-3 所示，menu_id 为 2002 的 remark 值为“menu test”。

对象　sys_menu @friday (local) - 表

开始事务　文本　筛选　排序　导入　导出

menu_id	menu_name	parent_id	order_num	path	remark	component	is_frame	menu_type	visible	perms	icon	create_by
1027	字典新增	105	2	#			1	F	0	system:dict:add	#	admin
1028	字典修改	105	3	#			1	F	0	system:dict:edit	#	admin
1029	字典删除	105	4	#			1	F	0	system:dict:remove	#	admin
1030	字典导出	105	5	#			1	F	0	system:dict:export	#	admin
1031	参数查询	106	1	#			1	F	0	system:config:query	#	admin
1032	参数新增	106	2	#			1	F	0	system:config:add	#	admin
1033	参数修改	106	3	#			1	F	0	system:config:edit	#	admin
1034	参数删除	106	4	#			1	F	0	system:config:remove	#	admin
1035	参数导出	106	5	#			1	F	0	system:config:export	#	admin
1036	公告查询	107	1	#			1	F	0	system:notice:query	#	admin
1037	公告新增	107	2	#			1	F	0	system:notice:add	#	admin
1038	公告修改	107	3	#			1	F	0	system:notice:edit	#	admin
1039	公告删除	107	4	#			1	F	0	system:notice:remove	#	admin
1040	操作查询	500	1	#			1	F	0	monitor:operlog:query	#	admin
1041	操作删除	500	2	#			1	F	0	monitor:operlog:remove	#	admin
1042	日志导出	500	4	#			1	F	0	monitor:operlog:export	#	admin
1043	登录查询	501	1	#			1	F	0	monitor:logininfor:query	#	admin
1044	登录删除	501	2	#			1	F	0	monitor:logininfor:remov	#	admin
1045	日志导出	501	3	#			1	F	0	monitor:logininfor:export	#	admin
1046	在线查询	109	1	#			1	F	0	monitor:online:query	#	admin
1047	批量强退	109	2	#			1	F	0	monitor:online:batchLogo	#	admin
1048	单条强退	109	3	#			1	F	0	monitor:online:forceLogo	#	admin
1049	任务查询	110	1	#			1	F	0	monitor:job:query	#	admin
1050	任务新增	110	2	#			1	F	0	monitor:job:add	#	admin
1051	任务修改	110	3	#			1	F	0	monitor:job:edit	#	admin
1052	任务删除	110	4	#			1	F	0	monitor:job:remove	#	admin
1053	状态修改	110	5	#			1	F	0	monitor:job:changeStatus	#	admin
1054	任务导出	110	7	#			1	F	0	monitor:job:export	#	admin
1055	生成查询	114	1	#			1	F	0	tool:gen:query	#	admin
1056	生成修改	114	2	#			1	F	0	tool:gen:edit	#	admin
1057	生成删除	114	3	#			1	F	0	tool:gen:remove	#	admin
1058	导入代码	114	2	#			1	F	0	tool:gen:import	#	admin
1059	预览代码	114	4	#			1	F	0	tool:gen:preview	#	admin
1060	生成代码	114	5	#			1	F	0	tool:gen:code	#	admin
2000	test	0	0	system		(Null)	0	M	1		theme	admin
2001	erter	2000	0	user		system/user/index	1	C	0	system:user:list	404	admin
2002	menu test	(Null)	(Null)	(Null)	menu test	(Null)	(Null)	(Null)	(Null)	(Null)	(Null)	system

图 6-3　sys_menu 表的数据

Postman 里输入 http://localhost:8080/system/menu，选择“PUT”模式，选择“body”→“RAW”，输入 JSON 数据，然后单击“Send”按钮，返回成功信息，如图 6-4 所示。

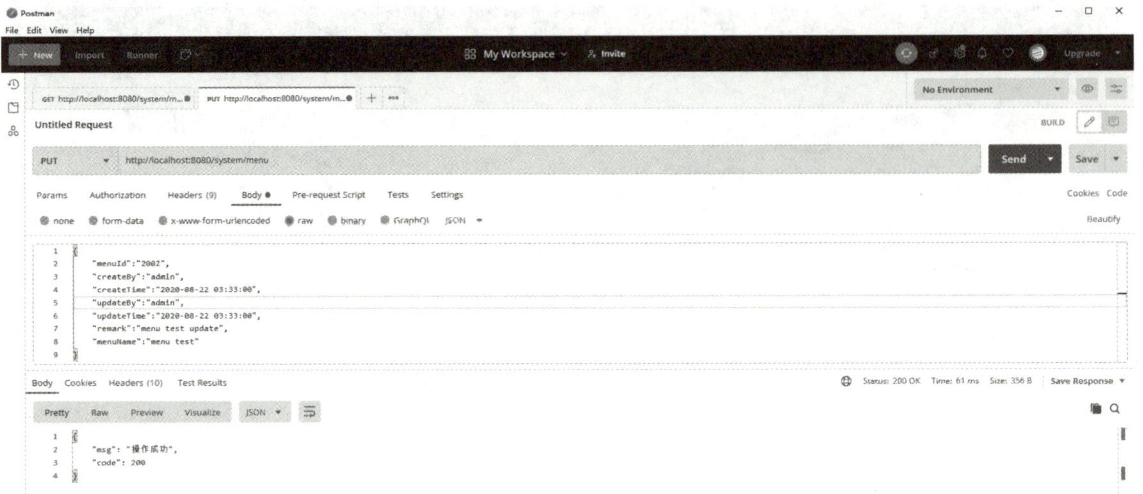

图 6-4　调用更新接口

查看数据表 sys_menu，menu_id 为 2002 的 remark 值被修改为“menu test update”，如图 6-5 所示。

menu_i	menu_name	parer	orc	path	componen	is_fram	menu_ty	visible	perms	icon	create_b	create_time	update_b	update_time	remark
1040	操作查询	500	1	#		1	F	0	monitor:operlc	#	admin	2018-03-16 11:	ry	2018-03-16 11:	
1041	操作删除	500	2	#		1	F	0	monitor:operlc	#	admin	2018-03-16 11:	ry	2018-03-16 11:	
1042	日志导出	500	4	#		1	F	0	monitor:operlc	#	admin	2018-03-16 11:	ry	2018-03-16 11:	
1043	登录查询	501	1	#		1	F	0	monitor:loginii	#	admin	2018-03-16 11:	ry	2018-03-16 11:	
1044	登录删除	501	2	#		1	F	0	monitor:loginii	#	admin	2018-03-16 11:	ry	2018-03-16 11:	
1045	日志导出	501	3	#		1	F	0	monitor:loginii	#	admin	2018-03-16 11:	ry	2018-03-16 11:	
1046	在线查询	109	1	#		1	F	0	monitor:online	#	admin	2018-03-16 11:	ry	2018-03-16 11:	
1047	批量强退	109	2	#		1	F	0	monitor:online	#	admin	2018-03-16 11:	ry	2018-03-16 11:	
1048	单条强退	109	3	#		1	F	0	monitor:online	#	admin	2018-03-16 11:	ry	2018-03-16 11:	
1049	任务查询	110	1	#		1	F	0	monitor:job:qu	#	admin	2018-03-16 11:	ry	2018-03-16 11:	
1050	任务新增	110	2	#		1	F	0	monitor:job:ad	#	admin	2018-03-16 11:	ry	2018-03-16 11:	
1051	任务修改	110	3	#		1	F	0	monitor:job:ed	#	admin	2018-03-16 11:	ry	2018-03-16 11:	
1052	任务删除	110	4	#		1	F	0	monitor:job:rei	#	admin	2018-03-16 11:	ry	2018-03-16 11:	
1053	状态修改	110	5	#		1	F	0	monitor:job:ch	#	admin	2018-03-16 11:	ry	2018-03-16 11:	
1054	任务导出	110	7	#		1	F	0	monitor:job:ex	#	admin	2018-03-16 11:	ry	2018-03-16 11:	
1055	生成查询	114	1	#		1	F	0	tool:gen:query	#	admin	2018-03-16 11:	ry	2018-03-16 11:	
1056	生成修改	114	2	#		1	F	0	tool:gen:edit	#	admin	2018-03-16 11:	ry	2018-03-16 11:	
1057	生成删除	114	3	#		1	F	0	tool:gen:remov	#	admin	2018-03-16 11:	ry	2018-03-16 11:	
1058	导入代码	114	2	#		1	F	0	tool:gen:impor	#	admin	2018-03-16 11:	ry	2018-03-16 11:	
1059	预览代码	114	4	#		1	F	0	tool:gen:previe	#	admin	2018-03-16 11:	ry	2018-03-16 11:	
1060	生成代码	114	5	#		1	F	0	tool:gen:code	#	admin	2018-03-16 11:	ry	2018-03-16 11:	
2000	test	0	0	system	(Null)	0	M	1		theme	admin	2020-03-14 16:	system	2020-04-07 00:	
2001	erter	2000	0	user	system/us	1	C	0	system:user:lis	404	admin	2020-03-14 16:	system	2020-04-07 00:	
2002	menu test	(Null)	ıll)	(Null)	(Null)	(Null)	(Null)	(Null)	(Null)	(Null)	admin	2021-08-30 19:	system	2021-08-30 19:	menu test update

图 6-5　remark 值为 menu test update

步骤 4　验证删除菜单数据接口

Postman 下测试删除菜单数据接口，输入 http://localhost:8080/system/menu/2002，选择“DELETE”模式，然后单击“Send”按钮，返回成功信息，过程如图 6-6 所示，然后查看

sys_menu 表，menu_id=2002 数据已被删除。

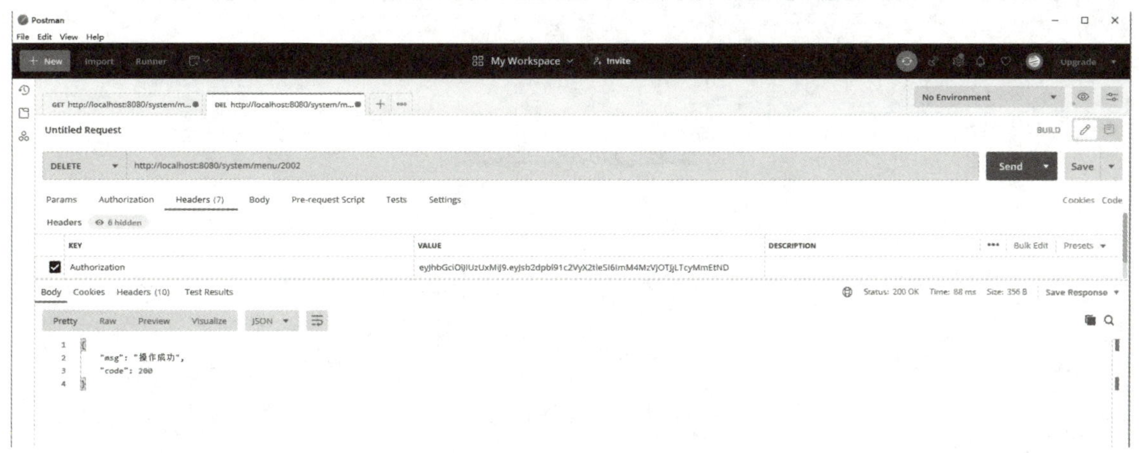

图 6-6　调用删除菜单数据接口

步骤 5　验证根据菜单编号获取详细信息接口

Postman 中测试根据菜单编号获取详细信息接口，选择“GET”模式，输入 http://localhost:8080/system/menu/2000，单击“Send”按钮，如图 6-7 所示，返回当前菜单详细信息数据。

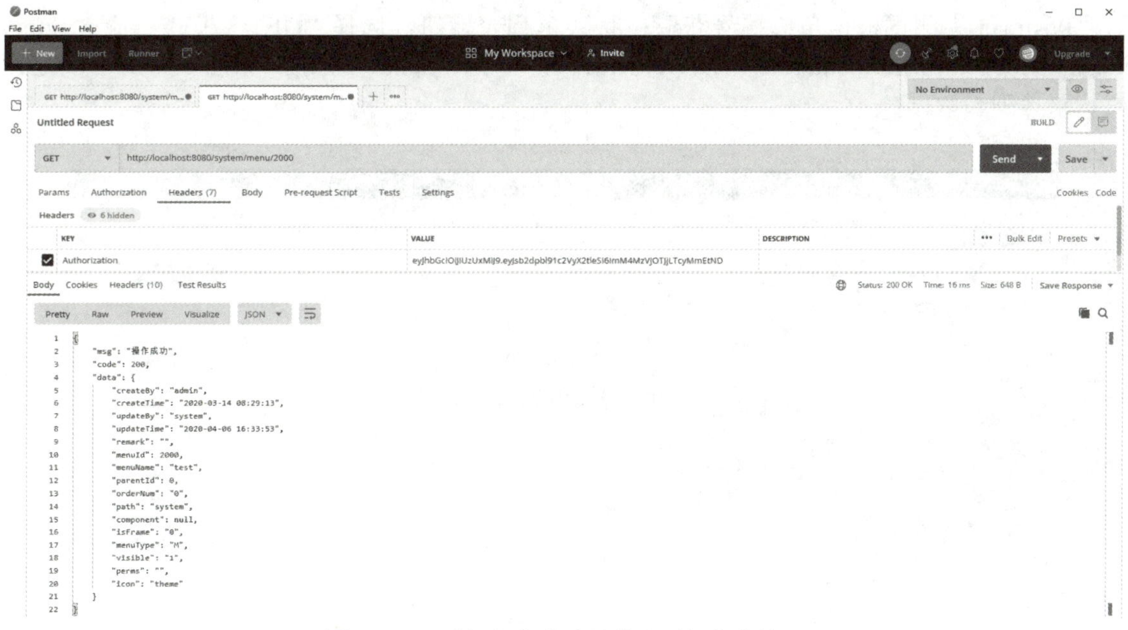

图 6-7　调用根据菜单编号获取详细信息接口

步骤 6　验证获取树状菜单列表接口

Postman 中测试获取树状菜单列表接口，选择“GET”模式，输入 http://localhost:8080/system/menu/treeselect，单击“Send”按钮，如图 6-8 所示，返回树状菜单列表数据。

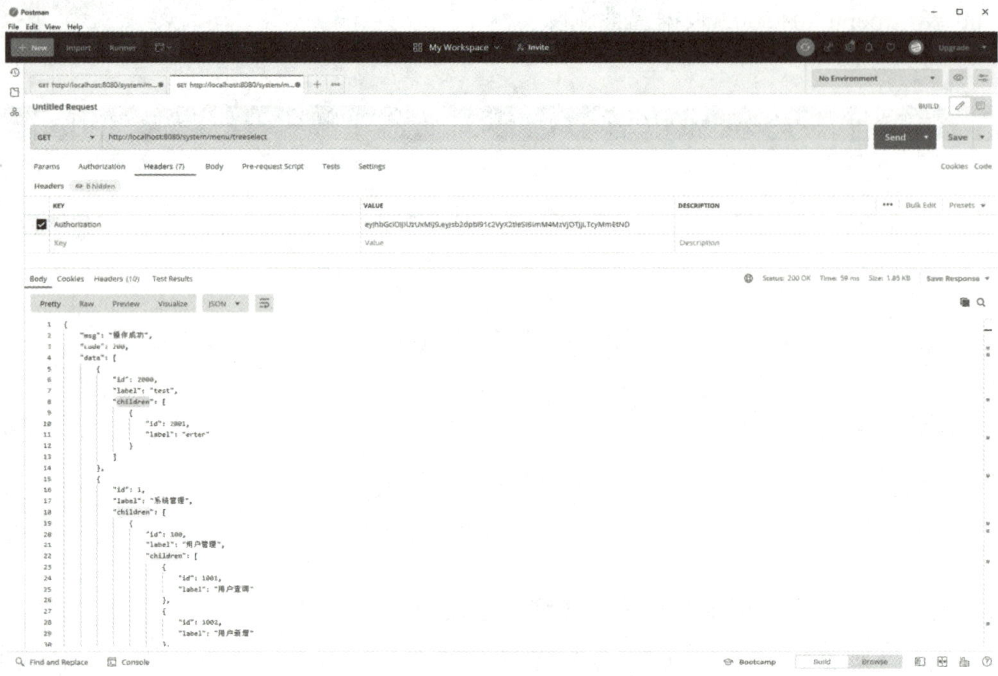

图 6-8　调用获取树状菜单列表接口

步骤 7　验证根据角色编号获取树状菜单列表接口

Postman 中测试根据角色编号获取树状菜单列表接口，选择“GET”模式，输入 http://localhost:8080/system/menu/roleMenuTreeselect/1，单击“Send”按钮，如图 6-9 所示，返回树状菜单列表数据。

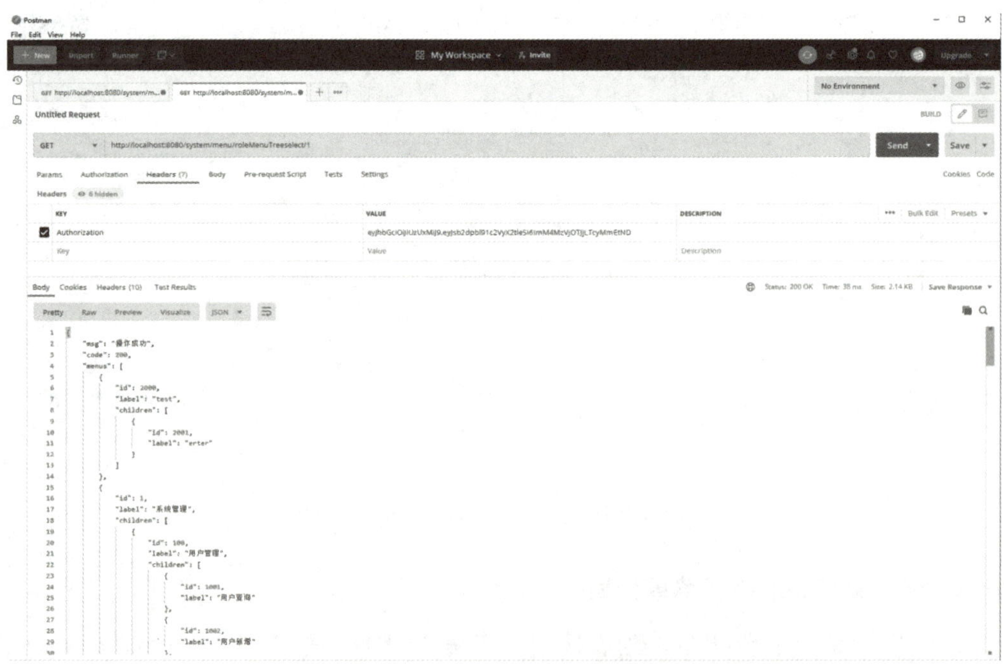

图 6-9　调用根据角色编号获取树状菜单列表接口

步骤 8 验证获取菜单路由信息接口

Postman 中测试获取菜单路由信息接口，选择“GET”模式，输入 http://localhost:8080/getRouters，单击“Send”按钮，如图 6-10 所示，返回菜单路由信息。

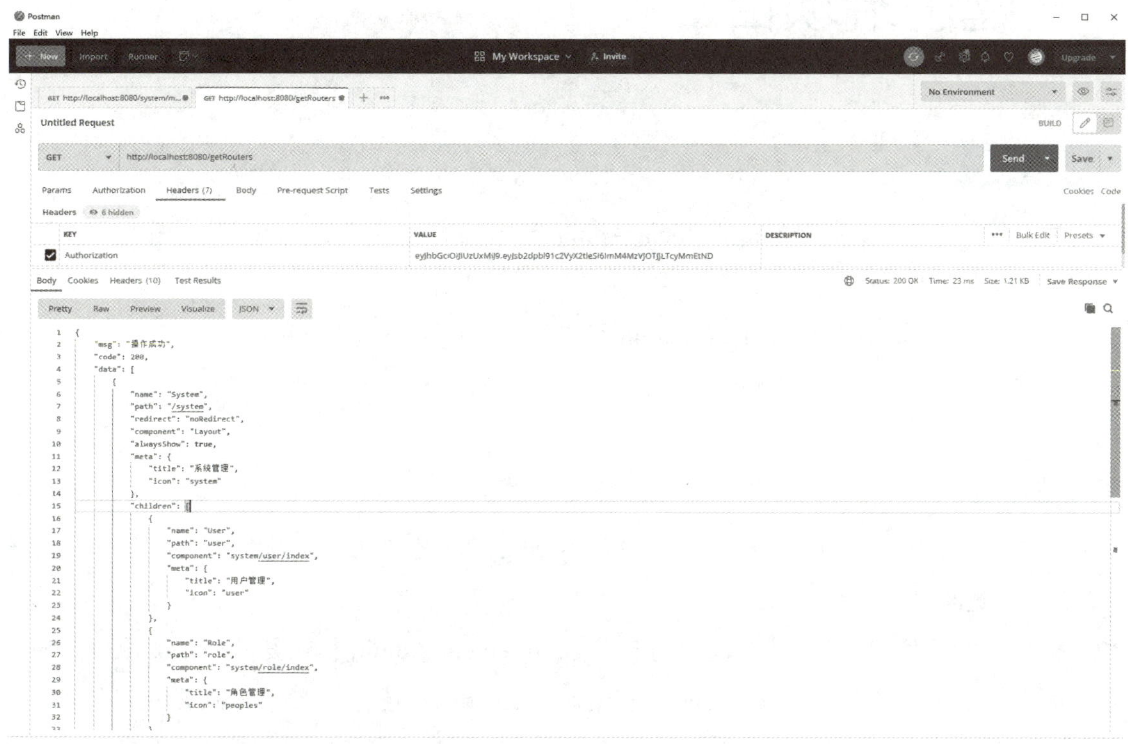

图 6-10　调用获取菜单路由信息接口

任务评价

技　能　点	知　识　点	自我评价（不熟悉 / 基本掌握 / 熟练掌握 / 灵活运用）
使用 Postman 验证后端服务接口	Postman 基础知识	

单元小结

本工作单元通过分层开发方法，专注于控制层、业务层和数据控制层的编码任务，高效实现菜单接口功能。控制层编写输入参数和返回值的接口定义；业务层根据常用面向对象设计原则选用合适的编程方法实现业务逻辑，如本工作单元中使用递归调用方法实现菜单的树型结构；数据控制层主要是根据业务层定义的逻辑操作数据表。通过完成本工作单元的任务和后面的实战强化，熟练掌握基于三层架构的分层开发的职业技能。

实战强化

按照本工作单元任务和步骤，编写实训项目“诚品书城”的图书菜单模块和图书推荐首页的后端接口。

工作单元 7

构建前端项目

职业能力

本工作单元主要是基于 Vue-Element-Admin 框架构建前端项目公共模块，最终希望学生达成如下职业能力目标：

1. 熟练掌握使用 Vue-Element-Admin 框架构建前端项目公共模块
2. 掌握 Vue-Element-Admin 框架的项目结构与公共工具类

任务情景

在前后端分离的项目中，开发人员使用 SpringBoot 框架快速构建后端项目，那么前端项目如何快速构建呢?

本工作单元的任务就是构建前端项目，Vue-Element-Admin 框架是一个主流的前后端分离系统的前端解决方案，它基于 VueJS 框架和 Element-UI 组件库实现，提供了动态路由，权限验证等丰富的功能组件，使用该框架可以帮助开发人员快速搭建企业级后台产品的前端项目。

前置知识

NPM 包管理器命令

NPM 全称 Node Package Manager，是 Node.js 的软件包管理工具，它是提供自动安装、配制、卸载和升级软件包的工具组合，在使用 Vue.js 前端框架时，使用 NPM 的包管理器命令就可以安装使用项目需要的模块包，非常便捷高效，常用的包管理器命令见表 7-1。

表 7-1　常用的包管理器命令

命令语法	命令作用
npm install <package name>	使用 npm install 安装依赖包的最新版
npm update <package name>	使用 npm update 更新包到最新版本
npm uninstall <package name>	使用 npm uninstall 下载依赖包
npm list/ls	使用 npm list/ls 查看已安装包的版本

任务 1 初始化前端项目

微课 7-1 初始化前端项目

步骤 1 下载 Vue-Element-Admin 框架

访问网址：https://github.com/PanJiaChen/vue-element-admin，单击“Code”→“Download ZIP”下载框架压缩包，然后将压缩包解压缩到“C:\friday”，解压名字为“friday-ui”，如图 7-1 所示。

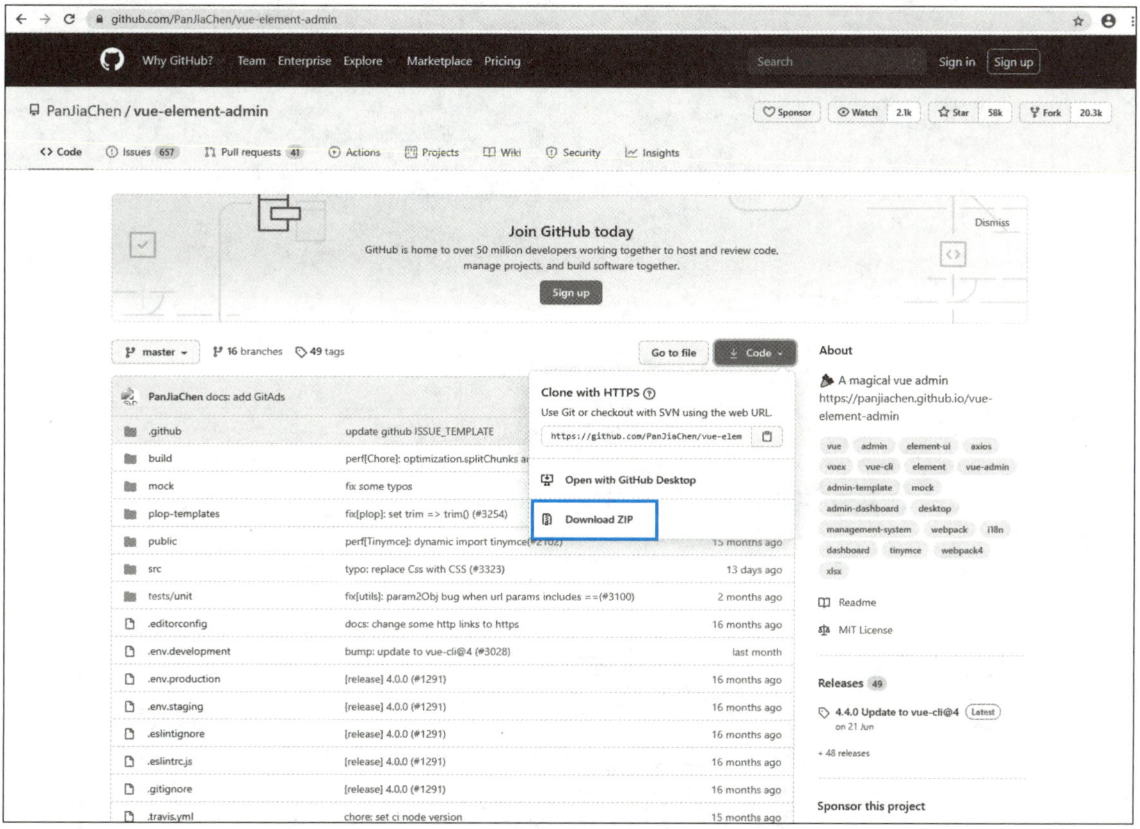

图 7-1 下载 Vue-Element-Admin 压缩包

步骤 2 导入项目到 Visual Studio Code 工具

使用 Visual Studio Code 导入前端项目，单击“Open folder…”，选择“C:\friday\friday-ui”文件夹，导入 friday-ui 项目，如图 7-2 所示。

步骤 3 安装项目依赖

按 <Ctrl+Shift+`> 组合键打开 TERMINAL 窗口，在 TERMINAL 中运行“npm install”命令安装项目依赖，NPM 包管理器下载的依赖模块都存储在“node_modules”目录下，过程如图 7-3 所示。

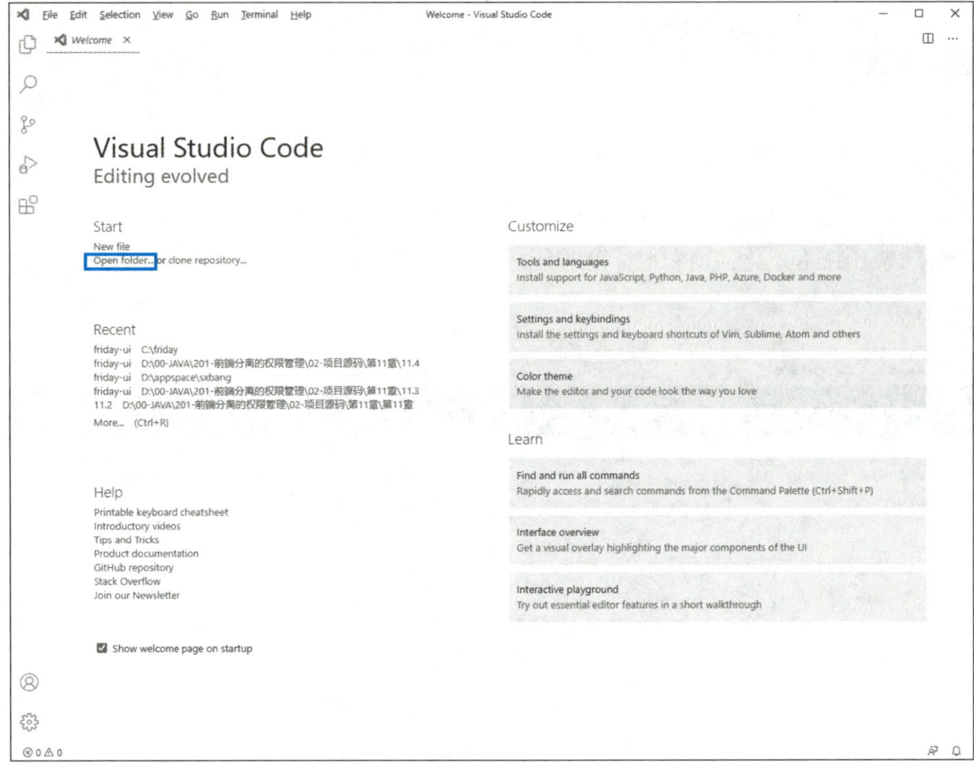

图 7-2　导入 friday-ui 项目

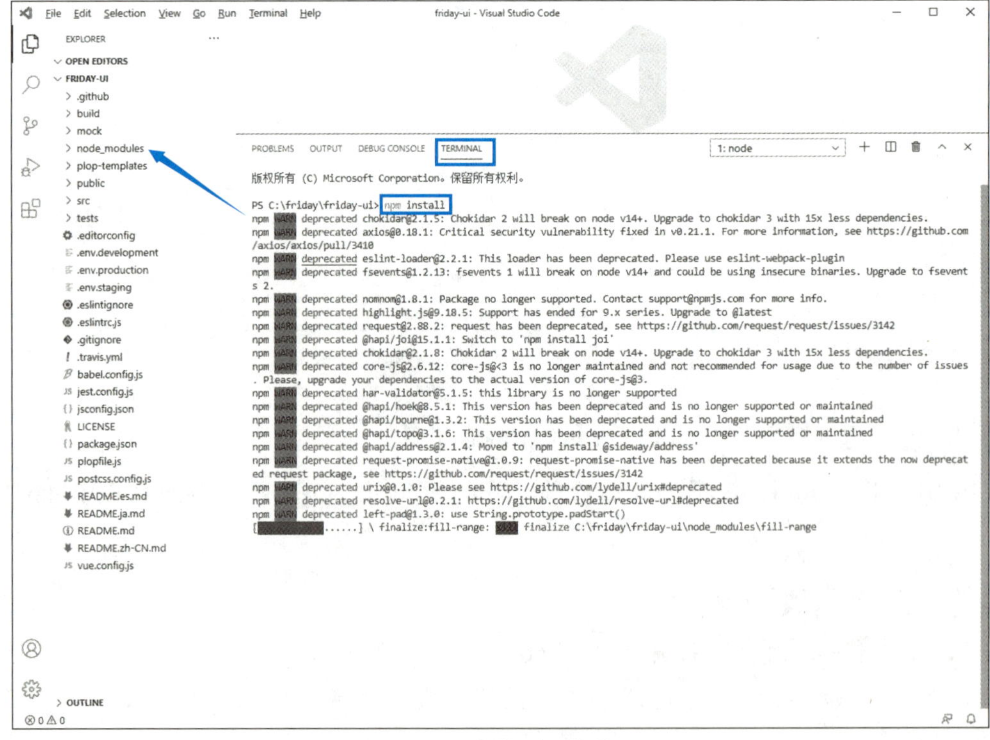

图 7-3　安装项目依赖

步骤 4　运行前端项目

在 TERMINAL 窗口执行“npm run dev”命令启动项目，启动完成后自动弹出图 7-4 所示界面，即为项目启动成功。

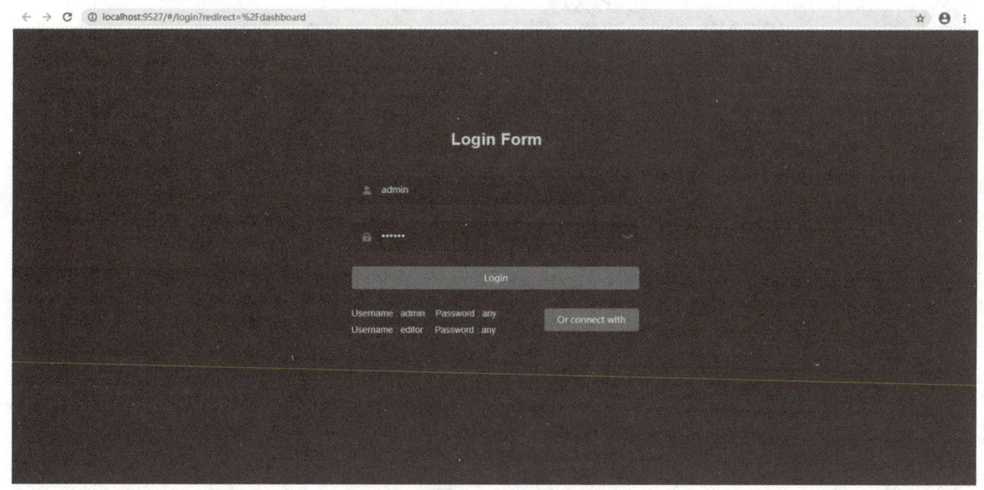

图 7-4　项目启动成功界面

任务评价

技　能　点	知　识　点	自我评价（不熟悉 / 基本掌握 / 熟练掌握 / 灵活运用）
使用 Vue-Element-Admin 框架构建前端项目	NPM 包管理器命令	

任务 2　裁剪前端项目结构

微课 7-2　裁剪前端项目结构

任务分析

Vue-Element-Admin 框架提供丰富的功能组件，但这些功能组件需要大量模块代码支撑，导致项目文件规模庞大。本任务就是根据权限管理系统的需要裁剪项目，删除不需要的功能组件，并增加必备的工具类。

任务实施

步骤 1　熟悉项目结构

Vue-Element-Admin 框架的项目结构中，各个文件夹的功能如图 7-5 所示。

如图 7-6 所示，views 目录下存放了非常多的页面文件，本权限管理系统仅需要保留登录页面、首页和异常页面，接下来对不需要的页面进行删减。

步骤 2　裁剪前端项目

1）删除“src/views”目录下的部分代码，保留如图 7-7 所示的 dashboard、error-page

和 login 目录以及它们的子文件。

图 7-5 项目结构

```
FRIDAY-UI
 > components
 > directive
 > filters
 > icons
 > layout
 > router
 > store
 > styles
 > utils
 > vendor
 v views
   > charts
   > clipboard
   > components-demo
   > dashboard
   > documentation
   > error-log
   > error-page
   > example
   > excel
   > guide
   > icons
   > login
   > nested
   > pdf
   > permission
   > profile
   > qiniu
   > redirect
   > tab
   > table
   > theme
   > zip
 App.vue
 main.js
```

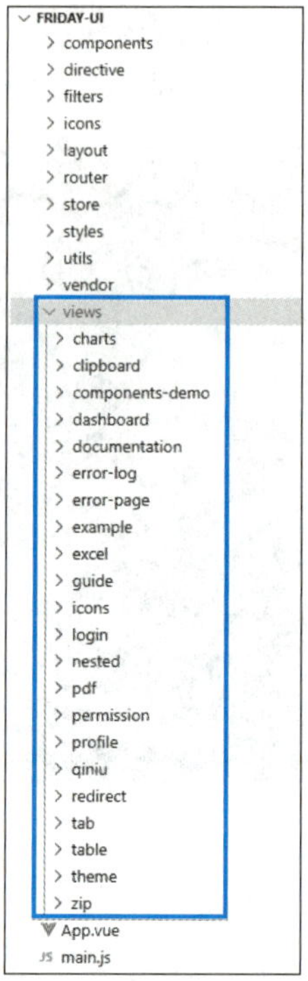

图 7-6 views 目录

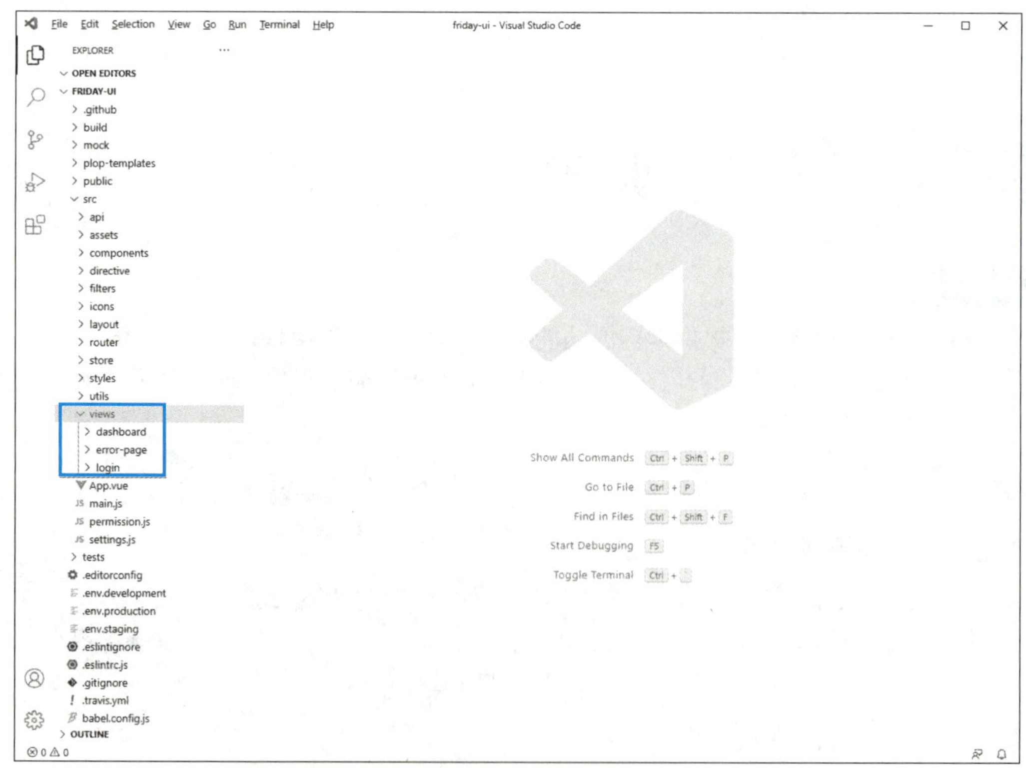

图 7-7 删除部分代码后的项目视图

2）修改“src/router”目录下的 index.js 文件，删除页面组件相应的路由配置，仅保留 dashboard、error-page 和 login 的路由配置。

```
import Vue from 'vue'
import Router from 'vue-router'
Vue.use(Router)

/* Layout */
import Layout from '@/layout'

export const constantRoutes = [
  {
    path: '/login',
    component: () => import('@/views/login/index'),
    hidden: true
  },
  {
    path: '/404',
    component: () => import('@/views/error-page/404'),
    hidden: true
  },
  {
    path: '/401',
    component: () => import('@/views/error-page/401'),
    hidden: true
  },
  {
    path: '/',
    component: Layout,
    redirect: '/dashboard',
    children: [
      {
       path: 'dashboard',
       component: () => import('@/views/dashboard/index'),
       name: 'Dashboard',
       meta: { title: 'Dashboard', icon: 'dashboard', affix: true }
      }
    ]
  }
]

/**
 * asyncRoutes
 * the routes that need to be dynamically loaded based on user roles
 */
export const asyncRoutes = [
  // 404 page must be placed at the end !!!
  { path: '*', redirect: '/404', hidden: true }
]

const createRouter = () => new Router({
  // mode: 'history', // require service support
  scrollBehavior: () => ({ y: 0 }),
  routes: constantRoutes
})

const router = createRouter()
```

```
export function resetRouter() {
  const newRouter = createRouter()
  router.matcher = newRouter.matcher // reset router
}
export default router
```

3）删除“src/router/modules”文件夹。

4）删除“src/verdor”文件夹。

步骤 3 增加通用工具类

1）在“/src/utils”目录下增加 commonUtil.js 工具类，用于实现项目的通用功能，通用功能具体如下：

- 根据表达式进行日期格式化的 parseTime 函数；
- 进行表单重置的 resetForm 函数；
- 回显数据字典的 selectDictLabel 函数；
- 通用下载方法的 download 函数；
- 将 undefined、null 等转化为 "" 的 praseStrEmpty 函数；
- 构造树形结构数据的 handleTree 函数。

```
const baseURL = process.env.VUE_APP_BASE_API
// 日期格式化
export function parseTime(time, pattern) {
  if (arguments.length === 0 || !time) {
    return null
  }
  const format = pattern || '{y}-{m}-{d} {h}:{i}:{s}'
  let date
  if (typeof time === 'object') {
    date = time
  } else {
    if ((typeof time === 'string') && (/^[0-9]+$/.test(time))) {
      time = parseInt(time)
    }
    if ((typeof time === 'number') && (time.toString().length === 10)) {
      time = time * 1000
    }
    date = new Date(time)
  }
  const formatObj = {
    y: date.getFullYear(),
    m: date.getMonth() + 1,
    d: date.getDate(),
    h: date.getHours(),
    i: date.getMinutes(),
    s: date.getSeconds(),
    a: date.getDay()
  }
  const time_str = format.replace(/{(y|m|d|h|i|s|a)+}/g, (result, key) => {
    let value = formatObj[key]
    // Note: getDay() returns 0 on Sunday
    if (key === 'a') { return ['日', '一', '二', '三', '四', '五', '六'][value] }
```

```
        if (result.length > 0 && value < 10) {
          value = '0' + value
        }
        return value || 0
      })
      return time_str
    }

    // 表单重置
    export function resetForm(refName) {
      if (this.$refs[refName]) {
        this.$refs[refName].resetFields()
      }
    }

    // 回显数据字典
    export function selectDictLabel(datas, value) {
      var actions = []
      Object.keys(datas).map((key) => {
        if (datas[key].dictValue === ('' + value)) {
          actions.push(datas[key].dictLabel)
          return false
        }
      })
      return actions.join('')
    }

    // 通用下载方法
    export function download(fileName) {
      window.location.href = baseURL + '/common/download?fileName=' + encodeURI(fileName) +
'&delete=' + true
    }

    // 转换字符串，将 undefined、null 等转化为 ""
    export function praseStrEmpty(str) {
      if (!str || str === 'undefined' || str === 'null') {
        return ''
      }
      return str
    }

    /**
     * 构造树型结构数据
     * @param {*} data 数据源
     * @param {*} id id 字段 默认 'id'
     * @param {*} parentId 父节点字段 默认 'parentId'
     * @param {*} children 子节点字段 默认 'children'
     * @param {*} rootId 根 Id 默认 0
     */
    export function handleTree(data, id, parentId, children, rootId) {
      id = id || 'id'
      parentId = parentId || 'parentId'
      children = children || 'children'
      rootId = rootId || 0
      // 对源数据深度克隆
      const cloneData = JSON.parse(JSON.stringify(data))
      // 循环所有项
      const treeData = cloneData.filter(father => {
```

```
        const branchArr = cloneData.filter(child => {
          // 返回每一项的子级数组
          return father[id] === child[parentId]
        })
        branchArr.length > 0 ? father.children = branchArr : ''
        // 返回第一层
        return father[parentId] === rootId
      })
      return treeData !== '' ? treeData : data
    }
```

2）修改“/src”目录下的 main.js 工具类，增加如下内容，将通用工具类的方法挂载为全局方法。

```
import { parseTime, resetForm, selectDictLabel, download, handleTree } from "@/utils/commonUtil";

// 全局方法挂载
Vue.prototype.parseTime = parseTime
Vue.prototype.resetForm = resetForm
Vue.prototype.selectDictLabel = selectDictLabel
Vue.prototype.download = download
Vue.prototype.handleTree = handleTree
```

步骤 4 运行前端项目

在 TERMINAL 窗口执行“npm run dev”命令启动项目，启动完成后单击“登录”按钮，弹出图 7-8 所示的界面效果，即为项目裁剪完成。

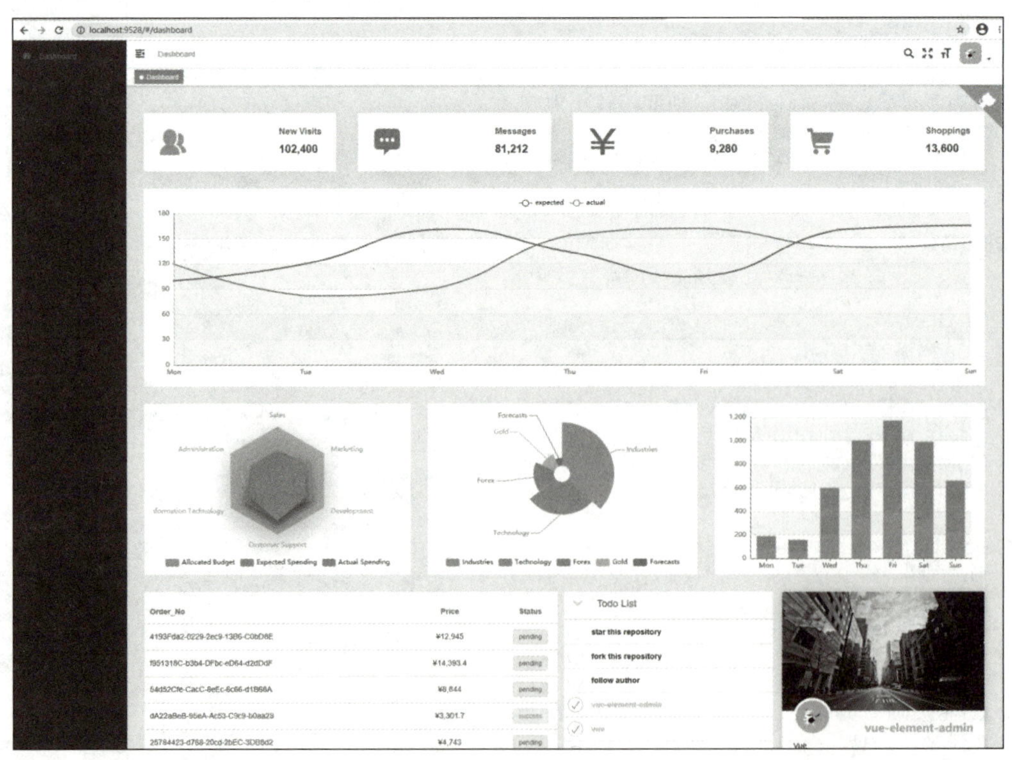

图 7-8　项目裁剪效果

知识小结

Vue-Element-Admin 框架提供了开发一个应用系统所必需的页面整体布局和基础工具类，在进行应用系统开发之前，应熟悉页面布局和这些工具类对应的代码位置。

1）页面整体布局，对应代码 @/layout。

2）路由和侧边栏，对应代码 @/router/index.js。

3）和服务端进行交互，主要用于调用后端接口，对应代码 @/utils/request.js。

@/utils/request.js 是基于 Axios 框架的封装，便于统一处理 POST、GET 等请求参数、请求头以及错误提示信息等。同时，它封装了全局 request 拦截器、response 拦截器、统一的错误处理、统一超时处理、baseURL 设置等。

任务评价

技能点	知识点	自我评价（不熟悉 / 基本掌握 / 熟练掌握 / 灵活运用）
根据项目需求裁剪 Vue-Element-Admin 框架构建的前端项目	裁剪 Vue-Element-Admin 框架	

单元小结

在前后端分离架构的项目中，前端项目的主要任务是展示页面数据、控制页面跳转和调用后端接口服务处理数据。Vue-Element-Admin 框架就是上述任务的前端解决方案，使用基于 VueJS 和 Element-UI 的 Vue-Element-Admin 框架可以快速构建前端项目。通过完成本工作单元的任务和实战强化，熟练掌握使用 Vue-Element-Admin 框架构建前端项目公共模块的职业技能。

实战强化

按照本工作单元任务实施步骤，构建实训项目“诚品书城”的前端项目。

工作单元 8

实现登录的前端功能

职业能力

本工作单元主要是基于 Vue-Element-Admin 框架的登录页面和公共工具类，实现完整的登录认证功能，最终希望学生达成如下职业能力目标：

1．掌握基于 Vue-Element-Admin 框架实现登录功能

2．掌握配置 Vue-Element-Admin 框架访问后端接口服务

3．掌握前端与服务端进行交互的代码实现

4．掌握根据登录用户权限生成动态菜单功能

任务情景

在权限管理系统里，登录与根据登录用户的权限生成首页的动态菜单是系统的必备功能。如何使用 Vue-Element-Admin 框架完成登录和生成动态菜单功能？如何调用后台接口服务来获取用户和权限数据？基于上述两个问题，本工作单元的具体任务如下：

1．基于 Vue-Element-Admin 框架编写调用后台接口函数完成用户登录功能

2．改写 Vue-Element-Admin 框架的 permission.js 工具类完成生成动态菜单功能

前置知识

任务 1 实现登录功能

微课 8-1 实现登录功能

任务分析

在前后端分离架构下，一个完整的前端 UI 交互到服务端处理流程如下：

1）UI 组件交互操作；
2）调用统一管理的 API 服务请求函数；
3）使用封装的 request.js 发送请求；
4）获取服务端返回的 data；
5）更新 data 到 UI 组件。

在 Vue-Element-Admin 框架中已经实现了 UI 组件交互操作的登录功能，以及更新 data 到 UI 组件。本任务要完成的是调用 API 服务的请求函数，便于从后端接口获取数据，另外由于 Vue-Element-Admin 是纯前端项目，所有的数据都是用 MockJS 模拟生成，所以本任务还需要移除本地 Mock 数据，并配置访问后端服务端的地址。

任务实施

步骤 1　编写 API 服务请求函数

在"\src\views\login\index.vue"中实现登录功能的 UI 组件，分别改写用于实现发送登录请求的 handleLogin() 方法以及 loginForm 的初始化值。

```
loginForm: {
        username: 'admin',
        password: '123456',
        code: '123',
        uuid: ''
      }

handleLogin() {
    this.$refs.loginForm.validate(valid => {
        if (valid) {
          this.loading = true
          this.$store.dispatch('user/login', this.loginForm)
            .then(() => {
            this.$router.push({ path: this.redirect || '/', query: this.otherQuery })
            this.loading = false
          })
            .catch(() => {
            this.loading = false
          })
        } else {
          console.log('error submit!!')
          return false
        }
    })
}
```

步骤 2　编写 VueX 的 actions 操作

编写 handleLogin() 中调用的名为"user/login"的 actions，用于更新用户登录后 token 值的变化，修改对应文件"\src\store\modules\user.js"中的 login({ commit }, userInfo) 方法，关于 VueX 的 actions 操作参见本单元知识小结。

```
login({ commit }, userInfo) {
    const username = userInfo.username.trim()
    const password = userInfo.password
    const uuid = userInfo.uuid
    return new Promise((resolve, reject) => {
      login(username, password, uuid).then(res => {
        setToken(res.token)
        commit('SET_TOKEN', res.token)
        resolve()
      }).catch(error => {
        reject(error)
      })
    })
}
```

步骤 3　编写接口调用方法

实现调用的 login(data) 方法，该方法用于发送请求到后端接口完成登录认证功能，编写文件“\src\api\user.js”，增加 login(data) 方法。

```
export function login(username, password, uuid, code) {
  const data = {
    username,
    password,
    code,
    uuid
  }
  return request({
    url: '/login',
    method: 'post',
    params: data
  })
}
```

步骤 4　移除 Mock 本地数据并配置后端服务端地址

1）修改“vue.config.js”文件，配置 devServer 并移除 Mock 服务，使前端不访问本地 Mock 数据，而是通过网络与后端服务端进行交互。

```
module.exports = {
  ...
  devServer: {
    host: '0.0.0.0',
    port: port,
    proxy: {
      [process.env.VUE_APP_BASE_API]: {
```

```
                target: `http://localhost:8080`,
                changeOrigin: true,
                pathRewrite: {
                    ['^' + process.env.VUE_APP_BASE_API]: ''
                }
            }
        },
        disableHostCheck: true
    }
    ...
}
```

2）修改“main.js”文件，移除 Mock 服务，将下面代码直接删除。

```
// if (process.env.NODE_ENV === 'production') {
//    const { mockXHR } = require('../mock')
//    mockXHR()
// }
```

步骤 5 改写封装的 request 工具类

修改文件“src\utils\request.js”，实现 Axios 框架的请求拦截器和响应拦截器功能。

➢ 请求拦截器用于前端向后端发送请求的时候，在 headers 里添加 token 值来验证该请求是否授权。

➢ 响应拦截器用于处理返回的数据，根据返回的 code 做相应的处理。

```
import axios from 'axios'
import { Notification, MessageBox, Message } from 'element-ui'
import store from '@/store'
import { getToken } from '@/utils/auth'

// create an axios instance
const service = axios.create({
  baseURL: process.env.VUE_APP_BASE_API,
  timeout: 5000 // request timeout
})

// request 拦截器
service.interceptors.request.use(
  config => {
    if (getToken()) {
      config.headers['Authorization'] = 'Bearer ' + getToken()
    }
    return config
  },
  error => {
    console.log(error)
```

```
        Promise.reject(error)
      }
    )

    // 响应拦截器
    service.interceptors.response.use(res => {
      const code = res.data.code
      if (code === 401) {
        MessageBox.confirm(
          '登录状态已过期，您可以继续留在该页面，或者重新登录',
          '系统提示',
          {
            confirmButtonText: '重新登录',
            cancelButtonText: '取消',
            type: 'warning'
          }
        ).then(() => {
          store.dispatch('LogOut').then(() => {
            location.reload() // 为了重新实例化 vue-router 对象避免 bug
          })
        })
      } else if (code !== 200) {
        Notification.error({
          title: res.data.msg
        })
        return Promise.reject('error')
      } else {
        return res.data
      }
    },
    error => {
      console.log('err' + error)
      Message({
        message: error.message,
        type: 'error',
        duration: 5 * 1000
      })
      return Promise.reject(error)
    }
    )
    export default service
```

步骤 6 运行调试

启动运行前端和后端服务，使用浏览器访问 URL 地址：http://localhost:9527/，按 <F12> 键打开控制台，选择“Network”，然后在登录页面单击“Login”按钮，如图 8-1 所示，登录请求的返回状态码是 200，表示登录验证已经成功，通过下一个任务来实现登录成功后的生成动态菜单功能，即可完成完整的登录与登录成功后的首页展示。

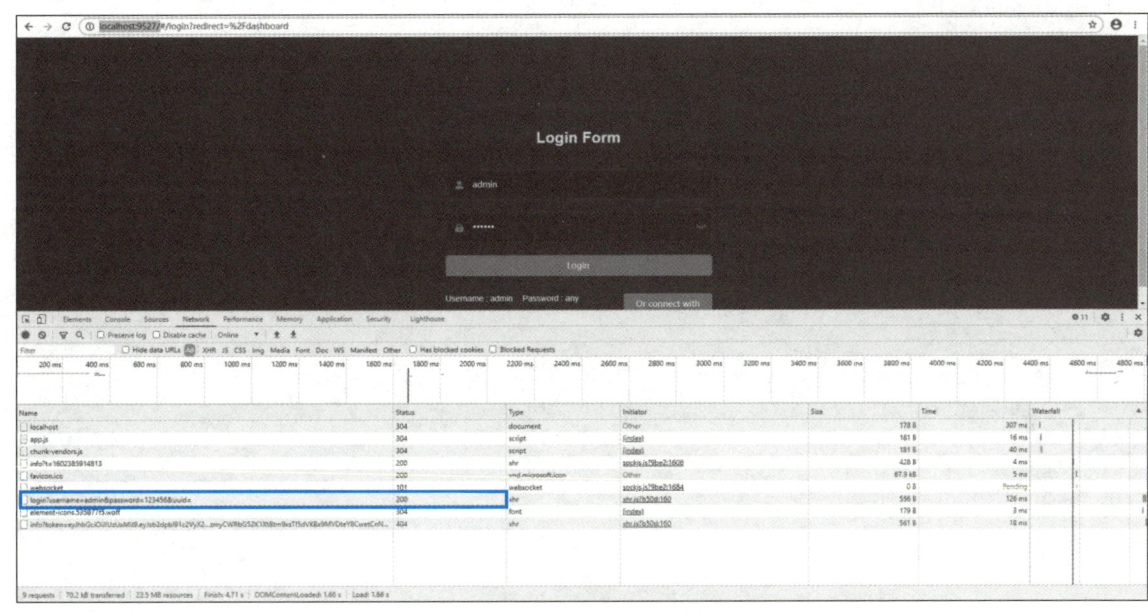

图 8-1　登录验证成功

知识小结

1. Vuex

Vuex 是一个专为 VueJS 应用程序开发的状态管理模式。在单页应用中，它主要用于管理组件之间的数据状态，例如，管理本任务中的登录状态 token 值。Vuex 的核心概念如下。

➢ State：Vuex 相当于一个仓库，里面存放很多对象，State 就是它的唯一数据源，每个应用仅包含一个 store 实例，Vue 组件从 store 实例中读取数据，若是 store 实例中的数据发生变化，依赖该数据的组件也会发生更新，例如，在本任务中 token 值作为数据保存在 State 中，代码如下。

```
const state = {
  token: getToken(),
}
```

➢ Mutation：更改 Vuex 的 store 中的状态，唯一方法是提交 Mutation，可通过如下代码定义 Mutation。

```
const mutations = {
  SET_TOKEN: (state, token) => {
     state.token = token
  }
}
```

➢ Action：Action 用于提交的是 Mutation，而 Action 不直接变更状态。

```
const actions = {
  // user login
  login({ commit }, userInfo) {
    const username = userInfo.username.trim()
    const password = userInfo.password
    const uuid = userInfo.uuid
    return new Promise((resolve, reject) => {
      login(username, password, uuid).then(res => {
        setToken(res.token)
        commit('SET_TOKEN', res.token)
        resolve()
      }).catch(error => {
        reject(error)
      })
    })
  }
}
```

2. Axios

Axios 是一个基于 promise 的易用、简洁且高效的 http 库，也是一个前端通信框架，因为 VueJS 更多地用于处理 DOM，并不具备通信功能，此时就需要额外使用一个通信框架与服务器交互，Axios 就是用来完成这个任务的。

3. Axios 的请求拦截器与响应拦截器

➢ 请求拦截器：请求拦截器的作用是在发送请求前进行一些操作，例如，在本任务中，在每个请求体的 header 里加上 token 值，统一做验证权限处理，如果后续要修改也非常容易。

➢ 响应拦截器：响应拦截器的作用是在接收到响应后进行一些操作，例如，在本任务中，服务器返回 code=402，表示登录状态已过期，需要重新登录。

任务评价

技 能 点	知 识 点	自我评价（不熟悉 / 基本掌握 / 熟练掌握 / 灵活运用）
使用 Vuex 集中管理共享数据	Vuex 的工作过程	
使用 Axios 发送请求后端接口数据	Axios 的工作过程	

任务 2 实现菜单动态生成功能

微课 8-2 实现菜单动态生成功能

任务分析

用户登录成功之后，系统会根据登录用户的权限生成首页动态菜单展示给该用户，本任务就是实现菜单动态生成功能。

任务实施

步骤 1 替换静态文件

从提供的代码资源中获取 assets 目录下的文件，替换项目中 assets 目录下的全部资源文件，结果如图 8-2 所示。

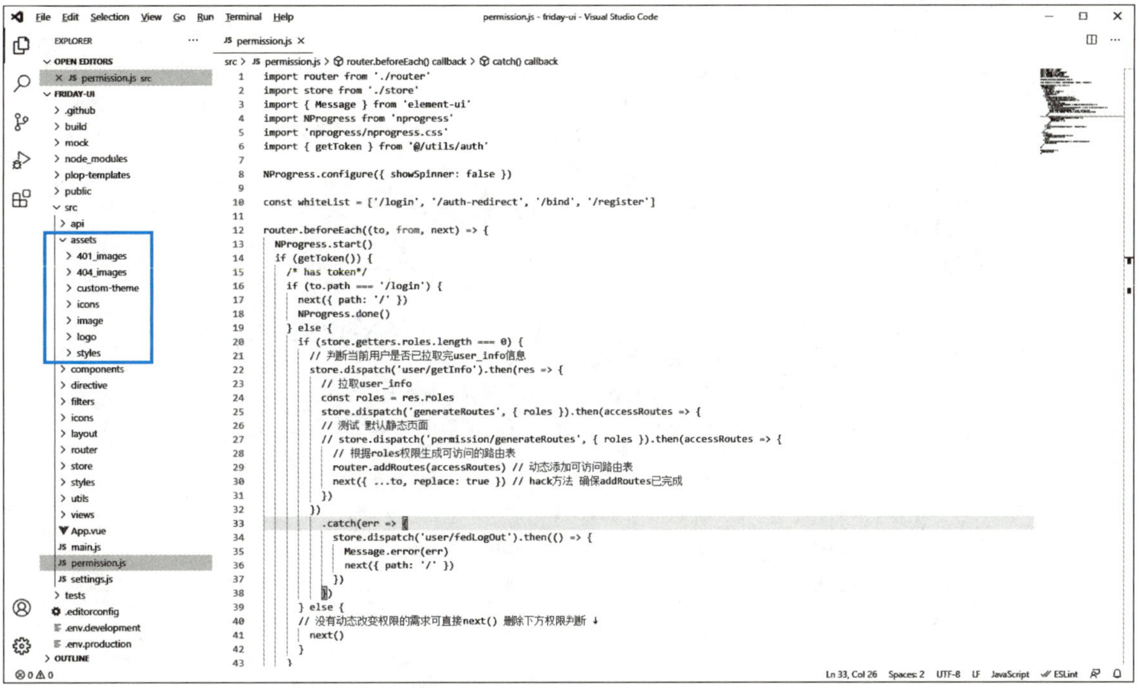

图 8-2 assets 目录文件

步骤 2 编写路由层权限控制工具类

修改“src\permission.js”文件，使用 router.beforeEach 注册一个全局前置守卫，当一个导航触发时，全局前置守卫按照下面代码逻辑顺序调用：首先判断是否有 token 值，如果有 token 值则调用 getInfo 方法获取用户信息，然后根据获取的用户信息调用 generateRoutes 方法获取有权限访问的路由表；如果没有 token 值，则跳转到登录页面。这里还引入了 NProgress 进度条插件，以优化用户体验。

```
import router from './router'
import store from './store'
import { Message } from 'element-ui'
import NProgress from 'nprogress'
import 'nprogress/nprogress.css'
import { getToken } from '@/utils/auth'

NProgress.configure({ showSpinner: false })
```

```
const whiteList = ['/login', '/auth-redirect', '/bind', '/register']

router.beforeEach((to, from, next) => {
  NProgress.start()
  if (getToken()) {
    /* has token*/
    if (to.path === '/login') {
      next({ path: '/' })
      NProgress.done()
    } else {
      if (store.getters.roles.length === 0) {
        // 判断当前用户是否已拉取完 user_info 信息
        store.dispatch('user/getInfo').then(res => {
          // 拉取 user_info
          const roles = res.roles
          store.dispatch('generateRoutes', { roles }).then(accessRoutes => {
          // 测试默认静态页面
          // store.dispatch('permission/generateRoutes', { roles }).then(accessRoutes => {
            // 根据 roles 权限生成可访问的路由表
            router.addRoutes(accessRoutes) // 动态添加可访问路由表
            next({ ...to, replace: true }) // hack 方法确保 addRoutes 已完成
          })
        })
          .catch(err => {
            store.dispatch('user/fedLogOut').then(() => {
              Message.error(err)
              next({ path: '/' })
            })
          })
      } else {
      // 没有动态改变权限的需求可直接使用 next() 删除下方权限判断
        next()
      }
    }
  } else {
    // 没有 token
    if (whiteList.indexOf(to.path) !== -1) {
      // 免登录白名单，可直接进入
      next()
    } else {
      next(`/login?redirect=${to.path}`) // 否则全部重定向到登录页
      NProgress.done()
    }
  }
})

router.afterEach(() => {
  NProgress.done()
})
```

步骤 3　编写 Vuex 的 actions 操作

1）修改“src\store\modules\user.js”文件，实现步骤 2 中 getInfo 与 fedLogOut 对应的 actions。

```
const state = {
  ...
  permissions: []
}

const mutations = {
  ...
  SET_PERMISSIONS: (state, permissions) => {
    state.permissions = permissions
  }
}

const actions = {
  ...
  getInfo({ commit, state }) {
    return new Promise((resolve, reject) => {
      getInfo(state.token).then(res => {
        const user = res.user
        const avatar = user.avatar === '' ? require('@/assets/image/profile.jpg') : user.avatar
        if (res.roles && res.roles.length > 0) { // 验证返回的 roles 是否是一个非空数组
          commit('SET_ROLES', res.roles)
          commit('SET_PERMISSIONS', res.permissions)
        } else {
          commit('SET_ROLES', ['ROLE_DEFAULT'])
        }
        commit('SET_NAME', user.username)
        commit('SET_AVATAR', avatar)
        resolve(res)
      }).catch(error => {
        reject(error)
      })
    })
  },

  fedLogOut({ commit }) {
    return new Promise(resolve => {
      commit('SET_TOKEN', '')
      removeToken()
      resolve()
    })
  },
  ...
}
```

2）修改“src\store\modules\permission.js”文件，实现步骤 2 中 generateRoutes 对应的 actions。

```
import { constantRoutes } from '@/router'
import { getRouters } from '@/api/menu'
import Layout from '@/layout/index'

const permission = {
  state: {
    routes: [],
    addRoutes: []
  },
  mutations: {
    SET_ROUTES: (state, routes) => {
      state.addRoutes = routes
      state.routes = constantRoutes.concat(routes)
    }
  },
  actions: {
    // 生成路由
    generateRoutes({
      commit
    }) {
      return new Promise(resolve => {
        // 向后端请求路由数据
        getRouters().then(res => {
          const accessedRoutes = filterAsyncRouter(res.data)
          accessedRoutes.push({
            path: '*',
            redirect: '/404',
            hidden: true
          })
          commit('SET_ROUTES', accessedRoutes)
          resolve(accessedRoutes)
        })
      })
    }
  }
}

// 遍历后台传来的路由字符串，转换为组件对象
function filterAsyncRouter(asyncRouterMap) {
  return asyncRouterMap.filter(route => {
    if (route.component) {
     // Layout 组件特殊处理
     if (route.component === 'Layout') {
       route.component = Layout
     } else {
```

```
          route.component = loadView(route.component)
        }
      }
      if (route.children != null && route.children && route.children.length) {
        route.children = filterAsyncRouter(route.children)
      }
      return true
    })
}

export const loadView = (view) => { // 路由懒加载
    return (resolve) => require([`@/views/${view}`], resolve)
}

export default permission
```

步骤 4 编写接口调用方法

1）修改“src\api\user.js”文件，实现步骤 3 中的 getInfo 接口调用方法。

```
// 获取用户详细信息
export function getInfo() {
  return request({
    url: '/getInfo',
    method: 'get'
  })
}
```

2）在“src\api”目录下新增“menu.js”文件，实现步骤 3 中的 getRouters 接口调用方法。

```
import request from '@/utils/request'
// 获取路由
export const getRouters = () => {
    return request({
      url: '/getRouters',
      method: 'get'
    })
}
```

步骤 5 编写首页页面组件

修改“src\views\dashboard\admin\index.vue”文件，处理掉首页中无用的组件，修改后的代码如下。

```
<template>
    <div class="dashboard-editor-container">
```

```
    <panel-group @handleSetLineChartData="handleSetLineChartData" />
    <el-row style="background:#fff;padding:16px 16px 0;margin-bottom:32px;">
      <line-chart :chart-data="lineChartData" />
    </el-row>
    <el-row :gutter="32">
      <el-col :xs="24" :sm="24" :lg="8">
        <div class="chart-wrapper">
          <raddar-chart />
        </div>
      </el-col>
      <el-col :xs="24" :sm="24" :lg="8">
        <div class="chart-wrapper">
          <pie-chart />
        </div>
      </el-col>
      <el-col :xs="24" :sm="24" :lg="8">
        <div class="chart-wrapper">
          <bar-chart />
        </div>
      </el-col>
    </el-row>
  </div>
</template>

<script>
import PanelGroup from './components/PanelGroup'
import LineChart from './components/LineChart'
import RaddarChart from './components/RaddarChart'
import PieChart from './components/PieChart'
import BarChart from './components/BarChart'

const lineChartData = {
  newVisitis: {
    expectedData: [100, 120, 161, 134, 105, 160, 165],
    actualData: [120, 82, 91, 154, 162, 140, 145]
  },
  messages: {
    expectedData: [200, 192, 120, 144, 160, 130, 140],
    actualData: [180, 160, 151, 106, 145, 150, 130]
  },
  purchases: {
    expectedData: [80, 100, 121, 104, 105, 90, 100],
    actualData: [120, 90, 100, 138, 142, 130, 130]
  },
  shoppings: {
    expectedData: [130, 140, 141, 142, 145, 150, 160],
    actualData: [120, 82, 91, 154, 162, 140, 130]
  }
```

```
}

export default {
  name: 'Index',
  components: {
    PanelGroup,
    LineChart,
    RaddarChart,
    PieChart,
    BarChart
  },
  data() {
    return {
      lineChartData: lineChartData.newVisitis
    }
  },
  methods: {
    handleSetLineChartData(type) {
      this.lineChartData = lineChartData[type]
    }
  }
}
</script>

<style lang="scss" scoped>
.dashboard-editor-container {
  padding: 32px;
  background-color: rgb(240, 242, 245);
  position: relative;
  .chart-wrapper {
    background: #fff;
    padding: 16px 16px 0;
    margin-bottom: 32px;
  }
}

@media (max-width:1024px) {
  .chart-wrapper {
    padding: 8px;
  }
}
</style>
```

步骤 6　运行调试

启动前端和后端服务，使用浏览器访问 URL 地址：http://localhost:9527/，在登录页面单击“Login”按钮跳转到首页页面，且左侧有生成的系统菜单即为成功，如图 8-3 所示。

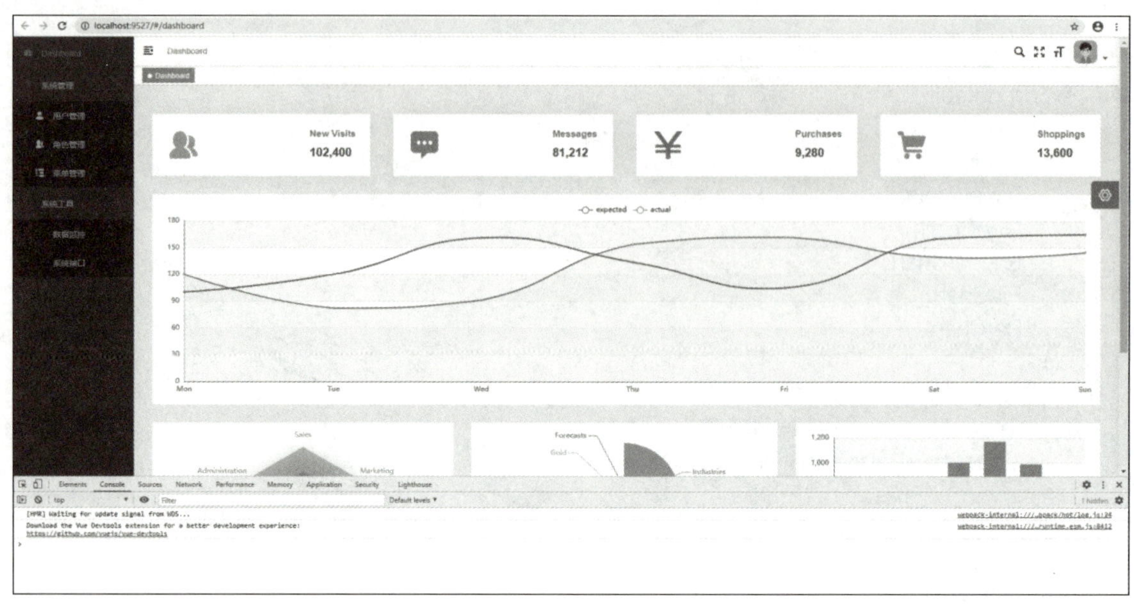

图 8-3　首页页面

知识小结

1. Vue Router 是 Vue.js 官方的路由管理器

Vue Router 和 Vue.js 的核心深度集成，让构建单页面应用变得易如反掌。使用 Vue.js 的组件来组成应用程序，单页面应用无法进行跳转动作。此时把 Vue Router 添加进来，将组件（components）映射到路由（routes），然后告诉 Vue Router 在哪里可以渲染它们，即能实现单页面应用跳转的功能。

2. Vue Router 的全局前置守卫

当一个导航触发时，全局前置守卫按照创建顺序调用。守卫是异步解析执行，此时导航在所有守卫解析完之前一直处于等待中。每个守卫方法接收如下 3 个参数。

➢ to: Route：即将要进入的目标路由对象。

➢ from: Route：当前导航正要离开的路由对象。

➢ next: Function：一定要调用该方法来处理这个钩子。执行效果依赖 next 方法的调用参数，效果见表 8-1。

表 8-1　next 方法的调用参数及执行效果

调 用 参 数	执 行 效 果
next()	进行管道中的下一个钩子，如果全部钩子执行完了，则导航的状态就是 confirmed（确认的）
next(false)	中断当前的导航。如果浏览器的 URL 改变了，那么 URL 地址会重置到 from 路由对应的地址
next('/') 或 next({path:'/'})	跳转到一个不同的地址。当前的导航被中断，然后进行一个新的导航
next(error)	传入 next 的参数是一个 Error 实例，则导航会被终止且该错误会被传递给 router.onError() 注册过的回调函数

任务评价

技 能 点	知 识 点	自我评价（不熟悉 / 基本掌握 / 熟练掌握 / 灵活运用）
使用 Vue Router 构建项目路由	Vue Router 的工作过程	

单元小结

在前后端分离架构的项目中，实现一个完整的前端 UI 交互到服务端处理业务流程的编码是核心基础技能。本工作单元主要是基于 Vue-Element-Admin 框架的登录页面和公共工具类，实现前后端联动的登录认证功能，以及基于 Vue Router 构建首页动态菜单功能。通过完成本工作单元的任务和实战强化，熟练掌握使用 Vue-Element-Admin 框架进行前后端全流程实现业务功能的职业技能。

实战强化

按照本工作单元任务实施步骤，构建实训项目“诚品书城”的登录和首页菜单功能。

工作单元 9

实现用户和角色管理的前端功能

职业能力

本工作单元实现项目的用户和角色管理的前端功能，具体实现用户的增、删、改、查页面，将角色的增、删、改、查页面设计成拓展练习，帮助学生巩固技能。最终希望学生达成的职业能力目标如下：

1. 掌握基于 Element UI 组件库实现增、删、改、查功能的页面效果
2. 掌握基于 Axios 向后台服务端发起 API 请求
3. 灵活使用 Vue-element-admin 框架的工具类

任务情景

权限管理系统里，增、删、改、查操作是页面的基础功能，如何围绕界面结构、操作流程来设计实现一个如图 9-1 所示的，简洁高效的增、删、改、查页面呢？

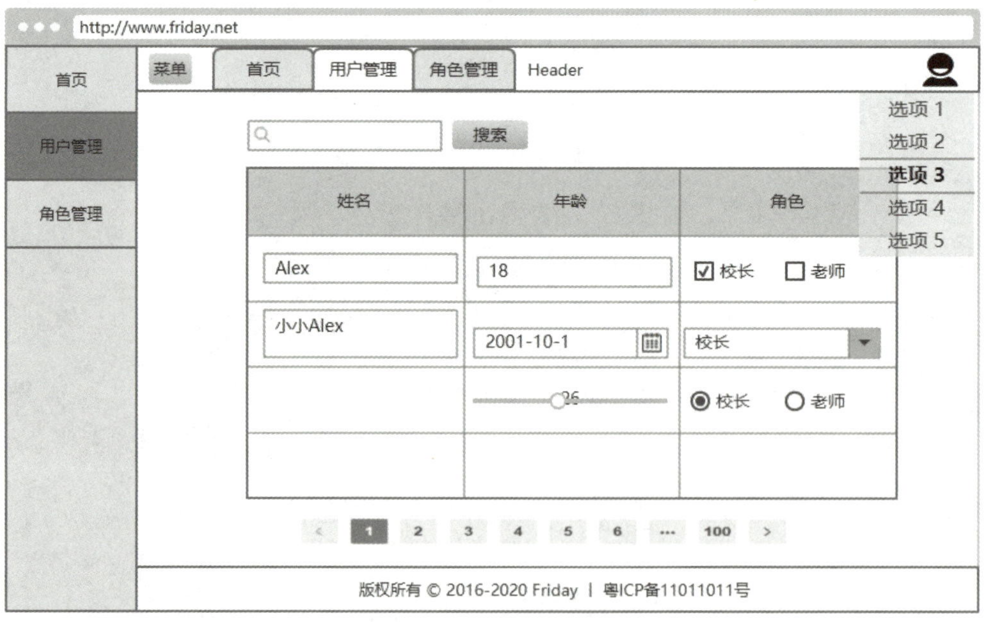

图 9-1　用户角色管理前端页面

网站快速成型工具 Element UI 可以快速搭建逻辑清晰、结构合理且高效易用的整体页面。本工作单元会使用 Element UI 完成如图 9-2 所示的用户角色管理前端页面。

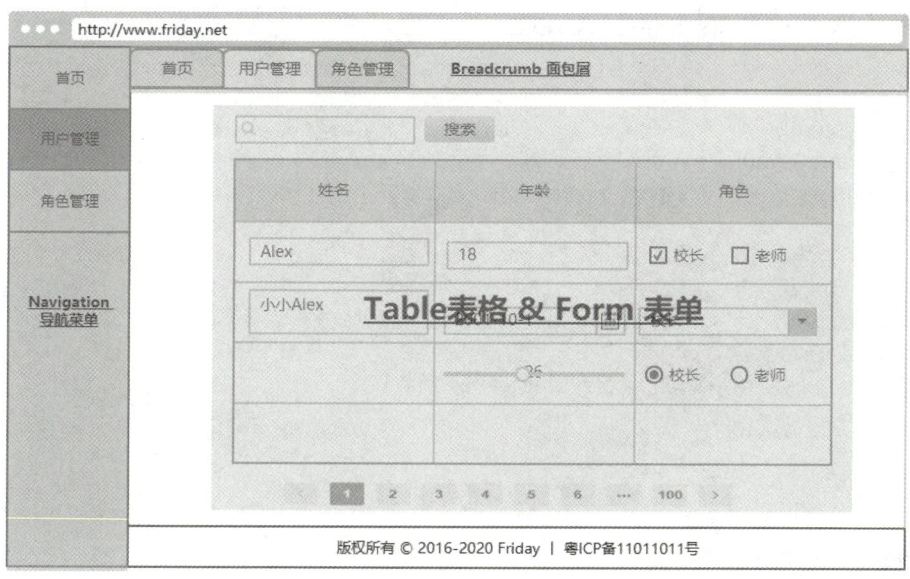

图 9-2　使用 Element UI 完成的用户角色管理前端页面

前置知识

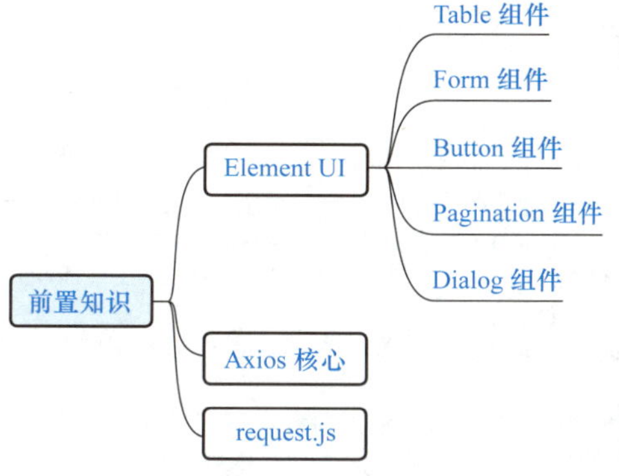

任务 1　实现显示用户列表页面

微课 9-1　实现显示用户列表页面

任务分析

本任务先配置显示用户列表页面的菜单路由，然后使用 Element UI 的 Table 组件构建用户列表页面，再使用封装 Axios 框架的 request.js 调用后端接口请求。

任务实施

步骤 1　确定路由映射文件

从下面的用户管理菜单路由中可以看出，访问路径 path 为“/system/user”，映射的组件 component 为“src/views/system/user/index”文件。

```
{
  path: '/system',
  component: Layout,
  meta: { title: ' 系统管理 ', icon: 'guide', affix: true },
  children: [
    {
     path: 'user',
     component: () => import('@/views/system/user/index'),
     name: 'user',
     meta: { title: ' 用户管理 ', icon: 'user', affix: true }
    }
  ]
}
```

步骤 2 编写页面代码

在“views”目录下创建“system/user/index.vue”组件，使用 Element table 组件实现用户列表展示。Element table 组件案例代码从下面的链接中获取：

https://element.faas.ele.me/#/zh-CN/component/table。

```
<template>
  <div class="app-container">
    <el-row :gutter="20">
      <!-- 用户数据 -->
      <el-col>
        <el-table :data="userList">
          <el-table-column type="selection" width="40" align="center" />
          <el-table-column label=" 用户编号 " align="center" prop="userId" />
          <el-table-column label=" 用户名称 " align="center" prop="userName" :show-overflow-tooltip="true" />
          <el-table-column label=" 用户昵称 " align="center" prop="nickName" :show-overflow-tooltip="true" />
          <el-table-column label=" 手机号码 " align="center" prop="phonenumber" />
          <el-table-column label=" 状态 " align="center">
            <template slot-scope="scope">
              <el-switch
                v-model="scope.row.status"
                active-value="0"
                inactive-value="1"
                @change="handleStatusChange(scope.row)"
              />
            </template>
          </el-table-column>
          <el-table-column label=" 创建日期 ">
            <template slot-scope="scope">
              <i class="el-icon-time" />
              <span>{{ scope.row.createDate }}</span>
            </template>
          </el-table-column>
          <el-table-column
            label=" 操作 "
```

```
                align="center"
                width="220"
                class-name="small-padding fixed-width"
              >
                <template slot-scope="scope">
                  <el-button
                    size="mini"
                    type="text"
                    icon="el-icon-edit"
                    @click="handleUpdate(scope.row)"
                  > 修改 </el-button>
                  <el-button
                    size="mini"
                    type="text"
                    icon="el-icon-delete"
                    @click="handleDelete(scope.row)"
                  > 删除 </el-button>
                </template>
              </el-table-column>
            </el-table>
          </el-col>
        </el-row>
      </div>
</template>
<script>
export default {
   data() {
      return {
        userList: [{
          userId: '12987122',
          userName: 'Alex',
          nickName: 'Alex Jordan',
          phonenumber: '13911223344',
          createDate: '2020-04-16'
        },
        {
          userId: '12956987',
          userName: 'Kelly',
          nickName: 'Jenson&Kyson',
          phonenumber: '13955669988',
          createDate: '2010-05-16'
        },
        {
          userId: '129569888',
          userName: 'Dalton',
          nickName: 'Lucas&Quintus',
          phonenumber: '13998745645',
          createDate: '2018-04-18'
        }
        ]
      }
   },
   methods: {
      handleUpdate(row) {
         console.log(row)
      },
      handleDelete(row) {
         console.log(row)
```

```
        },
        // 用户状态修改
        handleStatusChange(row) {
          console.log(row)
        }
      }
    }
    </script>
```

步骤 3　编写请求后端接口代码

在“src/api”目录下创建“system/uscr.js”文件，新增 listUser 函数，该函数使用 get 请求方法，请求后端接口“/system/user/list”，传递 query 参数，以便于获取用户列表数据，代码如下。

```
import request from '@/utils/request'

// 查询用户列表
export function listUser(query) {
  return request({
    url: '/system/user/list',
    method: 'get',
    params: query
  })
}
```

步骤 4　请求服务端数据，填充用户列表页面

在“/views/system/user/index.vue”文件中调用步骤 3 中的 listUser() 请求服务端数据，填充用户列表页面，代码细节分析如下：

➢　在 el-table 组件上绑定：data="userList"，用于渲染后端接口返回的用户列表数据，绑定 v-loading="listLoading" 用于后端返回到数据之前，显示加载中的效果，优化用户体验。

➢　在 created() 函数中调用 getList() 进行数据初始化，用于在模板渲染成 html 前，调用后端接口，返回用户列表数据。

```
<template>
  <div class="app-container">
    <el-row :gutter="20">
      <!-- 用户数据 -->
      <el-col>
        <el-table v-loading="listLoading" :data="userList">
          <el-table-column type="selection" width="40" align="center" />
          <el-table-column label=" 用户编号 " align="center" prop="userId" />
          <el-table-column label=" 用户名称 " align="center" prop="userName" :show-overflow-
tooltip="true" />
          <el-table-column label=" 用户昵称 " align="center" prop="nickName" :show-overflow-
tooltip="true" />
          <el-table-column label=" 手机号码 " align="center" prop="phonenumber" />
          <el-table-column label=" 状态 " align="center">
            <template slot-scope="scope">
              <el-switch
                v-model="scope.row.status"
```

```
              active-value="0"
              inactive-value="1"
              @change="handleStatusChange(scope.row)"
            />
          </template>
        </el-table-column>
        <el-table-column label=" 创建日期 ">
          <template slot-scope="scope">
            <i class="el-icon-time" />
            <span>{{ scope.row.createDate }}</span>
          </template>
        </el-table-column>
        <el-table-column
          label=" 操作 "
          align="center"
          width="220"
          class-name="small-padding fixed-width"
        >
          <template slot-scope="scope">
            <el-button
              size="mini"
              type="text"
              icon="el-icon-edit"
              @click="handleUpdate(scope.row)"
            > 修改 </el-button>
            <el-button
              size="mini"
              type="text"
              icon="el-icon-delete"
              @click="handleDelete(scope.row)"
            > 删除 </el-button>
          </template>
        </el-table-column>
      </el-table>
    </el-col>
  </el-row>
 </div>
</template>
<script>
import { listUser } from '@/api/system/user'
export default {
  data() {
    return {
      userList: null,
      listLoading: true
    }
  },
  created() {
    this.getList()
  },
  methods: {
    getList() {
      this.listLoading = true
      listUser().then(response => {
        this.userList = response.rows
        this.listLoading = false
      })
    },
```

```
        handleUpdate(row) {
          console.log(row)
        },
        handleDelete(row) {
          console.log(row)
        },
        // 用户状态修改
        handleStatusChange(row) {
          console.log(row)
        }
      }
    }
</script>
```

步骤 5 运行调试

分别启动运行前、后端服务，使用浏览器访问 URL 地址：http://localhost:9527/，登录成功后单击“系统管理”下的“用户管理”，显示用户列表，如图 9-3 所示。

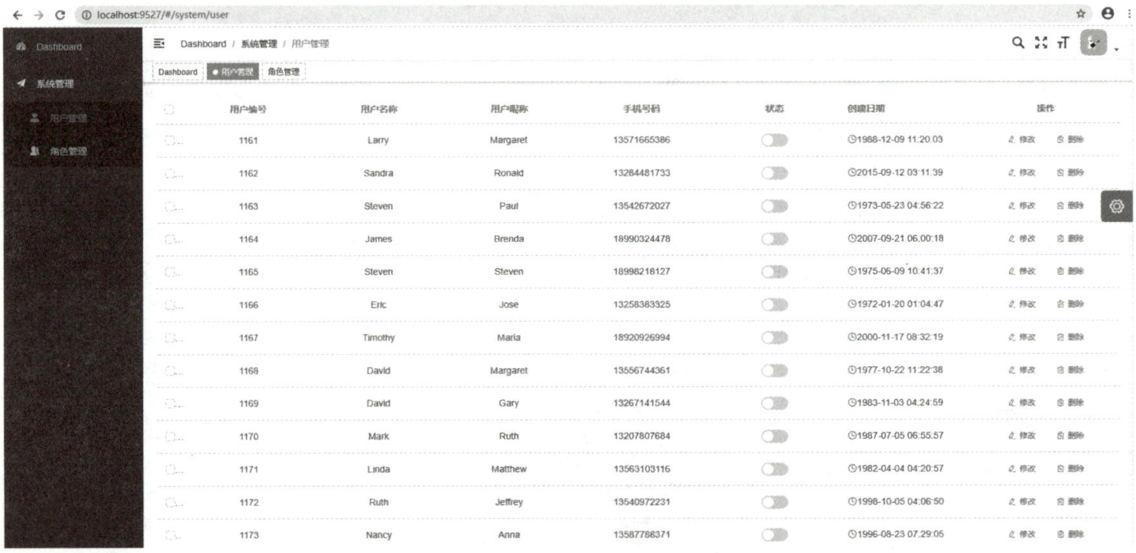

图 9-3　显示用户列表

任务评价

技　能　点	知　识　点	自我评价（不熟悉 / 基本掌握 / 熟练掌握 / 灵活运用）
使用 Element UI 的 Table 组件构建列表页面	Element UI 的 Table 组件的工作过程	
使用 el-table 组件填充后端接口返回的数据		

拓展练习

显示角色列表

模仿工作单元 9 任务 1 中的任务实施步骤，实现显示角色列表。显示角色列表前端结果如图 9-4 所示。

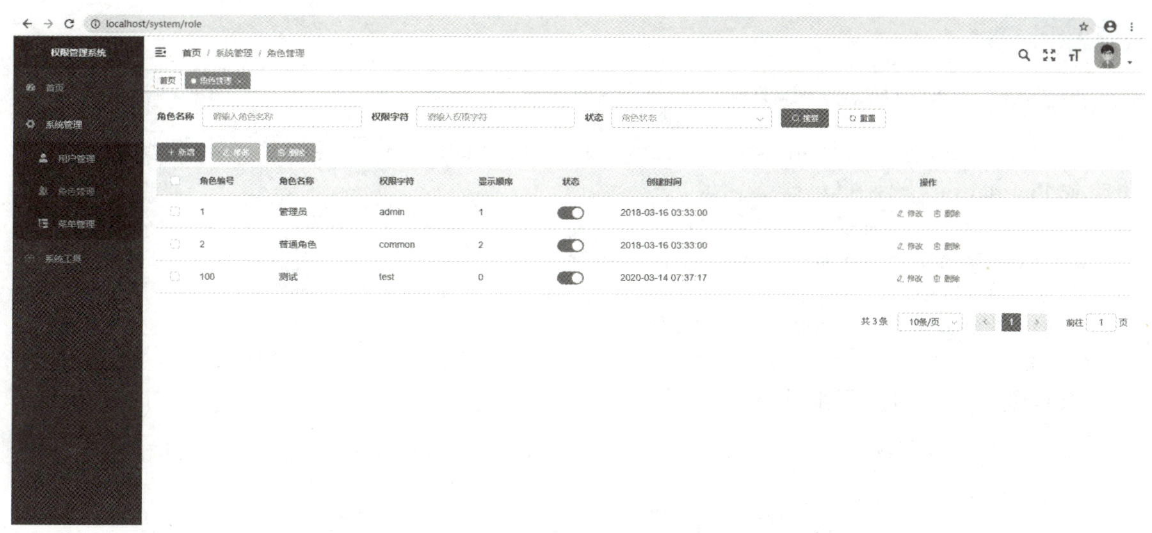

图 9-4 显示角色列表前端结果

任务 2 实现用户列表分页

微课 9-2 实现用户列表分页

任务分析

分页功能是网页中的一个常见功能，其作用就是将数据分割成多个页面来显示。当获取数据量达到一定规模的时候，使用分页来进行数据分割，可使数据展示形式更为齐整，实现分页的用户列表页面如图 9-5 所示。

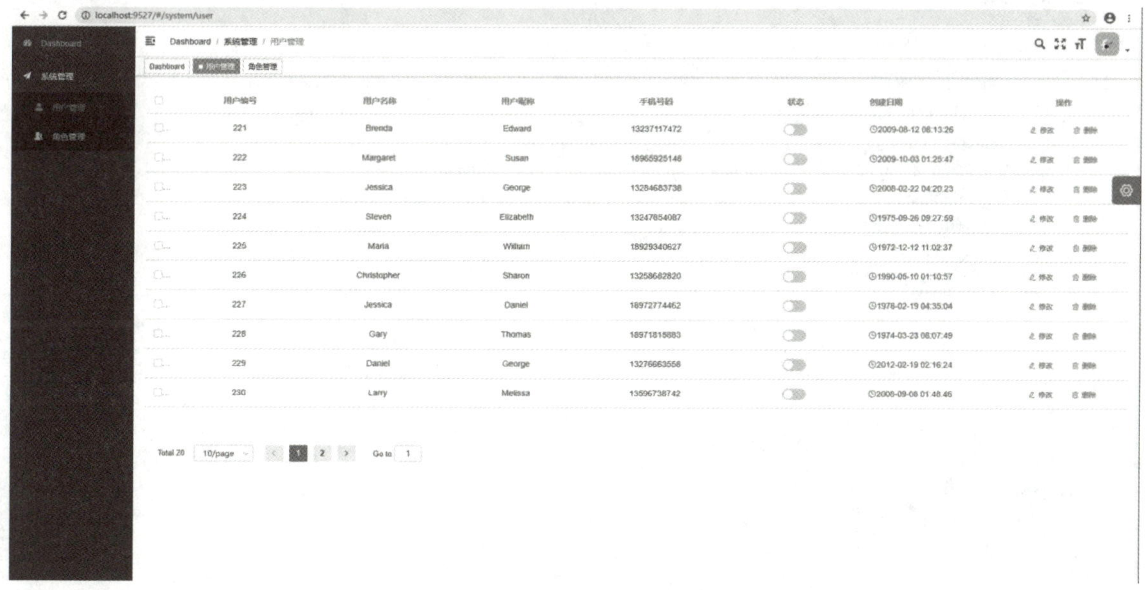

图 9-5 实现分页的用户列表页面

任务实施

步骤 1　全局挂载分页组件

使用 Vue-Element-Admin 框架的 /components/Pagination 组件实现分页，修改 main.js 文件挂载 Pagination 组件。

```
import Pagination from '@/components/Pagination'
// 全局组件挂载
Vue.component('Pagination', Pagination)
```

步骤 2　使用分页组件

1）修改“/views/system/user/index.vue”文件，在 <template> 标签里的“el-table”组件下增加 Pagination 组件。

```
<pagination
    v-show="total>0"
    :total="total"
    :page.sync="queryParams.page"
    :limit.sync="queryParams.size"
    @pagination="getList"
/>
```

2）Pagination 组件使用了 total、queryParams 两个变量，在 script 标签下的 data 中初始化该变量，同时把 queryParams 变量传入到 listUser(this.queryParams) 中，将返回的数据赋值给 total 变量，获取总数据数。

```
export default {
  data() {
    return {
      userList: null,
      listLoading: true,
      // 总条数
      total: 0,
      // 查询参数
      queryParams: {
        page: 1,
        size: 10
      }
    }
  },
  created() {
    this.getList()
  },
  methods: {
    getList() {
      this.listLoading = true
      listUser(this.queryParams).then(response => {
        this.total = response.data.total
```

```
            this.userList = response.data.items
            this.listLoading = false
          })
        },
        handleUpdate(row) {
          console.log(row)
        },
        handleDelete(row) {
          console.log(row)
        },
        // 用户状态修改
        handleStatusChange(row) {
          console.log(row)
        }
      }
    }
```

步骤 3 运行调试

运行调试步骤可参考本单元任务 1 中的实施步骤 5，最终测试运行效果如图 9-5 所示。

知识小结

Pagination 组件支持多种模式的分页形式。可以通过表 9-1 参数控制其形式。

表 9-1 Pagination 组件的参数列表

参 数	说 明	类 型	可 选 值	默 认 值
small	是否使用小型分页样式	boolean	—	false
background	是否为分页按钮添加背景色	boolean	—	false
page-size	每页显示条目个数，支持 .sync 修饰符	number	—	10
total	总条目数	number	—	—
page-count	总页数，在 total 和 page-count 两个参数中，设置任意一个参数就可以达到显示页码的功能；如果要支持 page-size 参数的更改，则需要使用 total 属性	number	—	—
pager-count	页码按钮的数量，当总页数超过该值时会折叠	number	大于等于 5 且小于等于 21 的奇数	7
current-page	当前页数，支持 .sync 修饰符	number	—	1
layout	组件布局，子组件名用逗号分隔	string	sizes, prev, pager, next, jumper, ->, total, slot	prev, pager, next, jumper, ->, total
page-sizes	每页显示个数选择器的选项设置	number[]	—	[10, 20, 30, 40, 50, 100]
popper-class	每页显示个数选择器的下拉框类名	string	—	—
prev-text	替代图标显示的上一页文字	string	—	—
next-text	替代图标显示的下一页文字	string	—	—
disabled	是否禁用	boolean	—	false

任务评价

技 能 点	知 识 点	自我评价（不熟悉 / 基本掌握 / 熟练掌握 / 灵活运用）
使用 pagination 分页组件构建页面分页的功能	分页组件的原理	

拓展练习

角色列表分页

模仿本任务的实施步骤，实现分页的角色列表页面结果如图 9-6 所示。

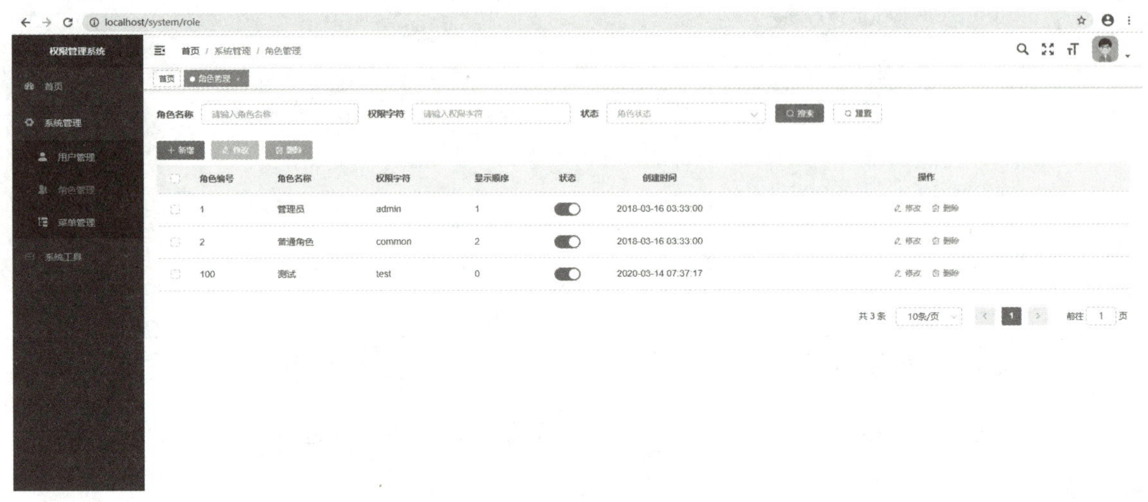

图 9-6　实现分页的角色列表页面

任务 3　实现用户查询功能

微课 9-3　实现用户查询功能

任务分析

在显示用户列表界面增加“新增”“修改”“删除”“搜索”“重置”按钮，实现按钮的事件绑定方法，通过输入用户名称、手机号码和状态来查询用户数据的功能。实现模糊查询功能的用户列表页面如图 9-7 所示。

用户名称 a　手机号码 请输入手机号码　状态 用户状态　搜索　重置

+ 新增　修改　删除

	用户编号	用户名称	用户昵称	手机号码	状态	创建时间	操作
	1	admin	若依	15888888888		2018-03-16 03:33:00	修改 重置
	109	asdf	asdd	13311122311		2020-04-02 00:02:48	修改 删除 重置
	110	asdf	asdd	13311122311		2020-04-02 00:02:48	修改 删除 重置
	111	asdf	asdd	13311122311		2020-04-02 00:02:48	修改 删除 重置
	112	asdf	asdd	13311122311		2020-04-02 00:02:48	修改 删除 重置
	113	asdas	asdas	13122321123		2020-04-02 00:03:27	修改 删除 重置
	114	asdasda	asdasda	13412311222		2020-04-02 00:06:55	修改 删除 重置
	115	del_flag	del_flag	15312112311		2020-04-02 00:08:59	修改 删除 重置

图 9-7　实现模糊查询功能的用户列表页面

任务实施

步骤 1　实现查询参数的展示

修改“/views/system/user/index.vue”文件，在 <template> 标签里使用 el-form 组件实现查询参数的展示功能，使用 el-button 组件实现按钮功能。

```
<el-row :gutter="20">
  <!-- 用户数据搜索与按钮 -->
  <el-col :span="20" :xs="24">
    <el-form ref="queryForm" :model="queryParams" :inline="true" label-width="68px">
      <el-form-item label=" 用户名称 " prop="userName">
        <el-input
          v-model="queryParams.userName"
          placeholder=" 请输入用户名称 "
          clearable
          size="small"
          style="width: 240px"
          @keyup.enter.native="handleQuery"
        />
      </el-form-item>
      <el-form-item label=" 手机号码 " prop="phonenumber">
        <el-input
          v-model="queryParams.phonenumber"
          placeholder=" 请输入手机号码 "
          clearable
          size="small"
          style="width: 240px"
          @keyup.enter.native="handleQuery"
        />
      </el-form-item>
      <el-form-item label=" 状态 " prop="status">
        <el-select
          v-model="queryParams.status"
          placeholder=" 用户状态 "
          clearable
          size="small"
          style="width: 240px"
        >
          <el-option
            v-for="dict in statusOptions"
            :key="dict.dictValue"
            :label="dict.dictLabel"
            :value="dict.dictValue"
          />
        </el-select>
      </el-form-item>
      <el-form-item>
        <el-button type="primary" icon="el-icon-search" size="mini" @click="handleQuery"> 搜索
</el-button>
        <el-button icon="el-icon-refresh" size="mini" @click="resetQuery"> 重置 </el-button>
```

```
      </el-form-item>
    </el-form>
    <el-row :gutter="10" class="mb8">
      <el-col :span="1.5">
        <el-button
          type="primary"
          icon="el-icon-plus"
          size="mini"
          @click="handleAdd"
        > 新增 </el-button>
      </el-col>
      <el-col :span="1.5">
        <el-button
          type="success"
          icon="el-icon-edit"
          size="mini"
          :disabled="single"
          @click="handleUpdate"
        > 修改 </el-button>
      </el-col>
      <el-col :span="1.5">
        <el-button
          type="danger"
          icon="el-icon-delete"
          size="mini"
          :disabled="multiple"
          @click="handleDelete"
        > 删除 </el-button>
      </el-col>
    </el-row>
  </el-col>
</el-row>
```

步骤 2 变量初始化

本任务使用了 queryParams.userName、queryParams.phonenumber、queryParams.status、statusOptions、single 以及 multiple 变量，在 <script> 标签下的 data 中初始化变量。

```
// 查询参数
 queryParams: {
   page: 1,
   size: 10,
   userName: undefined,
   phonenumber: undefined,
   status: undefined
 },
 statusOptions: [
   { dictLabel: ' 正常 ', dictValue: '0' },
   { dictLabel: ' 停用 ', dictValue: '1'  }
 ],
 single: true,
 multiple: true
```

步骤 3 / 实现按钮绑定事件的方法

```
// 搜索按钮操作
handleQuery() {
    this.queryParams.page = 1
    this.getList()
},
// 重置 Form 表单
resetQuery() {
    this.resetForm('queryForm')
    this.handleQuery()
},
// 新增用户
handleAdd() {
}
```

步骤 4 / 运行调试

分别运行前、后端服务，使用浏览器访问 URL 地址：http://localhost:9527/，登录成功后单击“系统管理”下的“用户管理”，在用户名称处输入“Barbara”，用户列表页面的查询结果如图 9-8 所示。

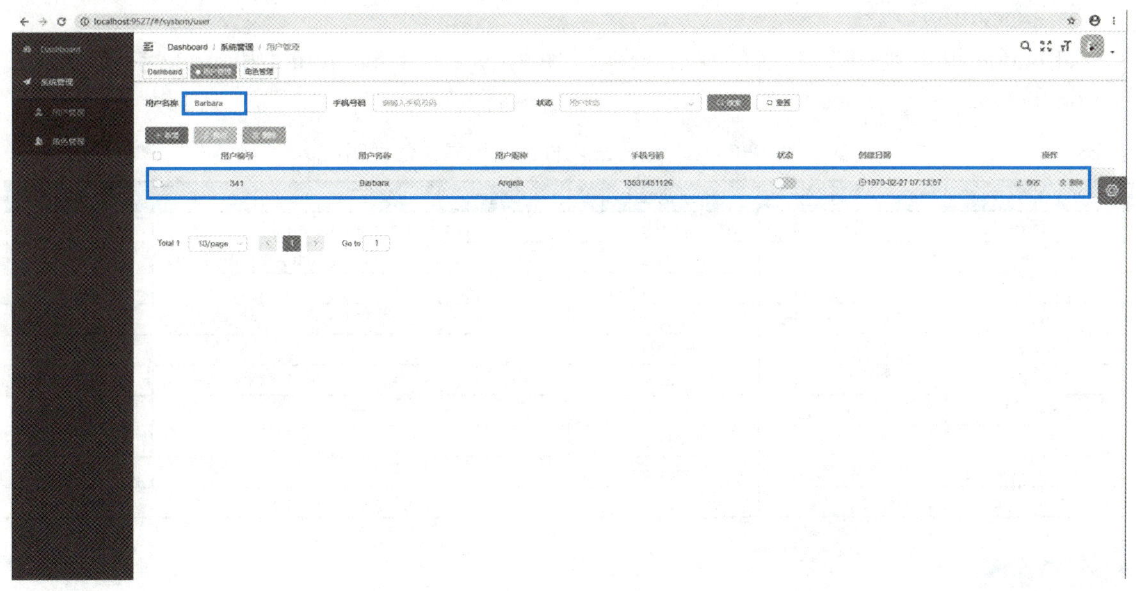

图 9-8　用户列表页面的查询结果

知识小结

1. el-form 组件

el-form 组件由输入框、选择器、单选框和多选框等控件组成，用以收集、校验和提交数据。一个典型表单包括各种表单项，如输入框、选择器、开关、单选框和多选框等。

在 Form 组件中，每一个表单域由一个 Form-Item 组件构成，表单域中可以放置各种类型的表单控件，包括 Input、Select、Checkbox、Radio、Switch、DatePicker 和 TimePicker。el-form 组件的常用属性值说明见表 9-2。

表 9-2　el-form 组件的常用属性值说明

参　数	说　明	类　型	可 选 值	默 认 值
model	表单数据对象	object	—	—
rules	表单验证规则	object	—	—
inline	行内表单模式	boolean	—	false
label-position	表单域标签的位置	string	right/left/top	right
label-width	表单域标签的宽度，作为 Form 直接子元素的 Form-Item 会继承该值	string	—	—
label-suffix	表单域标签的后缀	string	—	—
show-message	是否显示校验错误信息	boolean	—	true
inline-message	是否以行内形式展示校验信息	boolean	—	false
status-icon	是否在输入框中显示校验结果反馈图标	boolean	—	false
validate-on-rule-change	是否在 rules 属性改变后立即触发一次验证	boolean	—	true
size	用于控制该表单内组件的尺寸	string	medium / small / mini	—
disabled	是否禁用该表单内的所有组件。若设置为 true，则表单内的组件不再生效	boolean	—	false

2. el-button 组件

el-button 组件是一个常用的操作按钮，使用 type、plain、round 和 circle 参数来定义 Button 的 type 参数。el-button 组件的常用属性值见表 9-3。

表 9-3　el-button 组件的常用属性值说明

参　数	说　明	类　型	可 选 值	默 认 值
size	尺寸	string	medium / small / mini	—
type	类型	string	primary / success / warning / danger / info / text	—
plain	是否朴素按钮	boolean	—	false
round	是否圆角按钮	boolean	—	false
circle	是否圆形按钮	boolean	—	false
loading	是否加载中状态	boolean	—	false
disabled	是否禁用状态	boolean	—	false
icon	图标类名	string	—	—
autofocus	是否默认聚焦	boolean	—	false
native-type	原生 type 属性	string	button / submit / reset	button

任务评价

技 能 点	知 识 点	自我评价（不熟悉 / 基本掌握 / 熟练掌握 / 灵活运用）
使用 el-form 组件实现查询参数的表单提交	Element UI 的 el-form 和 el-button 组件的工作过程	
使用 el-button 组件实现按钮功能		

拓展练习

角色模糊查询

模仿本任务的实施步骤，实现角色列表的模糊查询结果如图 9-9 所示。

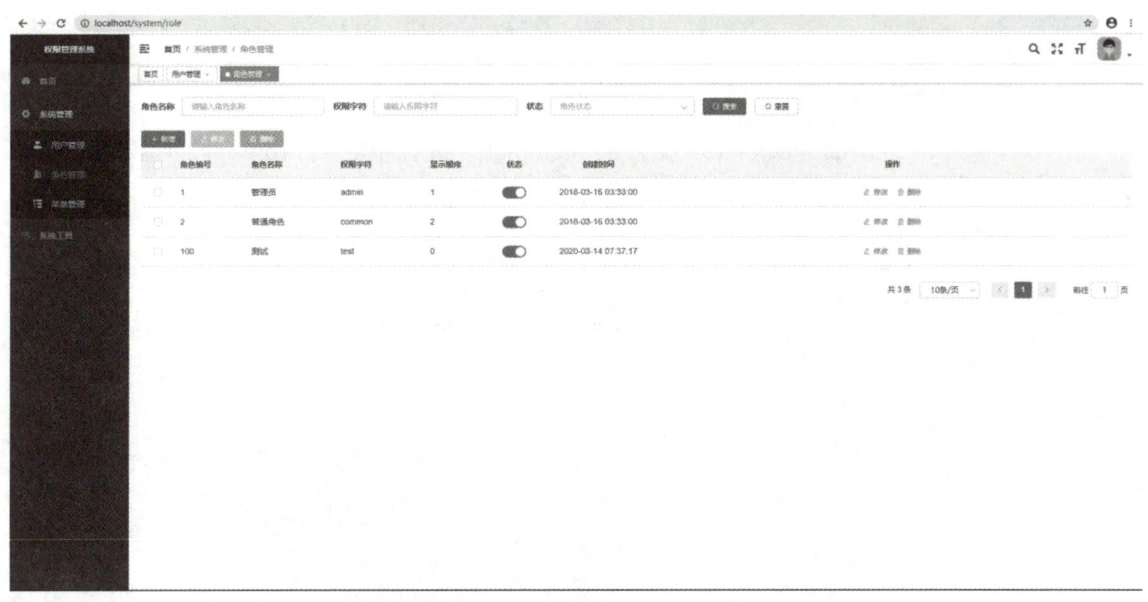

图 9-9　角色列表的模糊查询结果

任务 4　实现用户新增功能

微课 9-4　实现用户新增功能

任务分析

在用户管理界面，单击“新增”按钮弹出“添加用户”窗口，在窗口中录入数据，单击“确定”按钮，实现用户新增功能如图 9-10 所示。

添加用户

* 用户昵称　Athena_xiong　* 手机号码　13525452632

* 邮箱　487895@qq.com　* 用户名称　athena_xiong

* 用户密码　••••••　用户性别　女

角色　测试　状态　正常　停用

备注　test

确 定　取 消

图 9-10　用户新增功能

任务实施

步骤 1 编写新增页面代码

在“/views/system/user/index.vue”文件的 <template> 标签中的 <el-row> 下面增加 Element Dialog 组件，实现新增页面功能，组件的具体使用形式如下。

```
<!-- 添加或修改用户信息对话框 -->
    <el-dialog :title="title" :visible.sync="open" width="600px">
      <el-form ref="form" :model="form" :rules="rules" label-width="80px">
        <el-row>
          <el-col :span="12">
            <el-form-item label=" 用户昵称 " prop="nickName">
              <el-input v-model="form.nickName" placeholder=" 请输入用户昵称 " />
            </el-form-item>
          </el-col>
          <el-col :span="12">
            <el-form-item label=" 手机号码 " prop="phonenumber">
            <el-input v-model="form.phonenumber" placeholder=" 请输入手机号码 " maxlength="11" />
            </el-form-item>
          </el-col>
          <el-col :span="12">
            <el-form-item label=" 邮箱 " prop="email">
              <el-input v-model="form.email" placeholder=" 请输入邮箱 " maxlength="50" />
            </el-form-item>
          </el-col>
          <el-col :span="12">
            <el-form-item label=" 用户名称 " prop="userName">
              <el-input v-model="form.userName" placeholder=" 请输入用户名称 " />
            </el-form-item>
          </el-col>
          <el-col :span="12">
            <el-form-item v-if="form.userId == undefined" label=" 用户密码 " prop="password">
            <el-input v-model="form.password" placeholder=" 请输入用户密码 " type="password" />
            </el-form-item>
          </el-col>
          <el-col :span="12">
            <el-form-item label=" 用户性别 ">
              <el-select v-model="form.sex" placeholder=" 请选择 ">
                <el-option
                  v-for="dict in sexOptions"
                  :key="dict.dictValue"
                  :label="dict.dictLabel"
                  :value="dict.dictValue"
                />
              </el-select>
            </el-form-item>
          </el-col>
```

```
            <el-col :span="12">
              <el-form-item label=" 角色 ">
                <el-select v-model="form.roleIds" multiple placeholder=" 请选择 ">
                  <el-option
                    v-for="item in roleOptions"
                    :key="item.roleId"
                    :label="item.roleName"
                    :value="item.roleId"
                    :disabled="item.status == 1"
                  />
                </el-select>
              </el-form-item>
            </el-col>
            <el-col :span="12">
              <el-form-item label=" 状态 ">
                <el-radio-group v-model="form.status">
                  <el-radio
                    v-for="dict in statusOptions"
                    :key="dict.dictValue"
                    :label="dict.dictValue"
                  >{{ dict.dictLabel }}</el-radio>
                </el-radio-group>
              </el-form-item>
            </el-col>
            <el-col :span="24">
              <el-form-item label=" 备注 ">
                <el-input v-model="form.remark" type="textarea" placeholder=" 请输入内容 " />
              </el-form-item>
            </el-col>
          </el-row>
        </el-form>
        <div slot="footer" class="dialog-footer">
          <el-button type="primary" @click="submitForm"> 确 定 </el-button>
          <el-button @click="cancel"> 取 消 </el-button>
        </div>
      </el-dialog>
```

步骤 2 变量初始化

本任务使用了 title、sexOptions、roleOptions、initPassword、open、form 变量，在“/views/system/user/index.vue”文件的 <script> 标签下的 data 中初始化变量。

```
title: '',
sexOptions: [
{ dictLabel: ' 男 ', dictValue: '0' },
{ dictLabel: ' 女 ', dictValue: '1' }
],
roleOptions: [],
```

```
initPassword: '123456',
open: false,
form: {}
```

步骤 3 编写表单校验规则

本任务使用 rules 校验，在“/views/system/user/index.vue”文件的 <script> 标签下的 data 中定义校验规则。

```
// 表单校验
rules: {
    userName: [
      { required: true, message: ' 用户名称不能为空 ', trigger: 'blur' }
    ],
    nickName: [
      { required: true, message: ' 用户昵称不能为空 ', trigger: 'blur' }
    ],
    password: [
      { required: true, message: ' 用户密码不能为空 ', trigger: 'blur' }
    ],
    email: [
      { required: true, message: ' 邮箱地址不能为空 ', trigger: 'blur' },
      {
        type: 'email',
        message: ' 请输入正确的邮箱地址 ',
        trigger: ['blur', 'change']
      }
    ],
    phonenumber: [
      { required: true, message: ' 手机号码不能为空 ', trigger: 'blur' },
      {
        pattern: /^1[3|4|5|6|7|8|9][0-9]\d{8}$/,
        message: ' 请输入正确的手机号码 ',
        trigger: 'blur'
      }
    ]
}
```

步骤 4 编写表单处理方法

1）在“/views/system/user/index.vue”文件的 <script> 标签下的 methods 中实现对 form 表单的初始化 reset 方法。

```
reset() {
  this.form = {
    userId: undefined,
    nickName: undefined,
    userName: undefined,
    password: undefined,
```

```
      email: undefined,
      sex: undefined,
      status: 0,
      phonenumber: undefined,
      remark: undefined,
      roleIds: []
    }
    this.resetForm('form')
  }
```

2）在 “/views/system/user/index.vue” 的 <script> 标签下的 methods 中实现 handleAdd 方法用于初始化用户表单信息，其中 getUser() 方法可调用请求后端 API 接口数据。

```
import { listUser, getUser, addUser } from '@/api/system/user'

handleAdd() {
  this.reset()
  getUser().then(response => {
      this.roleOptions = response.roles
      this.open = true
      this.title = ' 添加用户 '
      this.form.password = this.initPassword
  })
}
```

3）在 “/views/system/user/index.vue” 的 <script> 标签下的 methods 中实现新增用户对话框的 “确定” 按钮绑定的 submitForm() 方法和 “取消” 按钮绑定的 cancel() 方法。

```
// 确定按钮
submitForm: function() {
  this.$refs['form'].validate(valid => {
    if (valid) {
      if (this.form.userId !== undefined) {
        console.log('update user')
      } else {
        addUser(this.form).then(response => {
          if (response.code === 200) {
            this.msgSuccess(' 新增成功 ')
            this.open = false
            this.getList()
          } else {
            this.msgError(response.msg)
          }
        })
      }
    }
  })
},
// 取消按钮
cancel() {
  this.open = false
```

```
    this.reset()
  }
}
```

步骤 5 编写请求后端接口代码

在“api/system/user.js”文件实现 getUser 函数和 addUser 函数，代码如下。

➢ getUser 函数使用 get 请求方法，请求后端接口“/system/user/”，传递 userId 参数，以便于获取用户数据和角色列表数据。

➢ addUser 函数使用 post 请求方法，请求后端接口“/system/user/”，传递页面输入的用户信息作为 data 参数，用于新增用户数据到数据库。

```
// 获取用户数据
export function getUser(userId) {
  return request({
    url: '/system/user/' + praseStrEmpty(userId),
    method: 'get'
  })
}
// 新增用户数据
export function addUser(data) {
  return request({
    url: '/system/user',
    method: 'post',
    data: data
  })
}
```

步骤 6 挂载全局消息的处理方法

在“main.js”文件中挂载整个项目中使用的 msgSuccess、msgError 以及 msgInfo 全局消息的处理方法。

```
Vue.prototype.msgSuccess = function(msg) {
  this.$message({ showClose: true, message: msg, type: 'success' })
}
Vue.prototype.msgError = function(msg) {
  this.$message({ showClose: true, message: msg, type: 'error' })
}
Vue.prototype.msgInfo = function(msg) {
  this.$message.info(msg)
}
```

步骤 7 运行调试

1）测试表单验证功能，所有字段全部为空，单击“确定”按钮提交空数据测试新增用户页面，如图 9-11 所示。

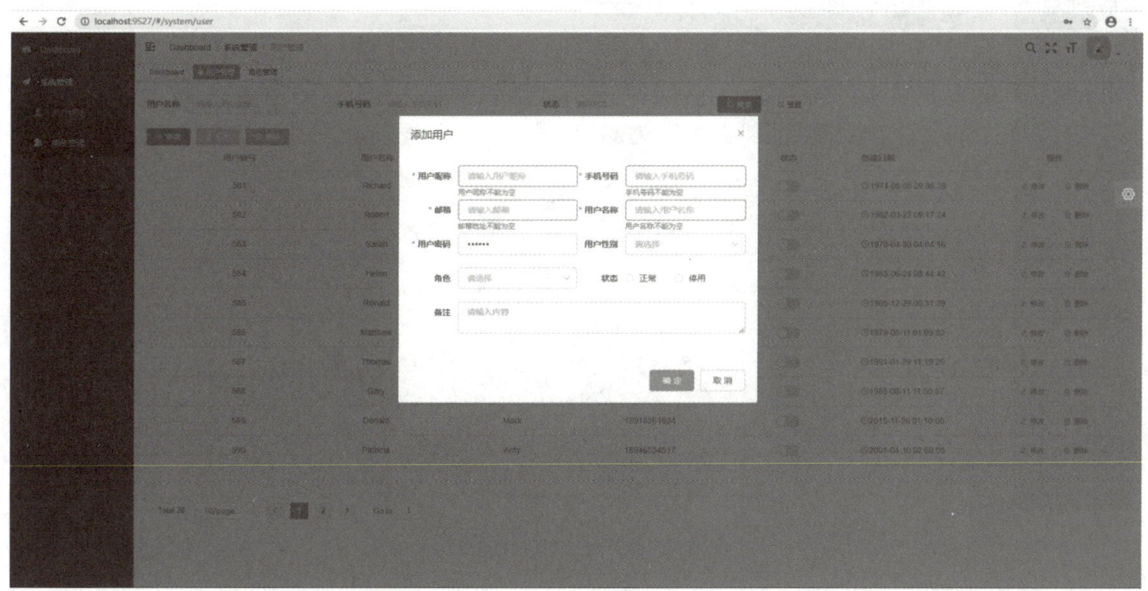

图 9-11　提交空数据测试新增用户页面

2）测试表单提交功能，如图 9-12 所示提交正常数据，新增用户成功界面如图 9-13 所示。

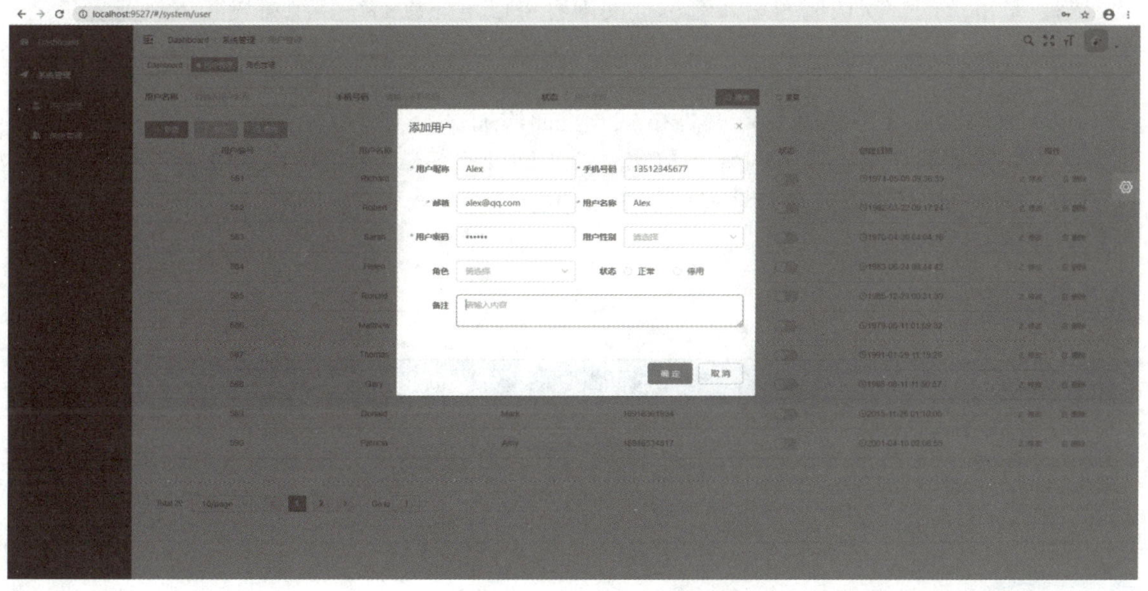

图 9-12　提交正常数据

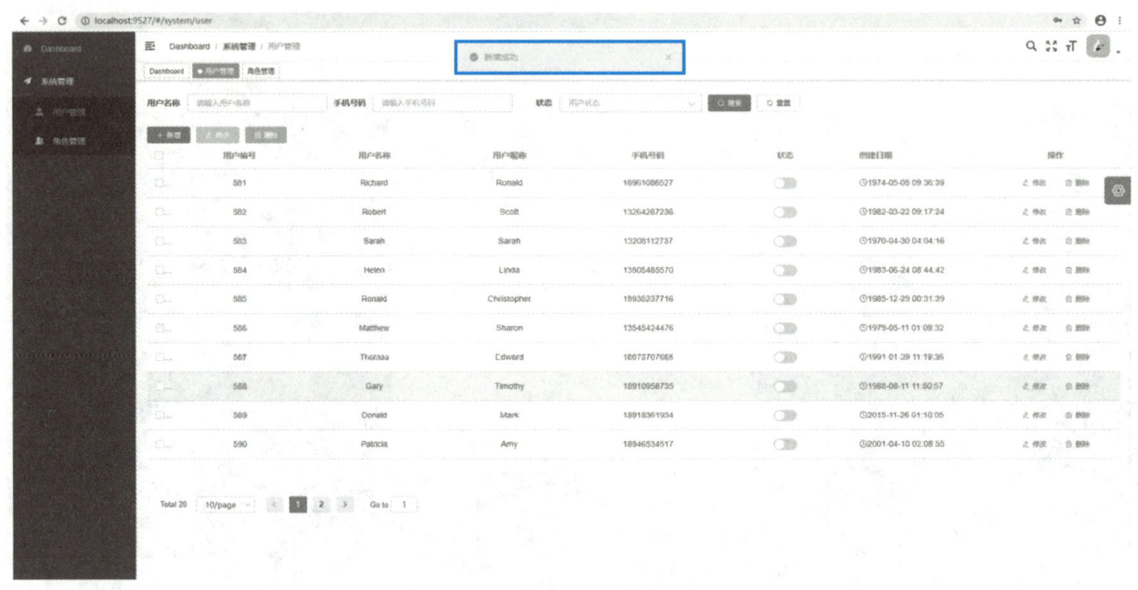

图 9-13　新增用户成功

知识小结

1. el-dialog 组件

el-dialog 组件在保留当前页面状态的情况下，弹出一个对话框，通知用户并承载相关操作。对话框分为两个部分：body 和 footer。footer 需要名为 footer 的 slot。title 属性用于定义标题，是可选的，默认值为空。对话框的内容可以是任意的，甚至可以是表格或表单。

2. rules 表单验证功能

rules 表单验证文件遵循 Json 文件格式，里面添加每一个需要验证的客户端组件名称，定义其需要验证的规范。使用 type 参数定义通用的校验类型，required 参数定义其是否是必要字段，message 参数定义其验证返回信息，trigger 参数定义验证的触发器，pattern 参数定义校验正则表达式。

任务评价

技　能　点	知　识　点	自我评价（不熟悉 / 基本掌握 / 熟练掌握 / 灵活运用）
使用 el-dialog 组件实现页面弹窗功能	Element UI 的 el-dialog 组件和 Element rules 表单验证规则的工作过程	
使用 Element rules 表单验证规则进行表单验证		

拓展练习

新增角色

模仿本任务的实施步骤实现新增角色功能，新增角色页面实现结果如图 9-14 所示。

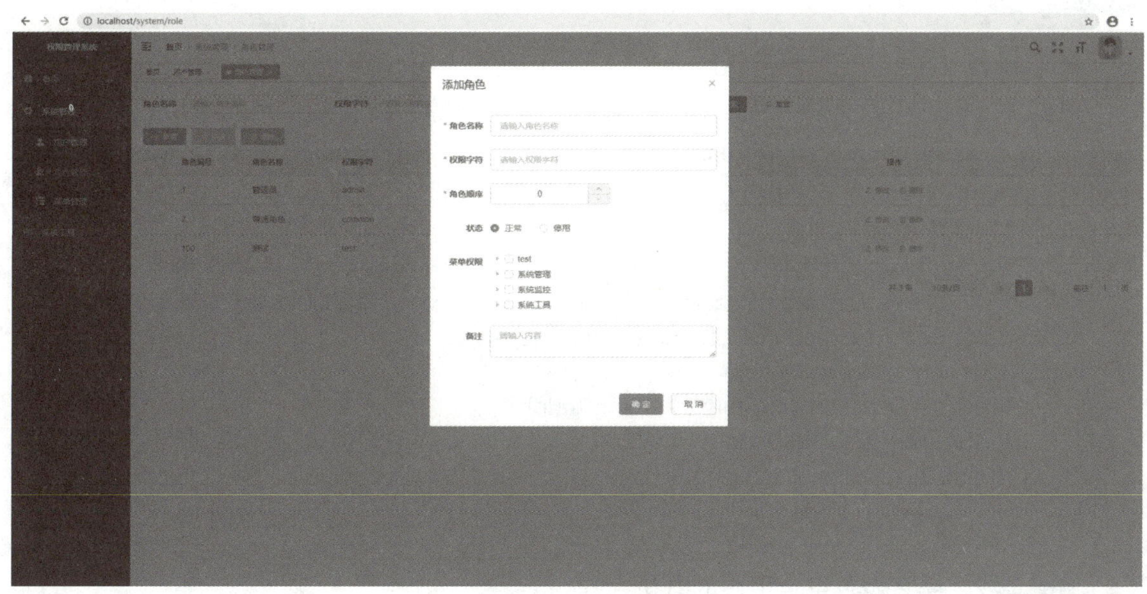

图 9-14　新增角色

任务 5　实现用户修改功能

微课 9-5　实现用户修改功能

任务分析

在用户管理界面，用户可以通过单击当前行的“修改”按钮或者选中多选框，然后单击头部的“修改”按钮来修改用户数据，页面效果如图 9-15 所示。

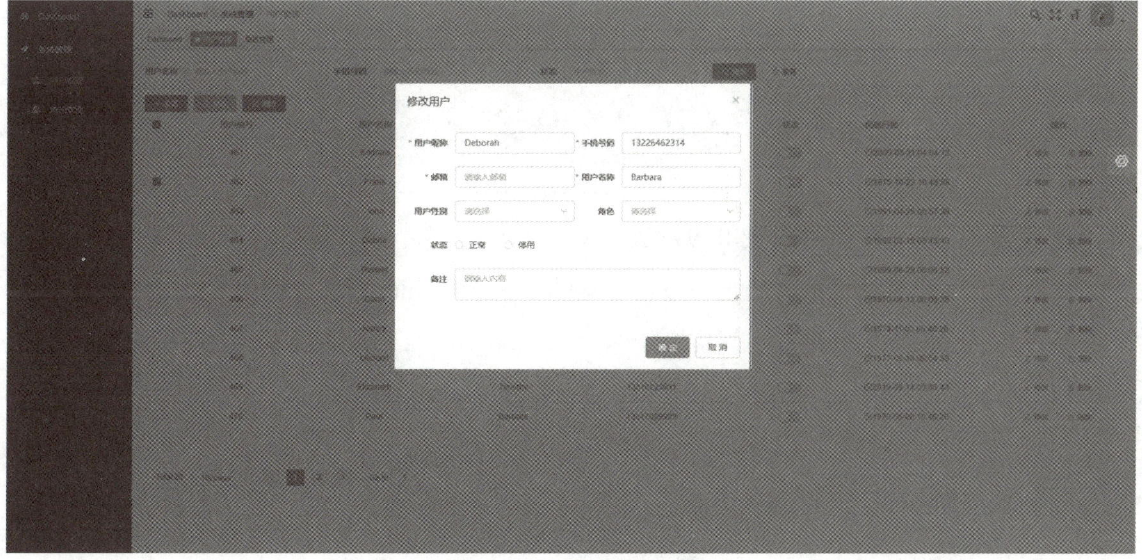

图 9-15　修改用户数据

任务实施

步骤 1 编写多选框实现代码

根据多选框选中数据项来控制顶部“修改”和“删除”按钮的启用和禁用。在 el-table 组件上使用 @selection-change="handleSelectionChange" 绑定 handleSelectionChange 方法。handleSelectionChange 方法实现的逻辑是当选中一条记录时，“修改”按钮启用，当选中多条记录时“删除”按钮启用。

```
<el-table v-loading="listLoading" :data="userList" @selection-change="handleSelectionChange">
// 多选框选中数据，method 下增加 handleSelectionChange
handleSelectionChange(selection) {
        this.ids = selection.map(item => item.userId);
        this.single = selection.length !== 1;
        this.multiple = !selection.length;
}
```

步骤 2 编写表单绑定用户的方法

实现事件绑定的 handleUpdate 方法，调用后端 API 接口获取待修改的用户数据，弹出“修改用户”窗口。

```
handleUpdate(row) {
  this.reset()
  const userId = row.userId || this.ids
  getUser(userId).then(response => {
    this.form = response.data
    this.roleOptions = response.roles
    this.form.roleIds = response.roleIds
    this.open = true
    this.title = ' 修改用户 '
    this.form.password = ''
  })
}
```

步骤 3 编写更新用户表单的方法

实现修改用户对话框的“确定”按钮绑定的 submitForm() 方法，调用 updateUser 函数，请求后端接口更新用户数据。

```
import { listUser, getUser, addUser, updateUser } from '@/api/system/user'
/** 确定按钮 */
submitForm: function() {
  this.$refs['form'].validate(valid => {
    if (valid) {
      if (this.form.userId !== undefined) {
```

```
        // 修改用户数据
        updateUser(this.form).then(response => {
          if (response.code === 200) {
            this.msgSuccess(' 修改成功 ')
            this.open = false
            this.getList()
          } else {
            this.msgError(response.msg)
          }
        })
      } else {
        // 新增用户数据
        addUser(this.form).then(response => {
          if (response.code === 200) {
            this.msgSuccess(' 新增成功 ')
            this.open = false
            this.getList()
          } else {
            this.msgError(response.msg)
          }
        })
      }
    }
  })
}
```

步骤 4　编写请求后端接口的代码

在“api/system/user.js”文件中增加 updateUser 函数，updateUser 函数使用 put 请求方法，请求后端接口“/system/user/”，传递页面输入的用户信息作为 data 参数，以便于更新用户数据到数据库。

```
// 修改用户
export function updateUser(data) {
  return request({
    url: '/system/user',
    method: 'put',
    data: data
  })
}
```

步骤 5　运行调试

单击“修改”按钮，如图 9-16 所示录入修改的用户数据，运行测试，修改用户数据成功，效果如图 9-17 所示。

图 9-16　录入修改的用户数据

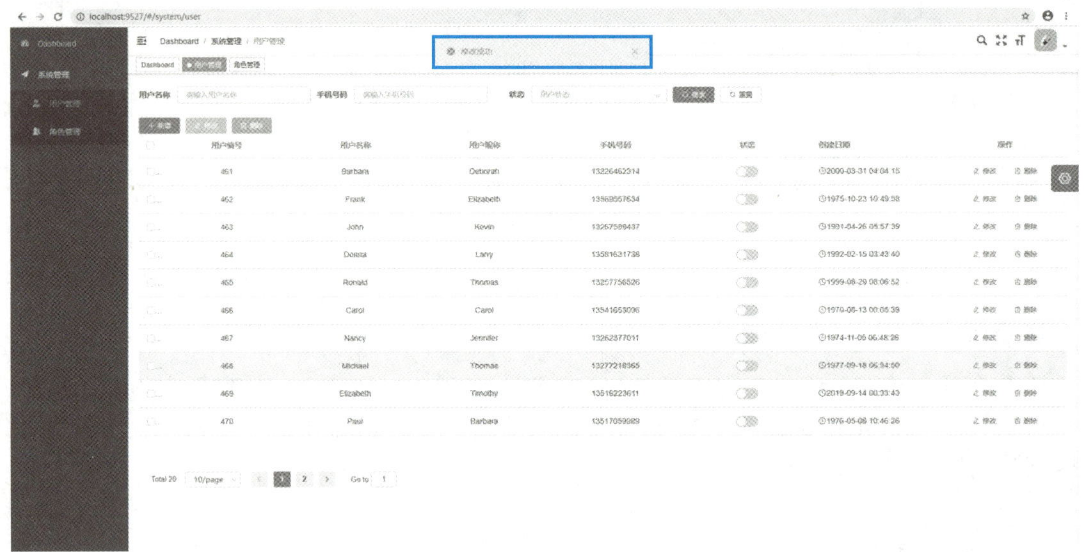

图 9-17　修改用户数据成功

任务评价

技　能　点	知　识　点	自我评价（不熟悉 / 基本掌握 / 熟练掌握 / 灵活运用）
使用 el-table 组件的 @selection-change 实现表格多选框选择数据	Element UI 的 el-table 组件的数据处理过程	

拓展练习

修改角色

模仿本任务的实施步骤实现修改角色功能，修改角色页面实现结果如图 9-18 所示。

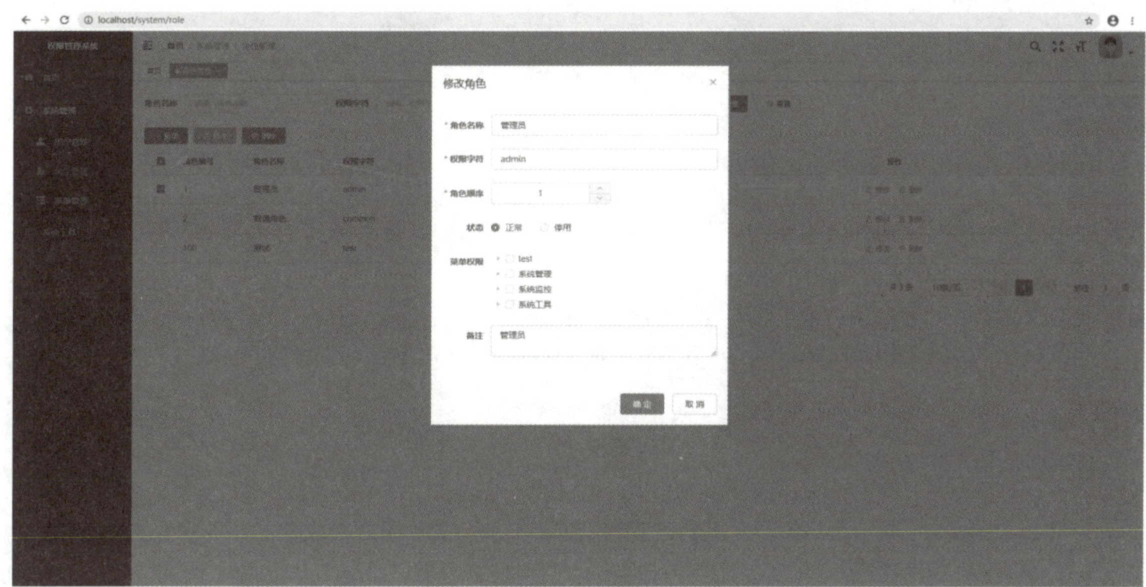

图 9-18　修改角色

任务 6　实现用户批量删除功能

微课 9-6　实现用户批量删除功能

任务分析

本任务实现删除单行用户数据和批量删除多个用户数据功能，单击“删除”按钮弹出警告窗口效果如图 9-19 所示。

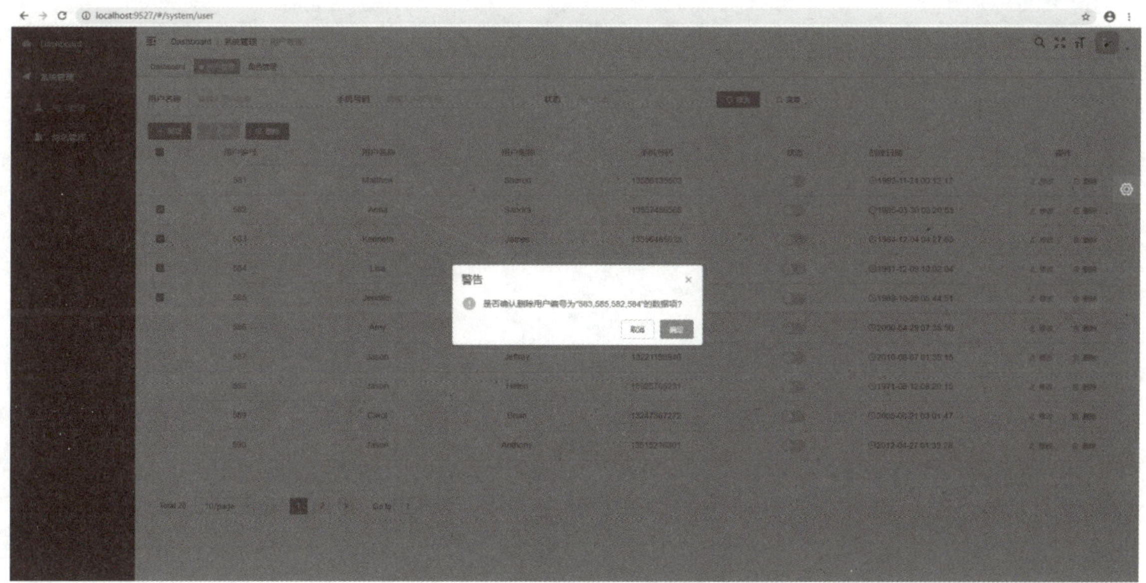

图 9-19　单击“删除”按钮弹出警告窗口

任务实施

步骤 1 编写删除用户数据代码

实现事件绑定的 handleDelete 方法，弹出警告窗口，如果用户单击“确定”按钮，则调用后端 API 接口删除该用户数据。

```
import { listUser, getUser, addUser, updateUser, delUser } from '@/api/system/user'

// 删除用户
handleDelete(row) {
  const userIds = row.userId || this.ids
  this.$confirm(' 是否确认删除用户编号为 "' + userIds + '" 的数据项 ?', ' 警告 ', {
    confirmButtonText: ' 确定 ',
    cancelButtonText: ' 取消 ',
    type: 'warning'
  }).then(function() {
    return delUser(userIds)
  }).then(() => {
    this.getList()
    this.msgSuccess(' 删除成功 ')
  }).catch(function() {})
}
```

步骤 2 编写请求后端接口代码

在“api/system/user.js”文件中增加 delUser 函数，delUser 函数使用 delete 请求方法，请求后端接口“/system/user/{userId}”，传递 userId 参数，以便于删除用户数据。

```
// 删除用户
export function delUser(userId) {
  return request({
    url: '/system/user/' + userId,
    method: 'delete'
  })
}
```

步骤 3 运行调试

如图 9-19 所示单击“删除”按钮后，测试运行结果，删除成功，效果如图 9-20 所示。

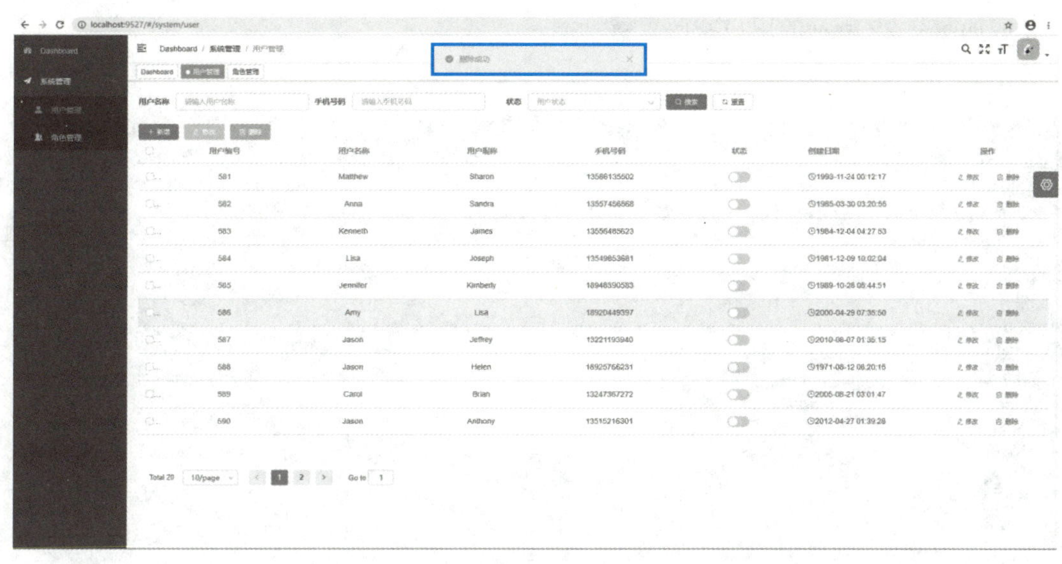

图 9-20　删除成功

任务评价

技　能　点	知　识　点	自我评价（不熟悉 / 基本掌握 / 熟练掌握 / 灵活运用）
使用 element ui 的 this.$confirm 实现确认框功能	Element UI 的 $confirm 的工作过程	

拓展练习

删除角色 & 重置密码

模仿本任务的实施步骤，实现删除角色，删除角色结果如图 9-21 所示。

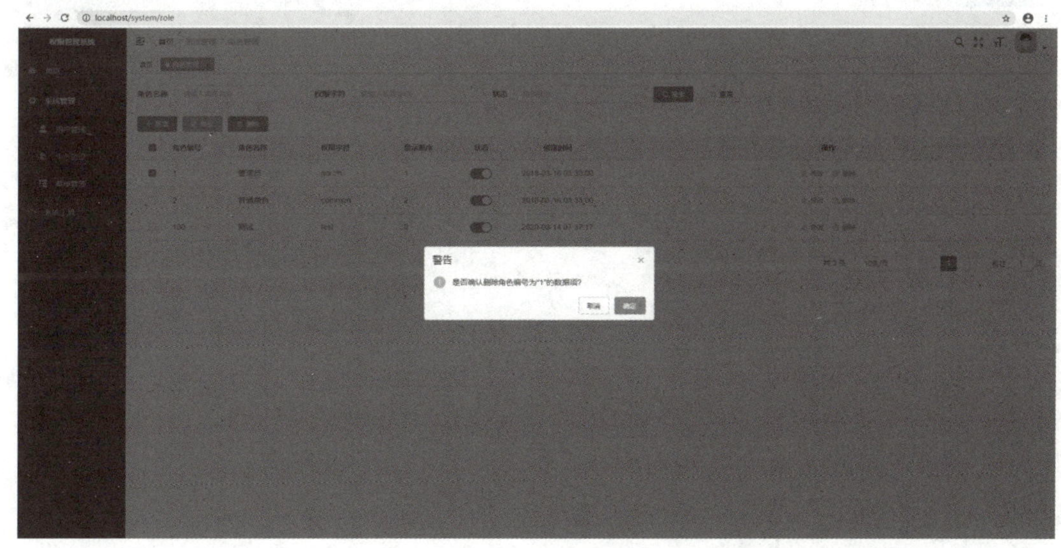

图 9-21　删除角色

完成如图 9-22 所示的用户密码重置功能。

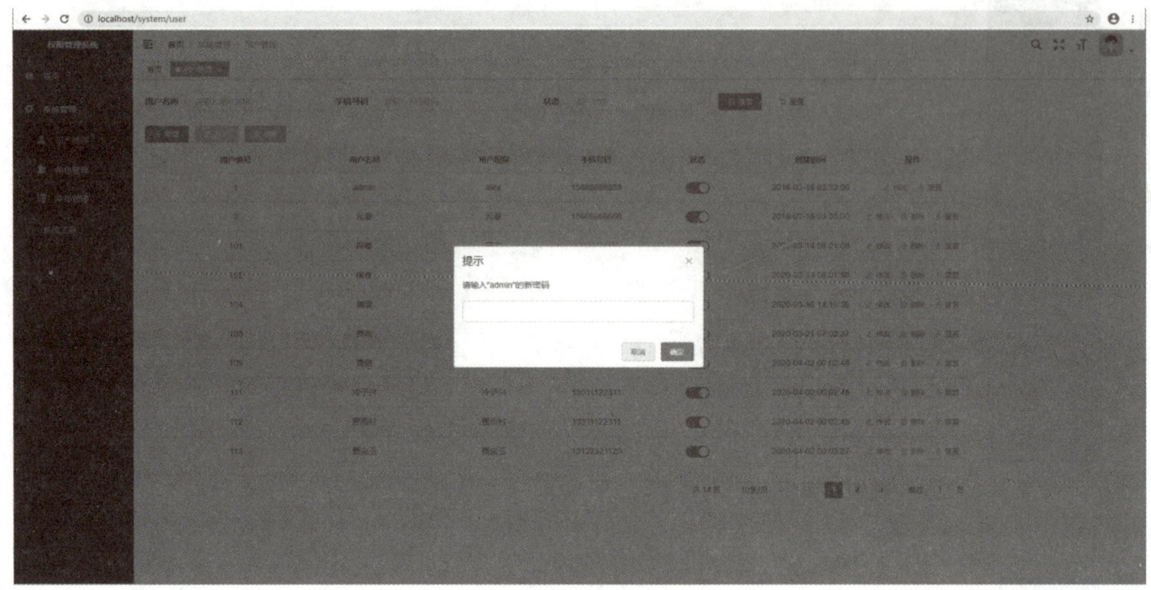

图 9-22　用户密码重置

单元小结

本工作单元实现了项目的用户和角色管理的前端功能，使用 Element UI 组件库实现增、删、改、查功能的页面效果，使用 Axios 向后台服务端发起 API 请求后台数据填充页面。通过学习并完成本工作单元的任务、拓展练习和实战强化，熟练掌握使用 Element UI 组件库实现常见的项目页面，以及使用 Axios 请求后端服务的职业技能。

实战强化

按照本工作单元任务实施步骤，实现实训项目“诚品书城”的商品管理页面。

工作单元 10

实现菜单前端管理功能

职业能力

本工作单元实现项目菜单前端管理功能，具体实现菜单的增、删、改、搜索以及树形显示功能，最终希望学生达成的职业能力目标是：

1．熟练掌握基于 Element UI 组件库实现菜单树形结构表格的页面效果

2．灵活使用 Vue-element-admin 框架的工具类

任务情景

在权限管理系统里，菜单用于定义和分配角色所拥有的菜单及其按钮的权限。本工作单元完成菜单管理的前端页面并与后端接口联调，实现完整的菜单管理功能，菜单前端管理界面如图 10-1 所示。

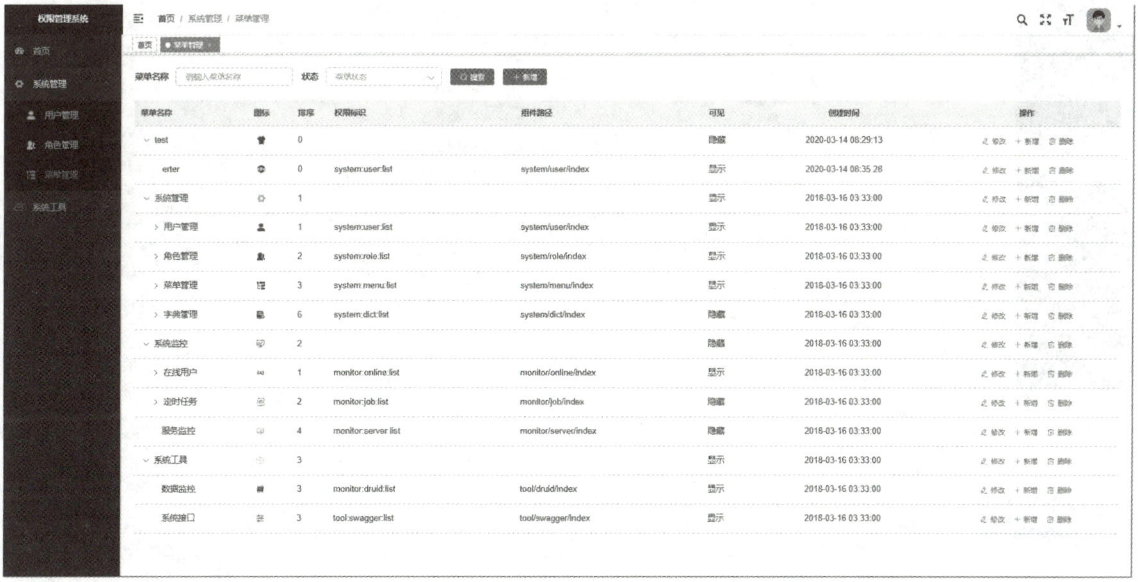

图 10-1　菜单前端管理界面

前置知识

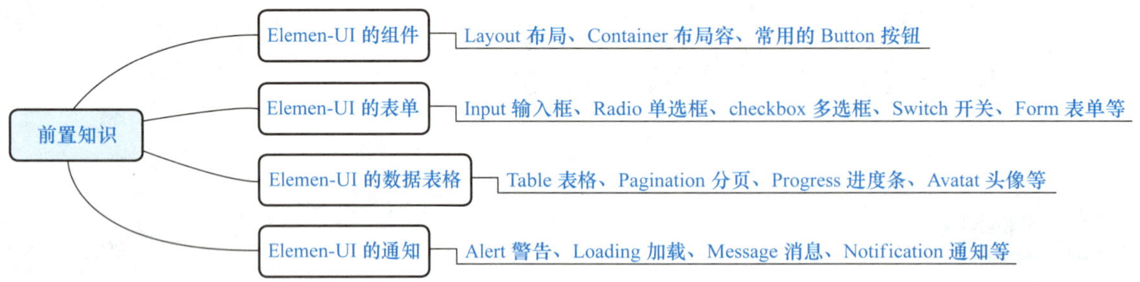

任务 1 实现菜单列表与查询页面

微课 10-1 实现菜单列表与查询页面

任务分析

本任务需要完成菜单的列表功能，通过调用后端接口获取菜单数据，并按照树形表格展示，实现效果如图 10-2 所示。

菜单名称	图标	排序	权限标识	组件路径	可见	创建时间	操作
test		0			隐藏	2020-03-14 08 29:13	修改 新增 删除
系统管理		1			显示	2018-03-16 03:33:00	修改 新增 删除
用户管理		1	system:user:list	system/user/index	显示	2018-03-16 03:33:00	修改 新增 删除
用户查询		1	system:user:query			2018-03-16 03:33:00	修改 新增 删除
用户新增		2	system:user:add			2018-03-16 03:33:00	修改 新增 删除
用户修改		3	system:user:edit			2018-03-16 03:33:00	修改 新增 删除
用户删除		4	system:user:remove			2018-03-16 03:33:00	修改 新增 删除
重置密码		7	system:user:resetPwd			2018-03-16 03:33:00	修改 新增 删除
角色管理		2	system:role:list	system/role/index	显示	2018-03-16 03:33:00	修改 新增 删除
菜单管理		3	system:menu:list	system/menu/index	显示	2018-03-16 03:33:00	修改 新增 删除
字典管理		6	system:dict:list	system/dict/index	隐藏	2018-03-16 03:33:00	修改 新增 删除
系统监控		2			隐藏	2018-03-16 03:33:00	修改 新增 删除
系统工具		3			显示	2018-03-16 03:33:00	修改 新增 删除

图 10-2　树形菜单列表

任务实施

步骤 1　编写菜单页面代码

在“views”目录下创建“system/menu/index.vue”组件，整体代码结构和用户管理前端代码高度相似，唯一不同就是使用 Element UI 的树形结构表格来展示菜单列表。

1）index.vue 页面的布局代码。

```
<template>
  <div class="app-container">
    <el-form :inline="true">
      <el-form-item label=" 菜单名称 ">
        <el-input v-model="queryParams.menuName" placeholder=" 请输入菜单名称 " clearable size="small" @keyup.enter.native="handleQuery" />
      </el-form-item>
      <el-form-item label=" 状态 ">
        <el-select v-model="queryParams.visible" placeholder=" 菜单状态 " clearable size="small">
          <el-option v-for="dict in visibleOptions" :key="dict.dictValue" :label="dict.dictLabel" :value="dict.dictValue" />
        </el-select>
      </el-form-item>
      <el-form-item>
        <el-button type="primary" icon="el-icon-search" size="mini" @click="handleQuery"> 搜索 </el-button>
        <el-button type="primary" icon="el-icon-plus" size="mini" @click="handleAdd"> 新增 </el-button>
      </el-form-item>
    </el-form>

    // 使用 Element UI 的树形结构表格，:tree-props="{children: 'children', hasChildren: 'hasChildren'}
    <el-table v-loading="loading" :data="menuList" row-key="menuId" :tree-props="{children: 'children', hasChildren: 'hasChildren'}">
      <el-table-column prop="menuName" label=" 菜单名称 " :show-overflow-tooltip="true" width="160" />
      <el-table-column prop="icon" label=" 图标 " align="center" width="100">
        <template slot-scope="scope">
          <svg-icon :icon-class="scope.row.icon" />
        </template>
      </el-table-column>
      <el-table-column prop="orderNum" label=" 排序 " width="60" />
      <el-table-column prop="perms" label=" 权限标识 " :show-overflow-tooltip="true" />
      <el-table-column prop="component" label=" 组件路径 " :show-overflow-tooltip="true" />
      <el-table-column prop="visible" label=" 可见 " :formatter="visibleFormat" width="80" />
      <el-table-column label=" 创建时间 " align="center" prop="createTime">
        <template slot-scope="scope">
          <span>{{ parseTime(scope.row.createTime) }}</span>
        </template>
      </el-table-column>
      <el-table-column label=" 操作 " align="center" class-name="small-padding fixed-width">
        <template slot-scope="scope">
          <el-button size="mini" type="text" icon="el-icon-edit" @click="handleUpdate(scope.
```

```
row)"> 修改 </el-button>
                <el-button size="mini" type="text" icon="el-icon-plus" @click="handleAdd(scope.
row)"> 新增 </el-button>
                <el-button size="mini" type="text" icon="el-icon-delete" @click="handleDelete(scope.
row)"> 删除 </el-button>
              </template>
            </el-table-column>
          </el-table>
        </div>
      </template>
```

2）编写初始化组件数据与页面组件绑定事件的代码。

```
<script>
import {
  listMenu
} from '@/api/system/menu'

export default {
  name: 'Menu',
  data() {
    return {
      // 遮罩层
      loading: true,
      // 菜单表格树数据
      menuList: [],
      // 菜单树选项
      menuOptions: [],
      // 弹出层标题
      title: '',
      // 是否显示弹出层
      open: false,
      // 菜单状态数据字典
      visibleOptions: [{
        'dictCode': '4',
        'dictSort': '1',
        'dictLabel': ' 显示 ',
        'dictValue': '0',
        'dictType': 'sys_show_hide',
        'cssClass': '',
        'listClass': 'primary',
        'isDefault': 'Y',
        'status': '0',
        'createBy': '1',
        'createTime': '0',
        'updateBy': '0',
        'updateTime': '0',
        'remark': ' 显示菜单 '
```

```
      }, {
        'dictCode': '5',
        'dictSort': '2',
        'dictLabel': ' 隐藏 ',
        'dictValue': '1',
        'dictType': 'sys_show_hide',
        'cssClass': '',
        'listClass': 'danger',
        'isDefault': 'N',
        'status': '0',
        'createBy': '0',
        'createTime': '0',
        'updateBy': '0',
        'updateTime': '0',
        'remark': ' 隐藏菜单 '
      }],
      // 查询参数
      queryParams: {
        menuName: undefined,
        visible: undefined
      },
      // 表单参数
      form: {},
      // 表单校验
      rules: {
        menuName: [{
          required: true,
          message: ' 菜单名称不能为空 ',
          trigger: 'blur'
        }],
        orderNum: [{
          required: true,
          message: ' 菜单顺序不能为空 ',
          trigger: 'blur'
        }]
      }
    }
  },
  created() {
    this.getList()
  },
  methods: {
    /** 查询菜单列表 */
    getList() {
      this.loading = true
      listMenu(this.queryParams).then(response => {
        this.menuList = this.handleTree(response.data, 'menuId')
```

```
        this.loading = false
      })
    },
    // 菜单显示状态字典翻译
    visibleFormat(row, column) {
      if (row.menuType === 'F') {
        return ''
      }
      return this.selectDictLabel(this.visibleOptions, row.visible)
    },
    /** 搜索按钮操作 */
    handleQuery() {
      this.getList()
    },
    /** 新增按钮操作 */
    handleAdd(row) {
    },
    /** 修改按钮操作 */
    handleUpdate(row) {
    },
    /** 删除按钮操作 */
    handleDelete(row) {
    }
  }
}
</script>
```

步骤 2 编写请求后端接口代码

在“api/system/menu.js”组件中增加 listMenu 函数，listMenu 函数使用 get 请求方法，请求后端接口“/system/menu/list”，传递页面输入的查询值作为 query 参数，以便于获取菜单列表数据。

```
// 查询菜单列表
export function listMenu(query) {
  return request({
    url: '/system/menu/list',
    method: 'get',
    params: query
  })
}
```

步骤 3 运行调试

分别启动运行前、后端服务，使用浏览器访问下面的 URL 地址：http://localhost:9527/，登录成功后单击左侧“系统管理”下的“菜单管理”，如图 10-3 所示，显示菜单列表即为成功。

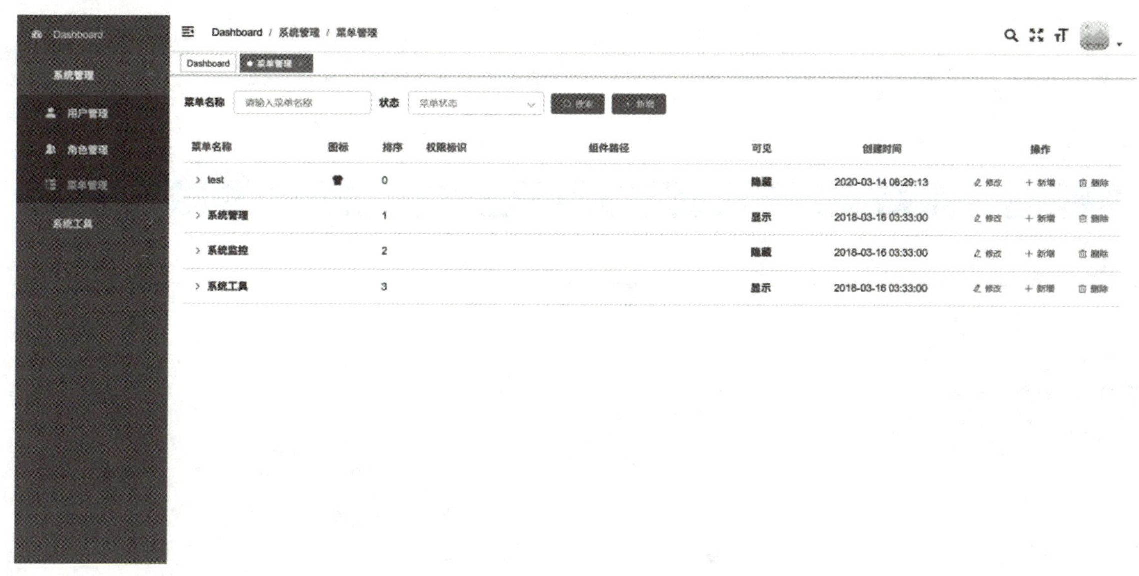

图 10-3　显示菜单列表

选择页面中状态的“显示”值，单击“查询”按钮，查询结果的菜单列表如图 10-4 所示。

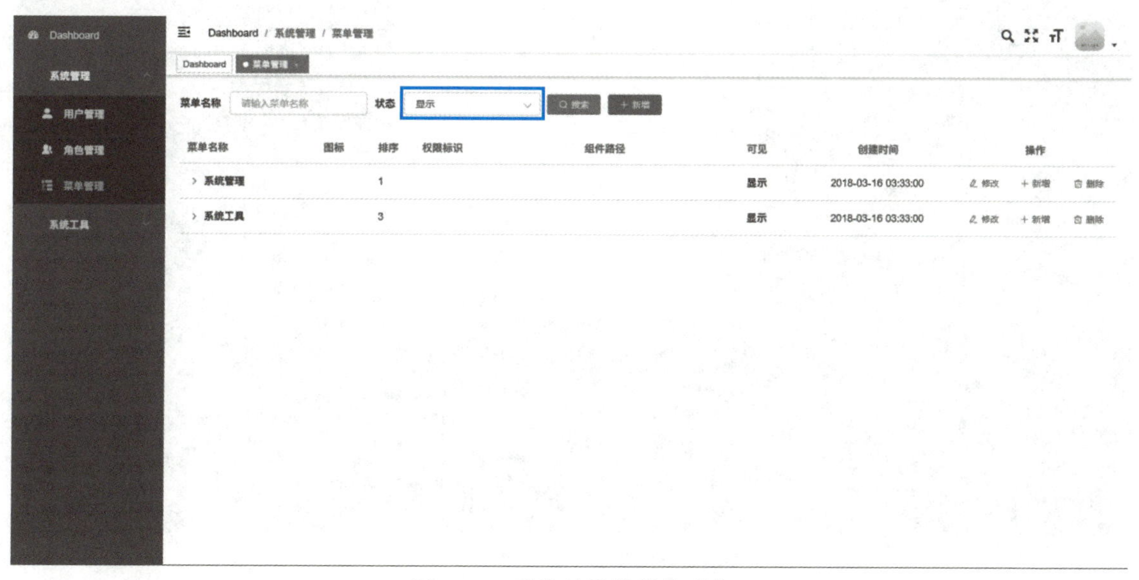

图 10-4　查询结果的菜单列表

知识小结

Element UI 的树形结构表格

在 Element UI 的 el-table 组件中，通过在 <el-table> 标签中增加 row-key="id" 和 :tree-props="{children: 'children', hasChildren: 'hasChildren'}，来支持树类型的数据显示。其中渲染树形数据时，必须要指定“row-key”参数，同时“tree-props”参数用于渲染嵌套数

据，通过指定数据行中的 hasChildren 字段来指定哪些行包含子节点，通过指定数据行中的 Children 字段来添加子节点的数据。

任务评价

技 能 点	知 识 点	自我评价（不熟悉/基本掌握/熟练掌握/灵活运用）
使用 Element UI 的树形表格实现菜单属性页面	Element UI 的树形表格组件的工作过程	

任务 2 实现菜单新增功能

微课 10-2 实现菜单新增功能

任务分析

本任务来实现新增菜单功能，在菜单管理界面，单击“新增”按钮弹出如图 10-5 所示的“添加菜单”窗口，在窗口中录入数据，单击“确定”按钮实现菜单的新增功能。

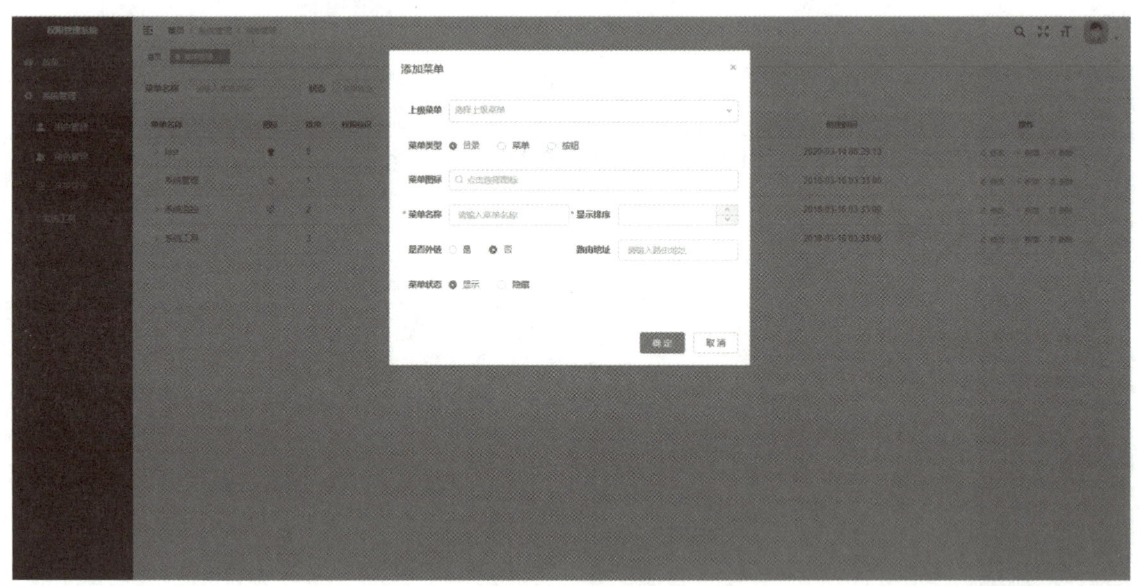

图 10-5 添加菜单

任务实施

步骤 1 编写新增按钮实现逻辑

修改 system/menu/index.vue 组件，实现“新增”按钮绑定的 handleAdd(row) 方法以及该方法内表单重置的 reset() 方法和获取菜单下拉树结构的 getTreeselect() 方法。

```
// 选择图标
selected(name) {
```

```
      this.form.icon = name;
    },
    /** 转换菜单数据结构 */
    normalizer(node) {
      if (node.children && !node.children.length) {
        delete node.children;
      }
      return {
        id: node.menuId,
        label: node.menuName,
        children: node.children
      };
    },
    /** 查询菜单下拉树结构 */
    getTreeselect() {
      listMenu().then(response => {
        this.menuOptions = [];
        const menu = {
          menuId: 0,
          menuName: ' 主类目 ',
          children: []
        };
        menu.children = this.handleTree(response.data, "menuId");
        this.menuOptions.push(menu);
      });
    },
    // 表单重置
    reset() {
      this.form = {
        menuId: undefined,
        parentId: 0,
        menuName: undefined,
        icon: undefined,
        menuType: "M",
        orderNum: undefined,
        isFrame: "1",
        visible: "0"
      };
      this.resetForm("form");
    },
    /** 新增按钮操作 */
      handleAdd(row) {
        this.reset();
        this.getTreeselect();
        if (row != null) {
          this.form.parentId = row.menuId;
    }
        this.open = true;
        this.title = " 添加菜单 ";
      }
```

步骤 2 增加添加对话框页面

修改 system/menu/index.vue 组件，使用 el-dialog 实现增加“添加或修改菜单对话框”效果，将 Treeselect 组件和 IconSelect 组件分别用于实现树形菜单选择和图标选择的效果。

```
<!-- 添加或修改菜单对话框 -->
<el-dialog :title="title" :visible.sync="open" width="600px">
  <el-form ref="form" :model="form" :rules="rules" label-width="80px">
    <el-row>
      <el-col :span="24">
        <el-form-item label=" 上级菜单 ">
          <treeselect v-model="form.parentId" :options="menuOptions" :normalizer="normalizer"
:show-count="true"
            placeholder=" 选择上级菜单 " />
        </el-form-item>
      </el-col>
      <el-col :span="24">
          <el-form-item label=" 菜单类型 " prop="menuType">
            <el-radio-group v-model="form.menuType">
              <el-radio label="M"> 目录 </el-radio>
              <el-radio label="C"> 菜单 </el-radio>
              <el-radio label="F"> 按钮 </el-radio>
            </el-radio-group>
          </el-form-item>
      </el-col>
      <el-col :span="24">
        <el-form-item v-if="form.menuType != 'F'" label=" 菜单图标 ">
          <el-popover placement="bottom-start" width="460" trigger="click" @
show="$refs['iconSelect'].reset()">
            <IconSelect ref="iconSelect" @selected="selected" />
            <el-input slot="reference" v-model="form.icon" placeholder=" 单击选择图标 "
readonly>
              <svg-icon v-if="form.icon" slot="prefix" :icon-class="form.icon" class="el-input__
icon" style="height: 32px;width: 16px;" />
              <i v-else slot="prefix" class="el-icon-search el-input__icon" />
            </el-input>
          </el-popover>
        </el-form-item>
      </el-col>
      <el-col :span="12">
          <el-form-item label=" 菜单名称 " prop="menuName">
            <el-input v-model="form.menuName" placeholder=" 请输入菜单名称 " />
          </el-form-item>
      </el-col>
      <el-col :span="12">
        <el-form-item label=" 显示排序 " prop="orderNum">
        <el-input-number v-model="form.orderNum" controls-position="right" :min="0" />
      </el-form-item>
      </el-col>
```

```
<el-col :span="12">
  <el-form-item v-if="form.menuType != 'F'" label=" 是否外链 ">
    <el-radio-group v-model="form.isFrame">
      <el-radio label="0"> 是 </el-radio>
      <el-radio label="1"> 否 </el-radio>
    </el-radio-group>
  </el-form-item>
 </el-col>
 <el-col :span="12">
   <el-form-item v-if="form.menuType != 'F'" label=" 路由地址 " prop="path">
     <el-input v-model="form.path" placeholder=" 请输入路由地址 " />
   </el-form-item>
  </el-col>
  <el-col :span="12" v-if="form.menuType == 'C'">
    <el-form-item label=" 组件路径 " prop="component">
      <el-input v-model="form.component" placeholder=" 请输入组件路径 " />
      </el-form-item>
  </el-col>
  <el-col :span="12">
    <el-form-item v-if="form.menuType != 'M'" label=" 权限标识 ">
  <el-input v-model="form.perms" placeholder=" 请权限标识 " maxlength="50" />
</el-form-item>
  </el-col>
  <el-col :span="24">
    <el-form-item v-if="form.menuType != 'F'" label=" 菜单状态 ">
      <el-radio-group v-model="form.visible">
    <el-radio v-for="dict in visibleOptions" :key="dict.dictValue" :label="dict.dictValue">{{dict.
dictLabel}}</el-radio>
      </el-radio-group>
    </el-form-item>
  </el-col>
</el-row>
</el-form>
<div slot="footer" class="dialog-footer">
 <el-button type="primary" @click="submitForm"> 确定 </el-button>
 <el-button @click="cancel"> 取消 </el-button>
</div>
</el-dialog>
```

步骤 3　实现菜单图标选择功能

1）创建“src\components\IconSelect\index.vue”文件，构建步骤 2 中使用的 IconSelect 组件的模板内容，用于实现菜单图标的选择功能。

```
<template>
  <div class="icon-body">
    <el-input v-model="name" style="position: relative;" clearable placeholder=" 请输入图标名称
" @clear="filterIcons" @input.native="filterIcons">
      <i slot="suffix" class="el-icon-search el-input__icon" />
    </el-input>
```

```
      <div class="icon-list">
        <div v-for="(item, index) in iconList" :key="index" @click="selectedIcon(item)">
          <svg-icon :icon-class="item" style="height: 30px;width: 16px;" />
          <span>{{ item }}</span>
        </div>
      </div>
    </div>
</template>

<script>
import icons from './requireIcons'
export default {
  name: 'IconSelect',
  data() {
    return {
      name: '',
      iconList: icons
    }
  },
  methods: {
    filterIcons() {
      if (this.name) {
        this.iconList = this.iconList.filter(item => item.includes(this.name))
      } else {
        this.iconList = icons
      }
    },
    selectedIcon(name) {
      this.$emit('selected', name)
      document.body.click()
    },
    reset() {
      this.name = ''
      this.iconList = icons
    }
  }
}
</script>

<style rel="stylesheet/scss" lang="scss" scoped>
  .icon-body {
    width: 100%;
    padding: 10px;
    .icon-list {
      height: 200px;
      overflow-y: scroll;
      div {
        height: 30px;
        line-height: 30px;
```

```
        margin-bottom: -5px;
        cursor: pointer;
        width: 33%;
        float: left;
      }
      span {
        display: inline-block;
        vertical-align: -0.15em;
        fill: currentColor;
        overflow: hidden;
      }
    }
  }
</style>
```

2）创建“src\components\IconSelect\requireIcons.js”文件，引入步骤 2 中使用的 IconSelect 组件的图标文件，用于实现菜单图标的选择功能 。

```
// requireIcons.js
const req = require.context('../../assets/icons/svg', false, /\.svg$/)
const requireAll = requireContext => requireContext.keys()
const re = /\.\/(.*)\.svg/

const icons = requireAll(req).map(i => {
  return i.match(re)[1]
})

export default icons
```

步骤 4　引入自定义组件

1）在“package.json”文件的 dependencies 下增加 Treeselect 依赖。

```
"@riophae/vue-treeselect": "0.4.0"
```

2）引入步骤 3 创建的 IconSelect 组件和 Treeselect 组件。

```
import Treeselect from "@riophae/vue-treeselect";
import "@riophae/vue-treeselect/dist/vue-treeselect.css";
import IconSelect from "@/components/IconSelect";
// export default 增加下面代码
components: {
  Treeselect,
  IconSelect
},
```

步骤 5　实现表单处理方法

1）实现步骤 2 中弹出框的 cancel 和 submitForm 方法。

```
// 取消按钮
cancel() {
  this.open = false;
  this.reset();
},
/** 提交按钮 */
submitForm: function() {
  this.$refs["form"].validate(valid => {
    if (valid) {
      addMenu(this.form).then(response => {
          if (response.code === 200) {
            this.msgSuccess(" 新增成功 ");
            this.open = false;
            this.getList();
          } else {
            this.msgError(response.msg);
    }
      });
}
  });
},
```

2）在 system/menu/index.vue 中引入步骤 5 中的 addMenu 方法。

```
import {
  listMenu,
  addMenu
} from '@/api/system/menu'
```

步骤 6 编写请求后端接口代码

在“api/system/menu.js”文件中增加 addMenu 函数，addMenu 函数使用 post 请求方法，请求后端接口“/system/menu/”，传递页面输入的菜单信息作为 data 参数，用于新增菜单数据到数据库。

```
// 新增菜单
export function addMenu(data) {
  return request({
    url: '/system/menu',
    method: 'post',
    data: data
  })
}
```

步骤 7 运行调试

测试新增菜单成功功能，提交正常数据的界面如图 10-6 所示，新增菜单成功界面如

图 10-7 所示。

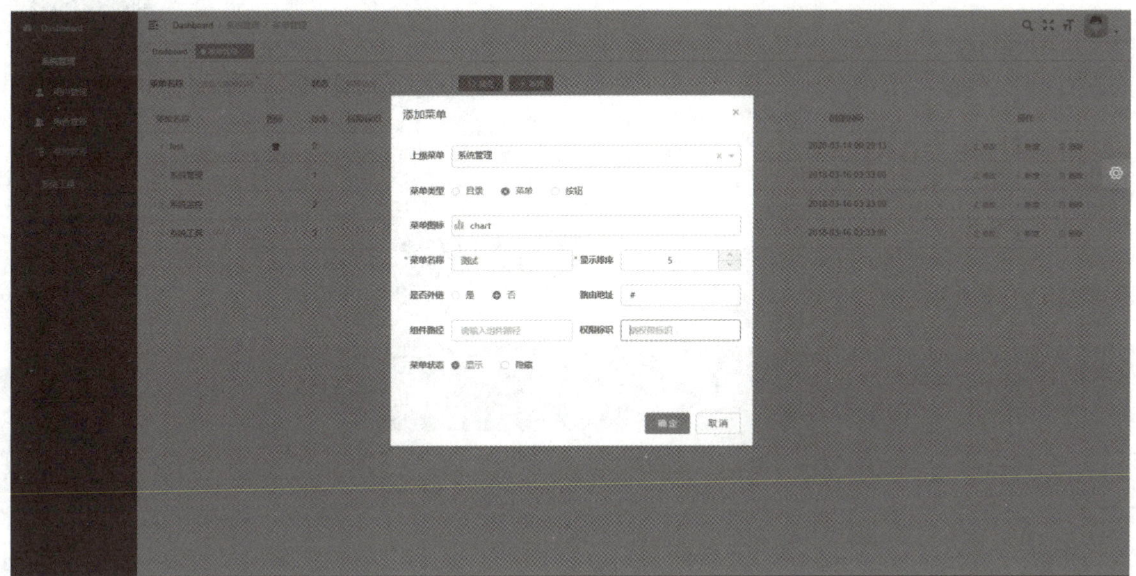

图 10-6　提交正常数据的界面

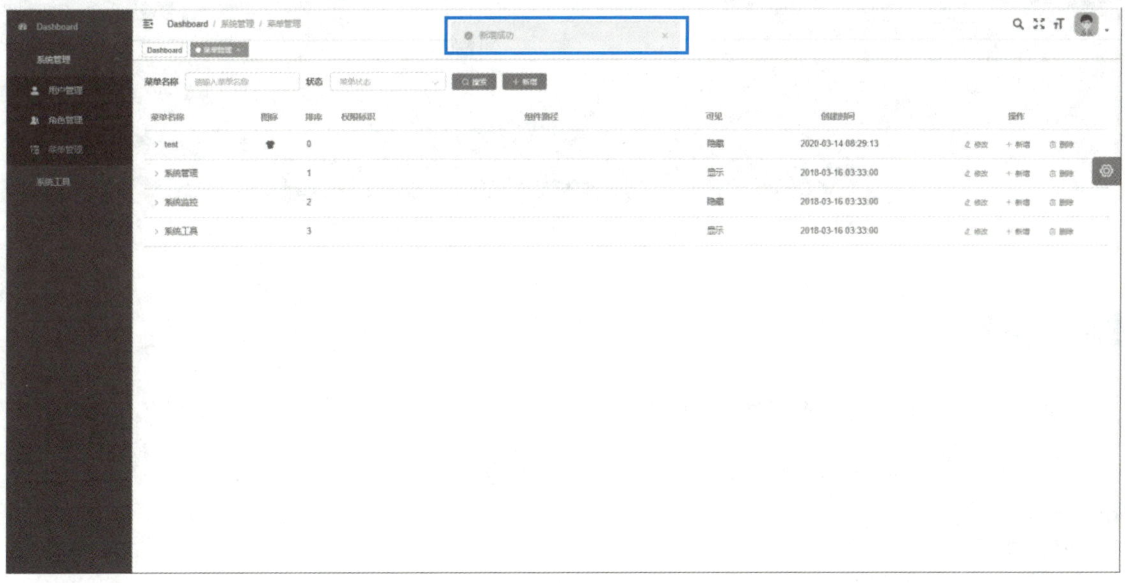

图 10-7　新增菜单成功界面

任务评价

技 能 点	知 识 点	自我评价（不熟悉 / 基本掌握 / 熟练掌握 / 灵活运用）
创建自定义组件实现图标的选择功能	创建自定义组件的工作过程 引入自定义组件的工作过程	
引入自定义组件 treeselect 实现树形菜单的选择功能		

任务 3 实现菜单修改功能

微课 10-3 实现菜单修改功能

任务分析

在菜单管理界面，用户可以通过单击当前行的“修改”按钮来修改菜单数据，页面效果如图 10-8 所示。

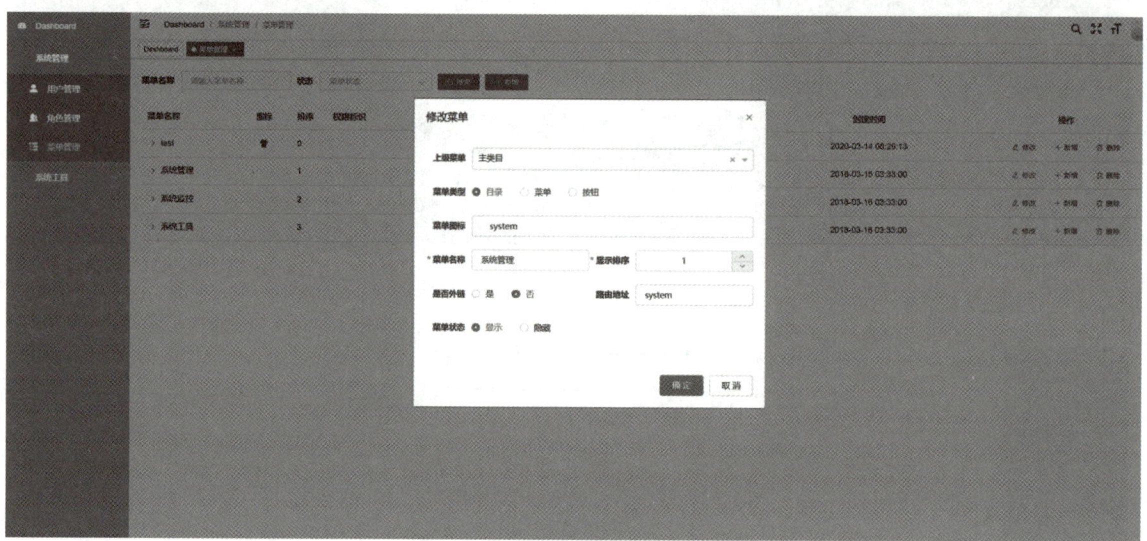

图 10-8 “修改”菜单

任务实施

步骤 1 实现修改菜单的事件

1）修改“system/menu/index.vue”组件，实现“修改”按钮绑定的 handleUpdate(row) 方法。

```
/** 修改按钮操作 */
  handleUpdate(row) {
    this.reset();
    this.getTreeselect();
    getMenu(row.menuId).then(response => {
      this.form = response.data;
      this.open = true;
      this.title = "修改菜单";
    });
  }
```

2）在“system/menu/index.vue”中导入 getMenu 方法和用来更新数据的 updateMenu 方法。

```
import {
    listMenu,
    addMenu,
    getMenu,
    updateMenu
} from "@/api/system/menu";
```

3）修改“system/menu/index.vue”组件的 submitForm 方法。

```
/** 提交按钮 */
 submitForm: function() {
  this.$refs["form"].validate(valid => {
    if (valid) {
     if (this.form.menuId != undefined) {
      updateMenu(this.form).then(response => {
        if (response.code === 200) {
         this.msgSuccess("修改成功");
         location.reload();
         this.open = false;
         this.getList();
        } else {
         this.msgError(response.msg);
        }
      });
     } else {
       addMenu(this.form).then(response => {
        if (response.code === 200) {
          this.msgSuccess("新增成功");
          this.open = false;
          this.getList();
        } else {
            this.msgError(response.msg);
          }
       });
     }
    }
  });
 }
```

步骤 2　编写请求后端接口代码

在“api/system/menu.js”组件中实现 getMenu 和 updateMenu 函数，代码如下。

➢ getMenu 函数使用 get 请求方法，请求后端接口“/system/menu/{menuId}”，传递 menuId 参数，以便于获取菜单详情数据。

➢ updateMenu 函数使用 put 请求方法，请求后端接口“/system/menu/”，传递页面输入的菜单信息作为 data 参数，用于更新菜单数据到数据库。

```
// 查询菜单详细数据
export function getMenu(menuId) {
  return request({
    url: '/system/menu/' + menuId,
    method: 'get'
  })
}
// 修改菜单
export function updateMenu(data) {
  return request({
    url: '/system/menu',
    method: 'put',
    data: data
  })
}
```

步骤 3 运行调试

单击“修改”按钮录入修改的菜单数据，如图 10-9 所示，运行测试成功，效果如图 10-10 所示。

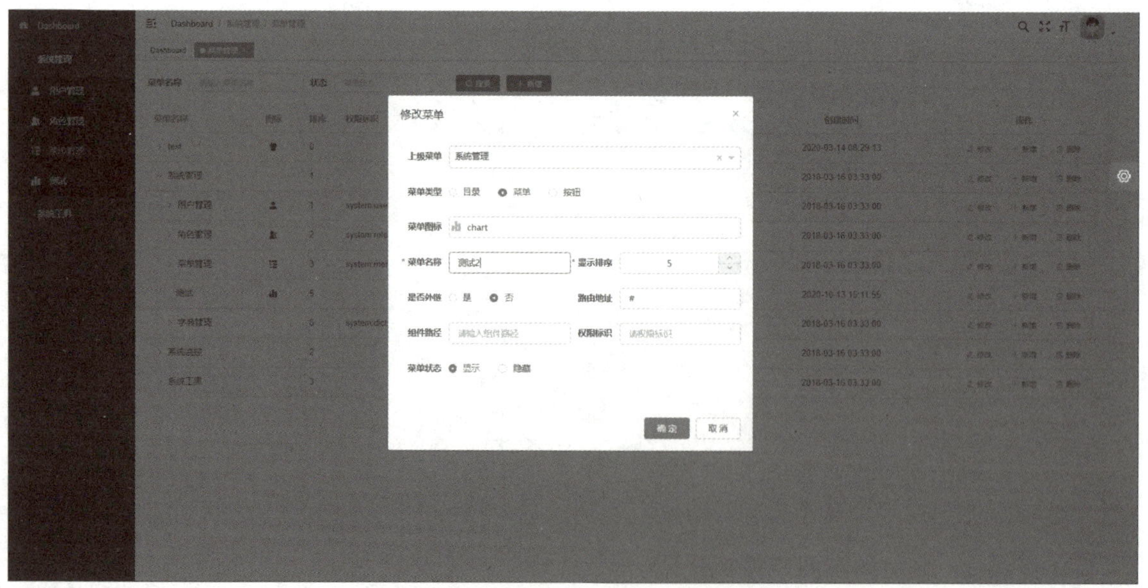

图 10-9　录入修改的菜单数据

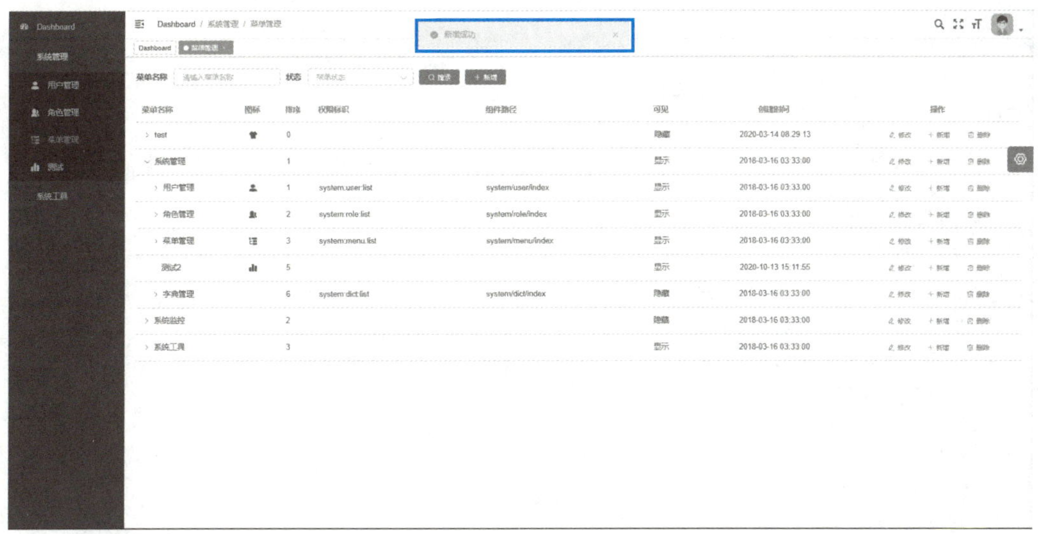

图 10-10　修改菜单数据成功

任务评价

技　能　点	知　识　点	自我评价（不熟悉 / 基本掌握 / 熟练掌握 / 灵活运用）
使用 Vue-Element-Admin 框架编写调用后台接口完成数据更新操作	前端 UI 调用服务端接口的处理流程	

任务 4　实现菜单删除功能

微课 10-4　实现菜单删除功能

任务分析

本任务实现菜单删除功能，通过单击“删除”按钮删除菜单数据，效果如图 10-11 所示。

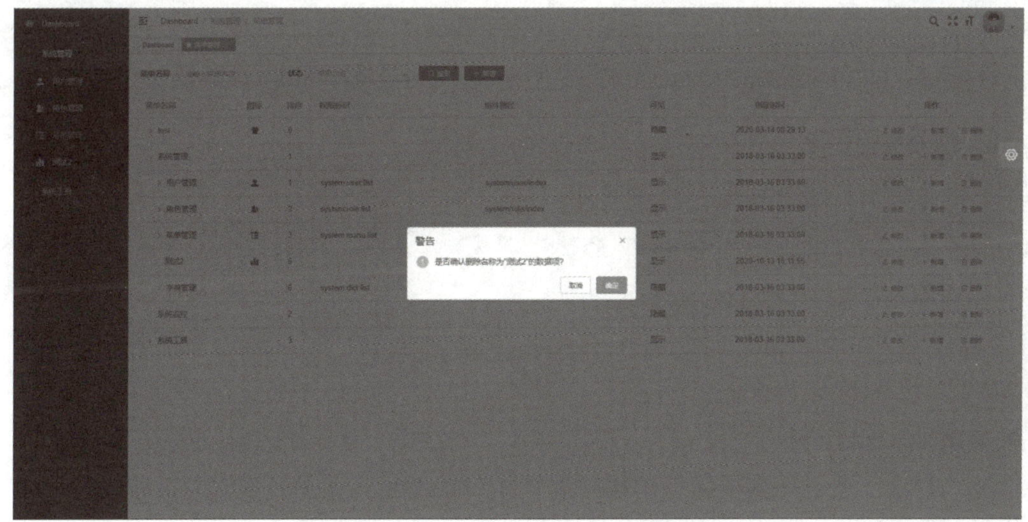

图 10-11　单击“删除”按钮弹出警告窗口

任务实施

步骤 1 编写删除菜单数据逻辑

1）修改“system/menu/index.vue”组件，实现“删除”按钮绑定的 handleDelete(row) 方法。

```
/** 删除按钮操作 */
  handleDelete(row) {
    this.$confirm('是否确认删除名称为"' + row.menuName + '"的数据项?', "警告", {
      confirmButtonText: "确定",
      cancelButtonText: "取消",
      type: "warning"
    }).then(function() {
      return delMenu(row.menuId);
    }).then(() => {
      this.getList();
      this.msgSuccess("删除成功");
    }).catch(function() {});
  }
```

2）在“system/menu/index.vue”中导入 delMenu 方法。

```
import {
    listMenu,
    addMenu,
    getMenu,
    updateMenu,
    delMenu
  } from "@/api/system/menu";
```

步骤 2 编写请求后端接口代码

在“api/system/menu.js”组件中增加 delMenu 函数，delMenu 函数使用 delete 请求方法，请求后端接口“/system/menu/{menuId}”，传递 menuId 参数，以便于删除菜单数据。

```
// 删除菜单数据
export function delMenu(menuId) {
  return request({
    url: '/system/menu/' + menuId,
    method: 'delete'
  })
}
```

步骤 3 运行调试

单击“删除”按钮后测试运行结果，如图 10-11 所示。菜单“测试 2”被成功删除，效果如图 10-12 所示。

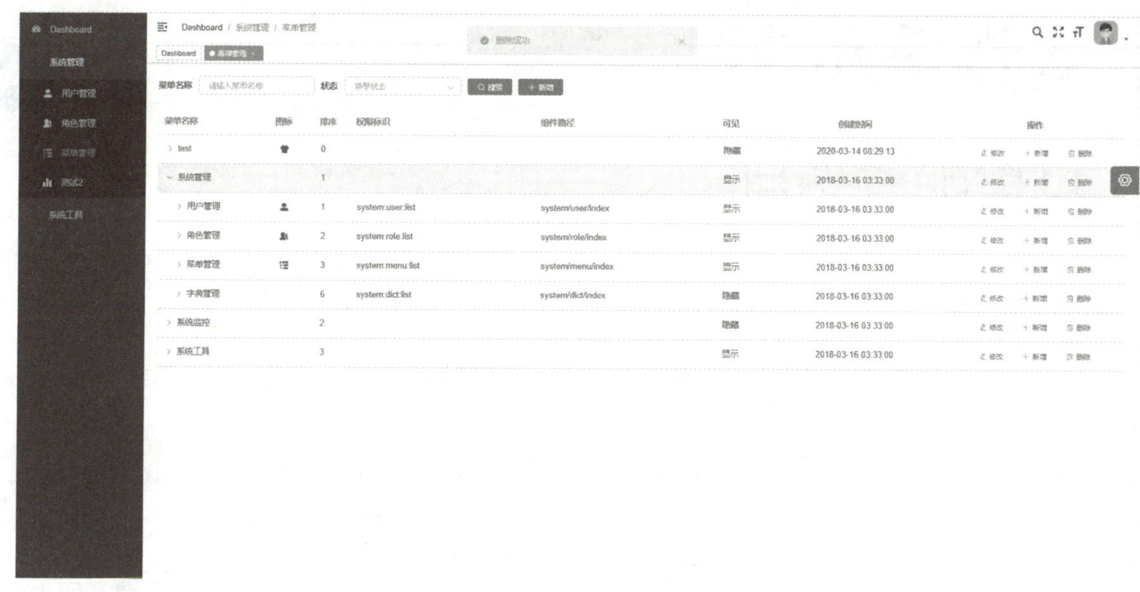

图 10-12　删除菜单“测试 2”

任务评价

技　能　点	知　识　点	自我评价（不熟悉 / 基本掌握 / 熟练掌握 / 灵活运用）
使用 Vue-Element-Admin 框架编写调用后台接口完成数据删除操作	前端 UI 调用服务端接口的处理流程	

单元小结

本工作单元通过运用 Element UI 中的组件、Axios 框架完成了菜单模块的增、删、改、查和树状菜单显示功能。通过完成本工作单元的任务和后面的实战强化，熟练掌握使用 Element UI 组件库和 Vue-element-admin 框架的工具类来实现前端页面诸多功能的职业技能。

实战强化

按照本工作单元任务实施步骤，实现实训项目“诚品书城”的订单管理页面。

工作单元 11

打包部署项目

职业能力

本工作单元主要完成在云服务器的 Linux 生产环境下打包和部署项目，最终希望学生达成如下职业能力目标：

1．熟练掌握在云服务器的 Linux 环境下安装与配置 JDK、MySQL、Redis 与 Ngnix 服务

2．熟练掌握后端项目的打包与部署

3．熟练掌握前端项目的打包与部署

任务情景

研发人员在开发机器上调试代码并通过测试后，需要将项目发布到外网服务器给用户提供服务，将面临下面三个问题：

1）外网服务器上如何安装项目需要的 Java 软件环境（如 JDK1.8、MySQL 等）？

2）后端项目如何打包部署到外网服务器？

3）前端项目如何打包部署到外网服务器？

基于上述三个问题，本工作单元的具体任务如下：

1）在云服务器上安装 JDK、MySQL、Redis 与 Ngnix 服务。

2）将后端项目打包部署到云服务器。

3）将前端项目打包部署到云服务器。

前置知识

Linux 系统的基础使用命令；Xshell 软件工具的使用；YUM 操作软件包常用命令的使用。

任务 1　搭建云服务器环境

微课 11-1　搭建云服务器环境

任务分析

JDK 在 Linux 系统中目前存在两种版本，一是 yum 提供的 OpenJDK，另一种是 Oracle

官网提供的 JDK 版本。其中 Oracle 提供的版本需要自己下载 rpm 文件进行安装，比较烦琐，因此本书使用 yum 提供的 OpenJDK 进行安装。

任务实施

步骤 1　安装 JDK 1.8

1）执行以下命令，查看可用的 Java 版本。

```
yum search java 或者 yum list java*
```

2）执行以下命令，使用 java-1.8.0-openjdk-devel.x86_64 版本安装 JDK。

```
yum install java-1.8.0-openjdk-devel.x86_64
```

安装过程中会询问“Is this OK?”输入“y”并按 <Enter> 键确认，显示安装完成，如图 11-1 所示。

```
Installed:
  java-1.8.0-openjdk-devel.x86_64 1:1.8.0.282.b08-1.el7_9

Dependency Updated:
  java-1.8.0-openjdk.x86_64 1:1.8.0.282.b08-1.el7_9                                java-1.8.0-open

Complete!
[root@VM_0_3_centos ~]# java  -version
```

图 11-1　JDK 安装完成

3）执行以下命令，验证 JDK 的版本，如果能正常输出版本号则说明安装成功，通过这种方式安装的 JDK 不需要自己再去配置环境变量，比较方便。

```
java -version
```

步骤 2　安装 MySQL 数据库

1）执行以下命令，下载并安装 MySQL。

```
// 下载并安装 MySQL 官方的 Yum Repository
wget https://repo.mysql.com//mysql80-community-release-el7-3.noarch.rpm
// 获得 rpm 文件后，通过命令添加镜像源
rpm -ivh mysql80-community-release-el7-3.noarch.rpm
// 安装 MySQL
yum install mysql-server
```

安装过程中会询问“Is this OK?”输入“y”并按 <Enter> 键确认，显示安装完成，如图 11-2 所示。

```
Installed:
  mysql-community-server.x86_64 0:8.0.23-1.el7

Dependency Installed:
  mysql-community-client.x86_64 0:8.0.23-1.el7
  mysql-community-client-plugins.x86_64 0:8.0.23-1.el7
  mysql-community-common.x86_64 0:8.0.23-1.el7
  mysql-community-libs.x86_64 0:8.0.23-1.el7

Complete!
[root@ecs-teacher ~]#
```

图 11-2　MySQL 安装完成

2）执行以下命令，启动 MySQL 数据库。

```
systemctl start mysqld.service
```

3）执行以下命令，查看 MySQL 运行状态，如图 11-3 所示，代表 MySQL 启动运行正常。

```
systemctl status mysqld.service
```

```
Complete!
[root@ecs-teacher ~]# systemctl start mysqld.service
[root@ecs-teacher ~]# systemctl status mysqld.service
● mysqld.service - SYSV: MySQL database server.
   Loaded: loaded (/etc/rc.d/init.d/mysqld; bad; vendor preset: disabled)
   Active: active (running) since Tue 2021-02-02 18:16:46 CST; 50s ago
     Docs: man:systemd-sysv-generator(8)
  Process: 13400 ExecStop=/etc/rc.d/init.d/mysqld stop (code=exited, status=0/SUCCESS)
  Process: 13457 ExecStart=/etc/rc.d/init.d/mysqld start (code=exited, status=0/SUCCESS)
 Main PID: 13701 (mysqld)
   CGroup: /system.slice/mysqld.service
           ├─13485 /bin/sh /usr/bin/mysqld_safe --datadir=/var/lib/mysql --socket=/var/lib/mysql...
           └─13701 /usr/sbin/mysqld --basedir=/usr --datadir=/var/lib/mysql --plugin-dir=/usr/li...

Feb 02 18:16:45 ecs-teacher systemd[1]: Starting SYSV: MySQL database server....
Feb 02 18:16:46 ecs-teacher mysqld[13457]: Starting mysqld:  [  OK  ]
Feb 02 18:16:46 ecs-teacher systemd[1]: mysqld.service: Supervising process 13701 which is no...ts.
Feb 02 18:16:46 ecs-teacher systemd[1]: Started SYSV: MySQL database server..
Hint: Some lines were ellipsized, use -l to show in full.
[root@ecs-teacher ~]#
```

图 11-3　MySQL 运行状态

4）执行以下命令，查看 MySQL 的初始密码，如图 11-4 所示，“.4s&D/9J?7XA”为查询到的密码。

```
grep "password" /var/log/mysqld.log
```

```
[root@iZwz97xzvgb8wvhl4hqhzxZ ~]# grep "password" /var/log/mysqld.log
2020-11-03T11:55:23.038189Z 6 [Note] [MY-010454] [Server] A temporary password is generated for root@localhost: .4s&D/9J?7XA
```

图 11-4　MySQL 的初始密码

5）执行以下命令，登录数据库，如图 11-5 所示。输入上一步骤查到的密码：.4s&D/9J?7XA，显示成功登录 MySQL 数据库。

```
mysql -uroot -p
```

```
   Status: "Server is operational"
   CGroup: /system.slice/mysqld.service
           └─1767 /usr/sbin/mysqld

Nov 03 19:55:18 iZwz97xzvgb8wvhl4hqhzxZ systemd[1]: Starting MySQL Server...
Nov 03 19:55:29 iZwz97xzvgb8wvhl4hqhzxZ systemd[1]: Started MySQL Server.
[root@iZwz97xzvgb8wvhl4hqhzxZ ~]# grep "password" /var/log/mysqld.log
2020-11-03T11:55:23.038189Z 6 [Note] [MY-010454] [Server] A temporary password is generated for
root@localhost: .4s&D/9J?7XA
[root@iZwz97xzvgb8wvhl4hqhzxZ ~]# mysql -uroot -p
Enter password:
Welcome to the MySQL monitor.  Commands end with ; or \g.
Your MySQL connection id is 8
Server version: 8.0.22

Copyright (c) 2000, 2020, Oracle and/or its affiliates. All rights reserved.

Oracle is a registered trademark of Oracle Corporation and/or its
affiliates. Other names may be trademarks of their respective
owners.

Type 'help;' or '\h' for help. Type '\c' to clear the current input statement.

mysql> _
```

图 11-5　成功登录 MySQL 数据库

6）执行以下命令，修改 MySQL 默认密码。

```
// 注意密码强度，需要大小写字母、数字和符号，否则会修改失败。
ALTER USER 'root' @ 'localhost' IDENTIFIED BY 'Apj123@pj'
```

7）执行以下命令，授予 root 用户远程管理权限。

默认只允许 root 用户在本地登录，如果要在其他机器上连接 MySQL，必须修改 root 权限并允许远程连接。

```
// 选择 'mysql'database
use mysql;
// 授权 root 用户的所有权限并设置远程访问
update user set host='%' where user ='root';
// 使修改生效
flush privileges;
```

8）执行以下命令，重新指定一个基于 mysql_native_password 的密码。

```
alter user 'root' @ '%' IDENTIFIED WITH mysql_native_password by 'Apj123@pj';
```

9）输入“exit”命令退出数据库，数据库配置完成。

步骤 3　安装 Redis

Redis 是一个开源的使用 ANSI C 语言编写、遵守 BSD 协议、支持网络、可基于内存亦可持久化的日志型、Key-Value 数据库，并提供多种语言的 API。

1）执行以下命令，安装 EPEL。

```
yum install epel-release
```

EPEL 的全称为 Extra Packages for Enterprise Linux。EPEL 是由 Fedora 社区打造，为 RHEL 及衍生发行版如 CentOS、Scientific Linux 等提供高质量软件包的项目。装上了 EPEL 之后，就相当于添加了一个第三方源。安装过程中会询问“Is this OK?”输入“y”并按 <Enter> 键确认，显示 EPEL 安装完成，如图 11-6 所示。

```
[root@iZwz97xzvgb8wvhl4hqhzxZ ~]# yum install epel-release
Loaded plugins: fastestmirror
Loading mirror speeds from cached hostfile
Resolving Dependencies
--> Running transaction check
---> Package epel-release.noarch 0:7-12 will be installed
--> Finished Dependency Resolution

Dependencies Resolved

================================================================================== Package
h                  Version               Repository          Size
==================================================================================Installing:
 epel-release                 noarch              7-12                  epel             15 k

Transaction Summary
==================================================================================Install  1 Package

Total download size: 15 k
Installed size: 24 k
Is this ok [y/d/N]: y
Downloading packages:
epel-release-7-12.noarch.rpm                                   |  15 kB  00:00:00
Running transaction check
Running transaction test
Transaction test succeeded
Running transaction
  Installing : epel-release-7-12.noarch                                             1/1
warning: /etc/yum.repos.d/epel.repo created as /etc/yum.repos.d/epel.repo.rpmnew
  Verifying  : epel-release-7-12.noarch                                             1/1

Installed:
  epel-release.noarch 0:7-12

Complete!
```

图 11-6　EPEL 安装完成

2）执行以下命令，安装 Redis 服务。

```
yum install redis
```

安装过程中会询问“Is this OK?”输入“y”并按 <Enter> 键确认，显示 Redis 安装完成，如图 11-7 所示。

3）执行以下命令，启动 Redis。

```
service redis start
```

4）执行下面命令，查看 Redis 目录。

```
whereis redis
```

5）配置 Redis 远程访问，打开 /etc/redis.conf 文件，按照下面内容修改并保存该文件。

```
// 注释掉 bind 127.0.0.1，否则无法访问
# bind 127.0.0.1
// 关闭保护模式，将 protected-mode 设置为 no
protected-mode no
// 设置 Redis 密码为 123456
requirepass 123456
```

```
[root@iZwz97xzvgb8wvhl4hqhzxZ ~]# yum install redis
Loaded plugins: fastestmirror
Loading mirror speeds from cached hostfile
Resolving Dependencies
--> Running transaction check
---> Package redis.x86_64 0:3.2.12-2.el7 will be installed
--> Processing Dependency: libjemalloc.so.1()(64bit) for package: redis-3.2.12-2.el7.x86_64
--> Running transaction check
---> Package jemalloc.x86_64 0:3.6.0-1.el7 will be installed
--> Finished Dependency Resolution

Dependencies Resolved

================================================================================================
 Package                   Arch               Version                 Repository          Size
================================================================================================
Installing:
 redis                     x86_64             3.2.12-2.el7            epel               544 k
Installing for dependencies:
 jemalloc                  x86_64             3.6.0-1.el7             epel               105 k

Transaction Summary
================================================================================================
Install  1 Package (+1 Dependent package)

Total download size: 648 k
Installed size: 1.7 M
Is this ok [y/d/N]: y
Downloading packages:
(1/2): jemalloc-3.6.0-1.el7.x86_64.rpm                              | 105 kB  00:00:00
(2/2): redis-3.2.12-2.el7.x86_64.rpm                                | 544 kB  00:00:00
------------------------------------------------------------------------------------------------
Total                                                     3.1 MB/s | 648 kB  00:00:00
Running transaction check
Running transaction test
Transaction test succeeded
Running transaction
  Installing : jemalloc-3.6.0-1.el7.x86_64                                                  1/2
  Installing : redis-3.2.12-2.el7.x86_64                                                    2/2
  Verifying  : redis-3.2.12-2.el7.x86_64                                                    1/2
  Verifying  : jemalloc-3.6.0-1.el7.x86_64                                                  2/2

Installed:
  redis.x86_64 0:3.2.12-2.el7

Dependency Installed:
  jemalloc.x86_64 0:3.6.0-1.el7

Complete!
```

图 11-7 Redis 安装完成

6）执行以下命令，重启 Redis。

```
service redis stop
service redis start
```

步骤 4 安装 Nginx

Nginx 是一款轻量级的 Web 服务器、反向代理服务器及电子邮件（IMAP/POP3）代理服务器，在 BSD-like 协议下发行。其特点是占有内存少，并发能力强，事实上 Nginx 的并发能力在同类型的网页服务器中表现较好。

1）执行以下命令，创建 Nginx 的 yum 源。

```
// 切换到 yum 安装源文件目录
cd /etc/yum.repos.d
// 创建一个新的 repo 文件来创建安装包源
vi nginx.repo
// 在 nginx.repo 文件中加入下列内容，复制、粘贴完成后按 <Esc> 键进入 vi 的命令模式，并输入 :wq
后按 <Enter> 键退出编辑模式
[nginx-stable]
name=nginx stable repo
baseurl=http://nginx.org/packages/centos/$releasever/$basearch/
gpgcheck=1
```

```
enabled=1
gpgkey=https://nginx.org/keys/nginx_signing.key
[nginx-mainline]
name=nginx mainline repo
baseurl=http://nginx.org/packages/mainline/centos/$releasever/$basearch/
gpgcheck=1
enabled=0
gpgkey=https://nginx.org/keys/nginx_signing.key
```

2）执行以下命令，查看当前可用的安装包源。

```
yum repolist
```

3）执行以下命令，安装 Nginx 服务。

```
yum install nginx
```

安装过程中会询问“Is this OK?”输入“y”并按 <Enter> 键确认，显示 Nginx 安装成功，如图 11-8 所示。

```
--------------------------------------------------------------------------------

Thanks for using nginx!

Please find the official documentation for nginx here:
* http://nginx.org/en/docs/

Please subscribe to nginx-announce mailing list to get
the most important news about nginx:
* http://nginx.org/en/support.html

Commercial subscriptions for nginx are available on:
* http://nginx.com/products/

--------------------------------------------------------------------------------
  Verifying  : 1:nginx-1.18.0-2.el7.ngx.x86_64                              1/1

Installed:
  nginx.x86_64 1:1.18.0-2.el7.ngx

Complete!
```

图 11-8　Nginx 安装成功

4）执行以下命令，启动 Nginx 服务。

```
service nginx start
```

5）Ngnix 的默认端口是 80，Nginx 要对外提供服务，需要配置云服务的访问安全组，可配置 80 端口用于控制对外开放的端口。

6）浏览器输入“云服务 ip:port”地址并按 <Enter> 键后，显示 Nginx 默认页面就表示安装完成，如图 11-9 所示。

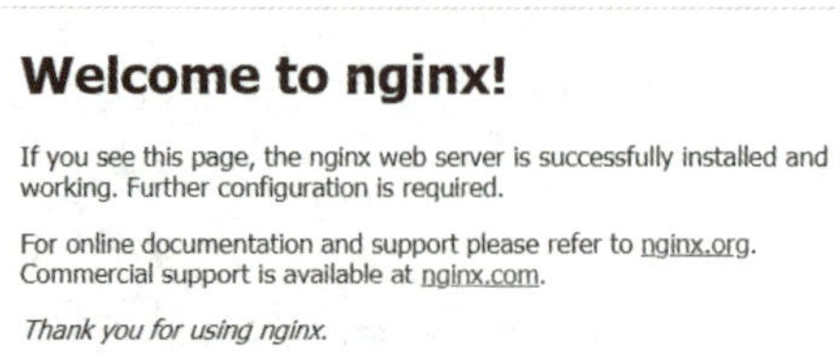

Welcome to nginx!

If you see this page, the nginx web server is successfully installed and working. Further configuration is required.

For online documentation and support please refer to nginx.org.
Commercial support is available at nginx.com.

Thank you for using nginx.

图 11-9　Nginx 默认页面

任务评价

技　能　点	知　识　点	自我评价（不熟悉 / 基本掌握 / 熟练掌握 / 灵活运用）
Linux 下安装 JDK8		
Linux 下安装 MySQL	Yum 安装原理	
Linux 下安装 Redis	Yum 命令	
Linux 下安装 Nginx		

任务 2　打包与部署项目后端

微课 11-2　打包与部署项目后端

从本书配套资源中获取项目源码，参考工作单元二任务 1 的内容导入项目数据到云服务器的 MySQL 中。

任务实施

步骤 1　打包项目后端

1）在 IDEA 右侧的 Maven 界面找到 Lifecycle 选项，双击 install 生成新的 jar 包，命令执行完毕后，在 /target 目录下会生成 jar 文件，如图 11-10 所示。

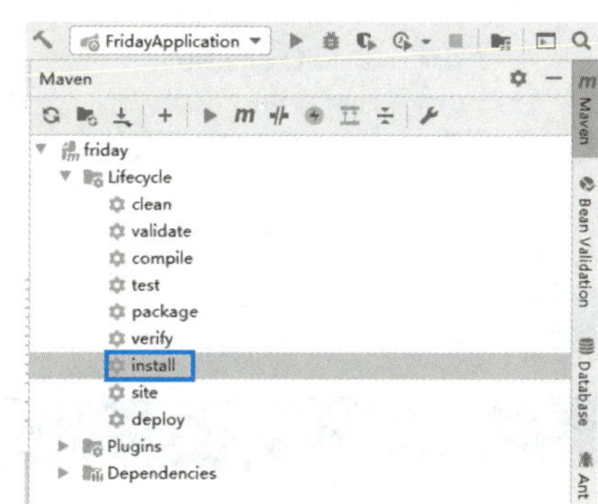

图 11-10　项目打包

2）使用 FTP 客户端把 jar 文件上传到服务器，如图 11-11 所示。

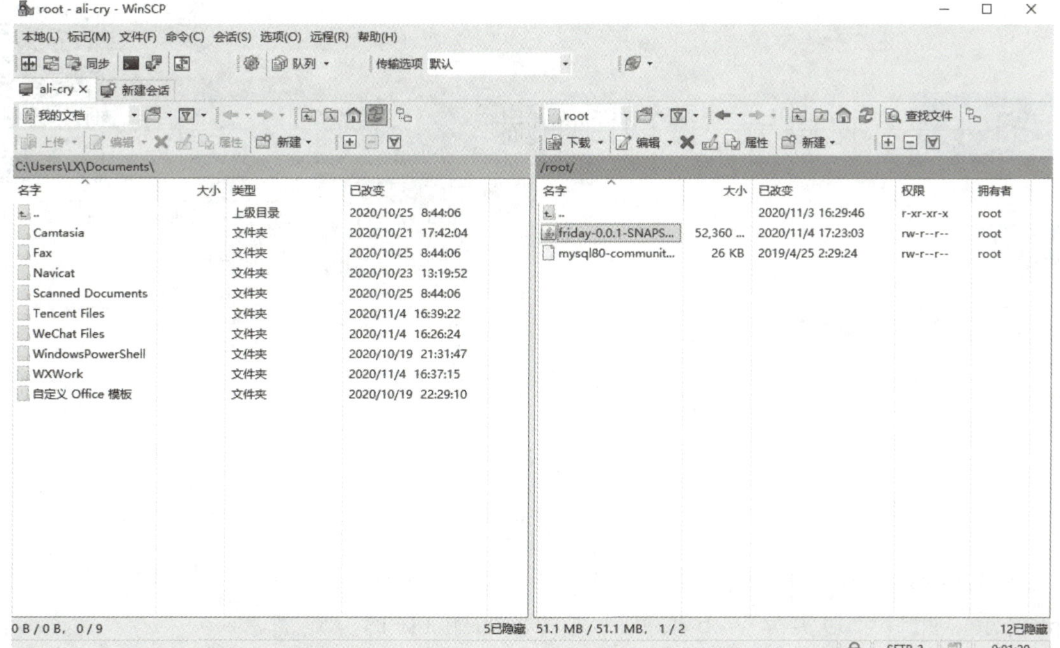

图 11-11　上传文件

步骤 2 部署运行项目后端

1）执行以下命令，启动 Friday 后端服务。

```
// nohup 指令可以保证在关掉命令行界面之后，后端服务不会停止
nohup java -jar friday-0.0.1-SNAPSHOT.jar &
```

2）启动之后通过 ls 指令可以看到在目录下生成了一个 nohup.out 文件，这个是后端项目输出的系统日志，然后通过 cat 指令观察项目的启动情况。

```
cat nohup.out
```

3）如果需要终止 Friday，可以执行以下命令。

```
// 查看 Friday 进程的 pid
ps -ef | grep friday
// 把查出的 pid 值替换下面的 pid 位置
kill -9 pid
```

任务评价

技能点	知识点	自我评价（不熟悉/基本掌握/熟练掌握/灵活运用）
打包项目后端	Maven 打包命令	
部署运行项目后端		

微课 11-3 打包与部署项目前端

任务 3 打包与部署项目前端

任务实施

步骤 1 打包与上传 Vue 项目

1）执行以下命令进行项目打包，打包文件存放在 dist 目录，如图 11-12 所示。

```
npm run build:prod
```

2）上传 dist 目录下的文件到服务器的 /usr/share/nginx/html 目录，此目录为实际 Web 项目的文件，如图 11-13 所示。

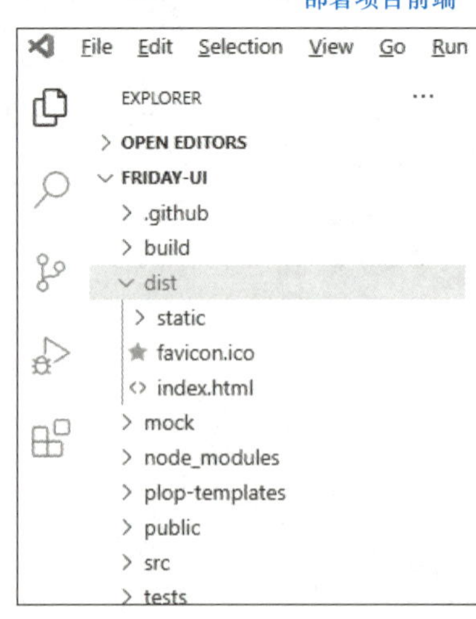

图 11-12 dist 目录

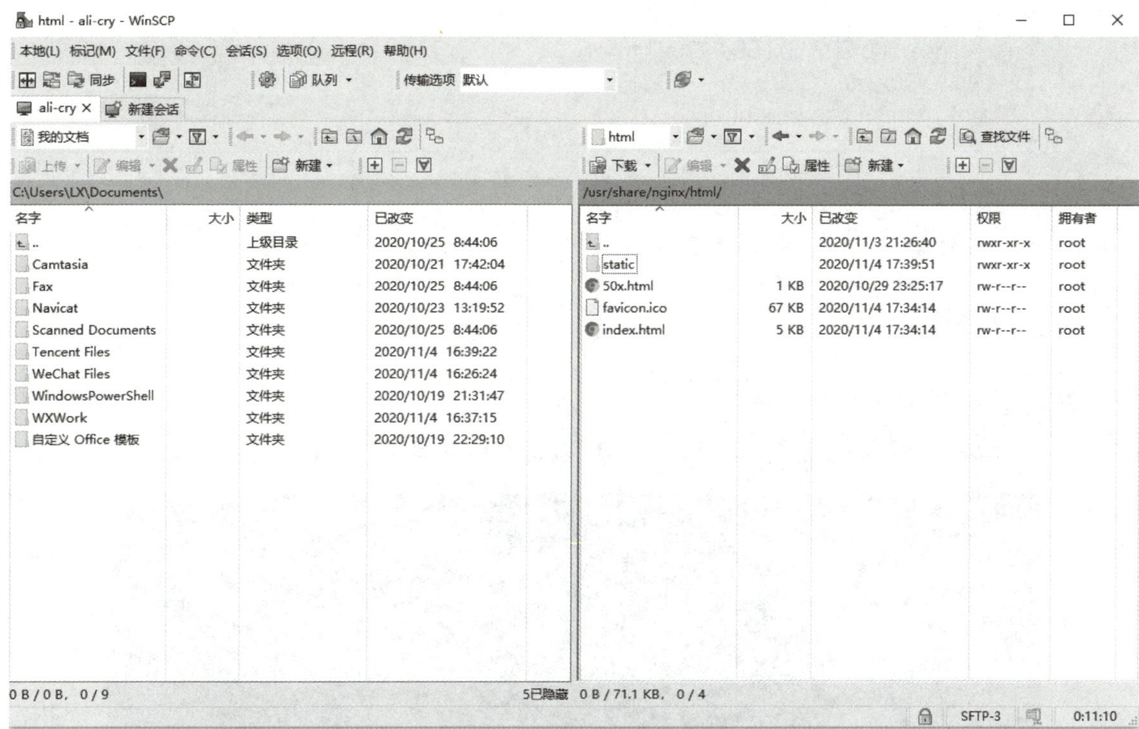

图 11-13 Web 项目目录

步骤 2 配置 Nginx 服务器

1）在 /etc/nginx/conf.d 目录下修改 default.conf，配置 Nginx 服务。

```
server {
    listen          80;                 # 监听端口
    server_name   localhost;            # 监听本地主机

    #charset koi8-r;
    #access_log  /var/log/nginx/host.access.log  main;

    location / {
        root   /usr/share/nginx/html;  # 根目录
        index  index.html index.htm;   # 设置默认页
        try_files $uri $uri/ /index.html;
    }

    location /prod-api/ {
        proxy_pass http://47.107.127.223:8080/;
        proxy_method POST;
        proxy_set_header X-Real-IP $remote_addr;
        proxy_set_header REMOTE-HOST $remote_addr;
        proxy_set_header X-Forwarded-For $proxy_add_x_forwarded_for;
    }
    #error_page  404              /404.html;

    # redirect server error pages to the static page /50x.html
```

```
    #
    error_page   500 502 503 504  /50x.html;
    location = /50x.html {
        root   /usr/share/nginx/html;
    }
}
```

2）执行以下命令，重新加载配置文件。

```
nginx -s reload
```

步骤 3 前后端联调验证

1）使用浏览器访问云服务器 IP 地址，项目正常访问登录页面，如图 11-14 所示。

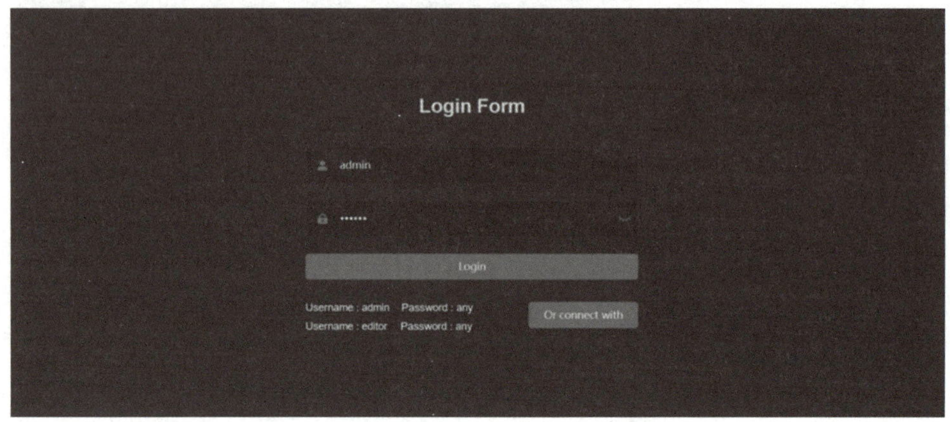

图 11-14 登录页面

2）单击“登录”按钮，进入项目的后台首页，如图 11-15 所示。

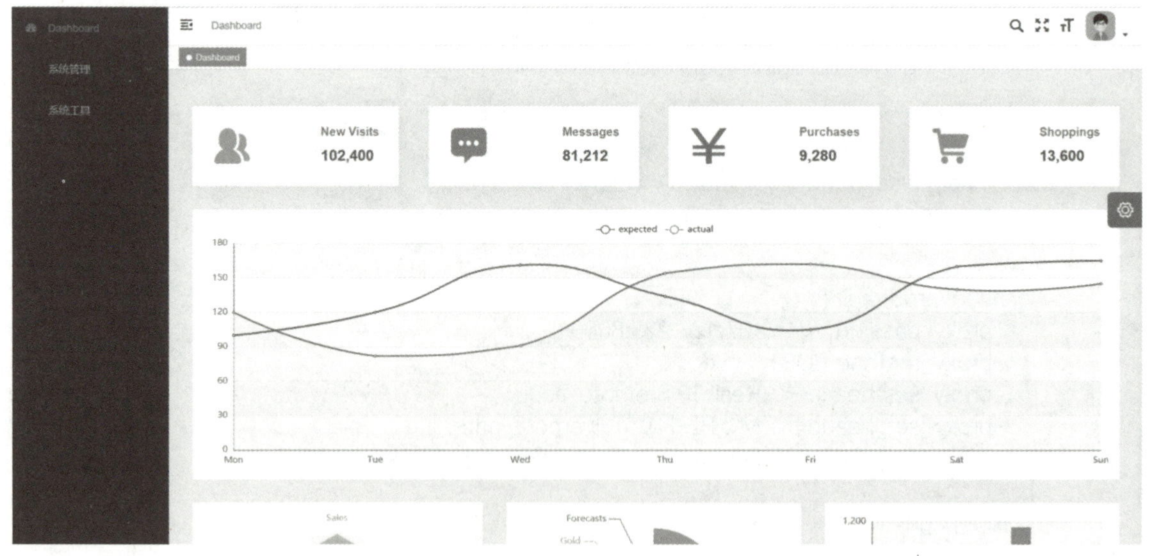

图 11-15 后台首页

知识小结

Nginx 的配置

在步骤 2 的配置文件中，只配置了一个 Nginx 服务器的 server 块和虚拟主机的相关参数，其实在一个 HTTP 中可以有多个 server 块。注意每个配置指令必须以分号结束。

配置文件中有一个配置选项 try_files，是 Nginx 中 http_core 核心模块所带的指令，主要是替代一些 rewrite 的指令，提高解析效率。参数定义了响应用户请求的文件搜索顺序，按指定的文件顺序查找存在的文件，并使用第一个找到的文件进行请求处理，如果给出的文件都没有匹配到，则重新请求最后一个参数给定的 URI，就是新的位置匹配。如果最后一个参数 =404，则最后返回 404 的响应码。

任务评价

技　能　点	知　识　点	自我评价（不熟悉 / 基本掌握 / 熟练掌握 / 灵活运用）
打包上传项目前端	Nginx 命令	
配置 Nginx 服务器		

单元小结

基于云服务器的项目环境搭建、项目前端和后端的打包部署、测试项目前后端联调是项目实施的重要环节。能够按照系统部署手册部署软件系统，并能够对系统进行测试和验证是 1+X 职业技能等级证书（Java 应用开发）的职业技能要求。通过完成本工作单元的任务和后面的实战强化，熟练掌握系统实施中系统部署与验证的职业技能。

实战强化

1．按照任务 2 的内容，在云服务器上部署实训项目“诚品书城”后端。

2．按照任务 3 的内容，在云服务器上部署实训项目“诚品书城”前端。

工作单元 12

自动打包部署项目

职业能力

本工作单元实现项目在生产环境Docker体系下的部署，具体实现了后端接口服务、前台应用的部署，将Docker环境的搭建和应用设计成拓展练习，帮助学生巩固技能。最终希望学生达成的职业能力目标是：

1．掌握基于Docker技术的基本镜像操作

2．掌握基于DockerFile文件和IDEA插件的后端项目自动化部署

3．掌握基于DockerFile文件和Docker Compose的前端项目自动化部署

任务情景

研发人员在开发机器上调试代码并通过测试后，正式将项目发布到外网服务器给用户提供服务之前，需要先在外网服务器和生产环境一致的条件下发布测试版本，供测试人员和少量用户进行测试，然后对发现的问题进行修复，修复后重新发布测试版本。等到生产环境下的问题修复得差不多之后，可以发布正式版供大量用户使用。

在这个过程中，需要解决两个问题：

1）如何减少重复发布新版本的工作量。

2）如何保证测试环境和生产服务器环境一致。

在本项目的权限管理系统里，将通过Docker技术实现自动化部署。本工作单元的具体任务如下：

1．安装Docker服务

2．自动发布后端项目到Docker容器

3．自动发布前端项目到Docker容器

前置知识

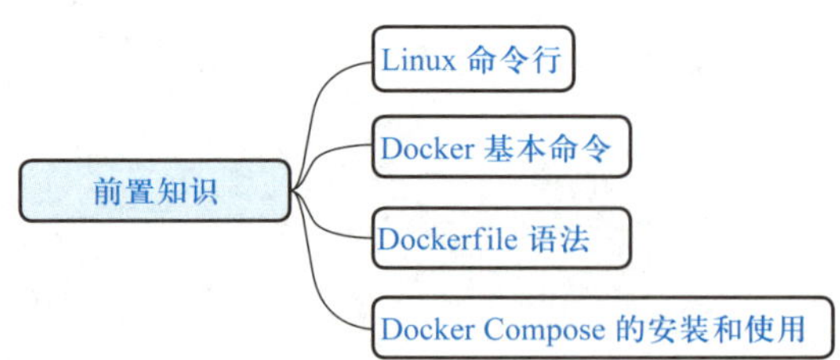

任务 1 安装 Docker 服务

微课 12-1　安装 Docker 服务

任务实施

步骤 1 检查服务器环境

1）打开 CMD 窗口，执行“ssh root@ 服务器的 ip”命令连接远程服务器，如图 12-1 所示。

```
命令提示符
Microsoft Windows [版本 10.0.17763.1577]
(c) 2018 Microsoft Corporation。保留所有权利。

C:\Users\liuxin>ssh root@47.107.127.223
root@47.107.127.223's password:
Last login: Thu Feb 25 10:50:07 2021 from 113.74.7.109

Welcome to Alibaba Cloud Elastic Compute Service !

[root@iZwz97xzvgb8wvhl4hqhzxZ ~]# Connection reset by 47.107.127.223 port 22
```

图 12-1　连接远程服务器

2）Docker 官方推荐其在 Linux 3.8 内核以上运行最佳，所以安装 Docker 前，执行“lsb_release -a”命令检查服务器环境，如图 12-2 所示，本系统环境为 CentOS 7.8。

```
[root@iZwz97xzvgb8wvhl4hqhzxZ ~]# lsb_release -a
LSB Version:    :core-4.1-amd64:core-4.1-noarch
Distributor ID: CentOS
Description:    CentOS Linux release 7.8.2003 (Core)
Release:        7.8.2003
Codename:       Core
```

图 12-2　系统环境为 CentOS 7.8

步骤 2 安装 Docker 服务

1）执行以下命令安装所需的软件包。

```
yum install -y yum-utils  device-mapper-persistent-data  lvm2
```

2）执行以下命令设置软件镜像。

```
yum-config-manager --add-repo https://download.docker.com/linux/centos/docker-ce.repo
```

3）执行以下命令安装 Docker 服务。

```
yum install docker-ce
```

4）执行以下命令启动 Docker 服务。

```
systemctl start docker
```

5）验证 Docker 服务，执行“docker run hello-world”命令运行 Hello-world 容器，图 12-3 中显示容器运行成功。

```
[root@iZwz97xzvgb8wvhl4hqhzxZ ~]# docker run hello-world
Unable to find image 'hello-world:latest' locally
latest: Pulling from library/hello-world
0e03bdcc26d7: Pull complete
Digest: sha256:7e02330c713f93b1d3e4c5003350d0dbe215ca269dd1d84a4abc577908344b30
Status: Downloaded newer image for hello-world:latest

Hello from Docker!
This message shows that your installation appears to be working correctly.

To generate this message, Docker took the following steps:
 1. The Docker client contacted the Docker daemon.
 2. The Docker daemon pulled the "hello-world" image from the Docker Hub.
    (amd64)
 3. The Docker daemon created a new container from that image which runs the
    executable that produces the output you are currently reading.
 4. The Docker daemon streamed that output to the Docker client, which sent it
    to your terminal.

To try something more ambitious, you can run an Ubuntu container with:
 $ docker run -it ubuntu bash

Share images, automate workflows, and more with a free Docker ID:
 https://hub.docker.com/

For more examples and ideas, visit:
 https://docs.docker.com/get-started/

[root@iZwz97xzvgb8wvhl4hqhzxZ ~]#
```

图 12-3　Hello-world 容器运行成功

步骤 3　开启 Docker 的远程连接

1）执行以下命令访问并编辑 Docker 配置文件。

```
vim /lib/systemd/system/docker.service
```

2）把文件中的“ExecStart=/usr/bin/dockerd-current”改为以下配置，其中“-H”选项表示客户端可以远程通过 23751 端口访问。

```
ExecStart=/usr/bin/dockerd-current -H tcp://0.0.0.0:23751 -H unix://var/run/docker.sock
```

3）修改结果如图 12-4 所示，按 <Esc> 键跳到命令模式，输入“:wq!”并按 <Enter> 键，强制保存文件并退出。

```
scription=Docker Application Container Engine
cumentation=http://docs.docker.com
ter=network.target
nts=docker-storage-setup.service
quires=docker-cleanup.timer

ervice]
pe=notify
tifyAccess=main
vironmentFile=-/run/containers/registries.conf
vironmentFile=-/etc/sysconfig/docker
vironmentFile=-/etc/sysconfig/docker-storage
vironmentFile=-/etc/sysconfig/docker-network
vironment=GOTRACEBACK=crash
vironment=DOCKER_HTTP_HOST_COMPAT=1
vironment=PATH=/usr/libexec/docker:/usr/bin:/usr/sbin
ecStart=/usr/bin/dockerd-current -H tcp://0.0.0.0:23751 -H unix://var/run/docker.sock \
        --add-runtime docker-runc=/usr/libexec/docker/docker-runc-current \
        --default-runtime=docker-runc \
        --exec-opt native.cgroupdriver=systemd \
        --userland-proxy-path=/usr/libexec/docker/docker-proxy-current \
        --init-path=/usr/libexec/docker/docker-init-current \
        --seccomp-profile=/etc/docker/seccomp.json \
        $OPTIONS \
        $DOCKER_STORAGE_OPTIONS \
        $DOCKER_NETWORK_OPTIONS \
        $ADD_REGISTRY \
        $BLOCK_REGISTRY \
        $INSECURE_REGISTRY \
        $REGISTRIES
ecReload=/bin/kill -s HUP $MAINPID
mitNOFILE=1048576
mitNPROC=1048576
mitCORE=infinity
meoutStartSec=0
start=on-abnormal
llMode=process

nstall]
ntedBy=multi-user.target
```

图 12-4　编辑 Docker 配置文件内容

4）执行以下命令重新加载并重启 Docker 服务，完成 Docker 远程连接配置。

```
systemctl daemon-reload && systemctl restart docker
```

任务评价

技　能　点	知　识　点	自我评价（不熟悉 / 基本掌握 / 熟练掌握 / 灵活运用）
Linux 下安装 Docker 服务	Yum 安装原理	
使用 Vim 编辑 Docker 文件	Vim 命令	

任务 2　自动发布后端项目到 Docker 容器

微课 12-2　自动发布后端项目到 Docker 容器

任务分析

通过配置 Dockerfile，使用 IDEA 的 Docker 插件构建 Docker 镜像并部署在远程服务器上。

任务实施

步骤 1　安装 IntelliJ IDEA 的 Docker 插件

以管理员身份运行 IDEA 工具，执行 File → Settings → Plugins → Install JetBrains plugin

命令进入插件安装界面，在搜索框中输入 docker，可以看到 Docker 选项，单击右边的 Install 按钮进行安装，安装后重启 IDEA，过程如图 12-5 所示。

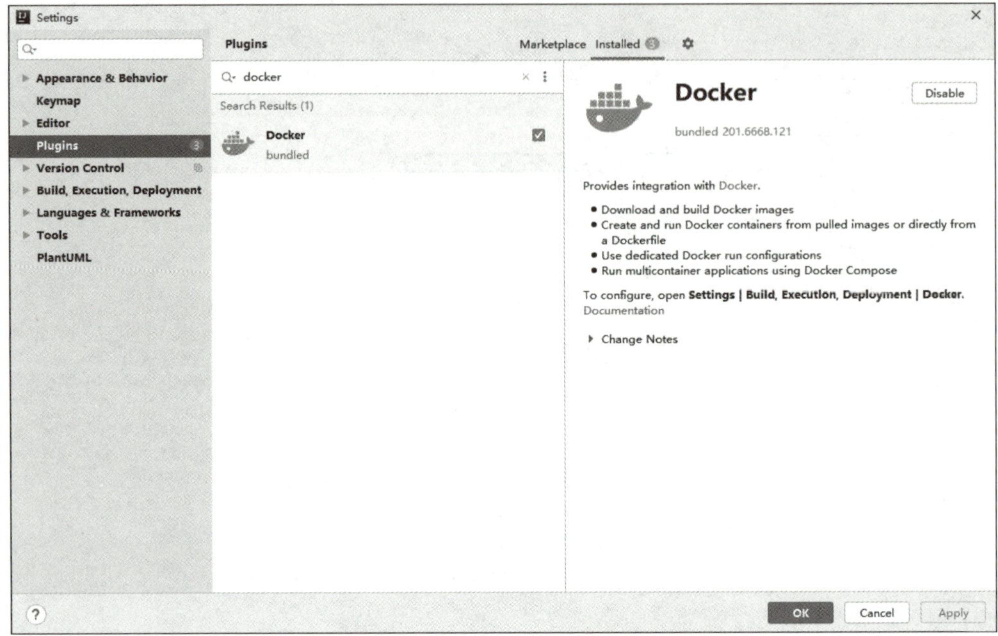

图 12-5　安装 Docker 插件

步骤 2　配置连接 Docker 服务

执行 File → Settings → Build,Execution,Deployment → Docker 命令打开配置界面。在设置页面，单击“+”按钮创建一个 Docker server 并进行设置，Engine API URL 中输入 Docker 服务的 IP 地址，连接成功页面上会提示 Connection successful，过程如图 12-6 所示。

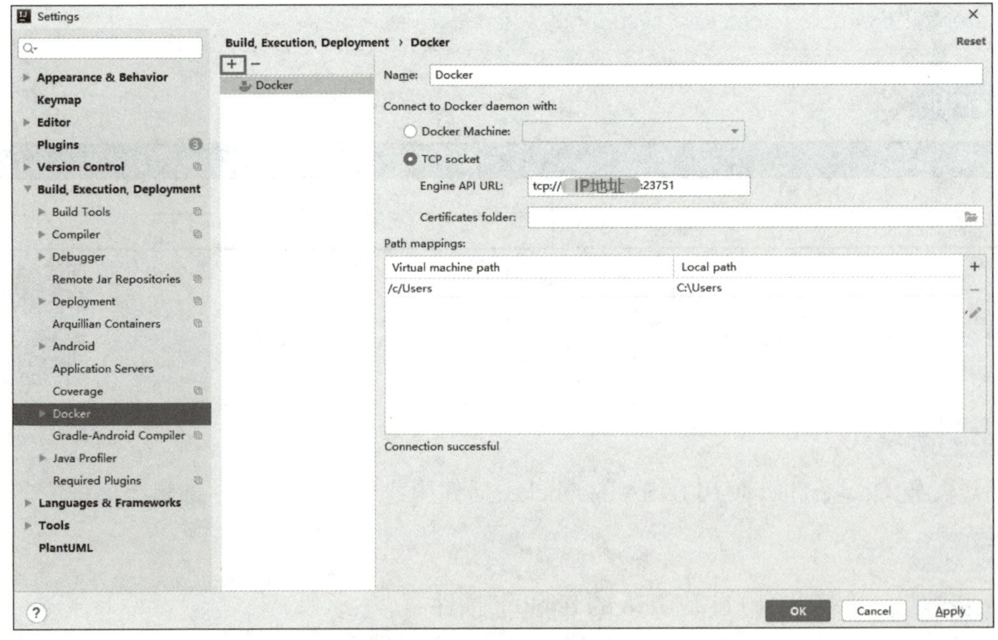

图 12-6　配置 Docker 服务

步骤 3　编写 Dockerfile 镜像文件

在 friday 项目的根目录下创建 Dockerfile 文件，文件内容如下：

```
# Docker image for springboot application
# VERSION 0.0.1
# Author: Alex
### 基础镜像，使用 alpine 操作系统，openjkd 使用 8u201
FROM openjdk:8u201-jdk-alpine3.9
# 作者
MAINTAINER Alex <503024749@qq.com>
# 系统编码
ENV LANG=C.UTF-8 LC_ALL=C.UTF-8
# 声明一个挂载点，容器内此路径会对应宿主机的某个文件夹
VOLUME /tmp
# 应用构建成功后 jar 文件被复制到镜像内，名字修改成 app.jar
ADD target/friday-0.0.1-SNAPSHOT.jar app.jar
# 启动容器器时的进程
ENTRYPOINT ["java","-jar","/app.jar"]
# 暴露 8080 端口
EXPOSE 8080
```

步骤 4　配置 Dockerfile

1）单击“Edit Configuration”选项，如图 12-7 所示。

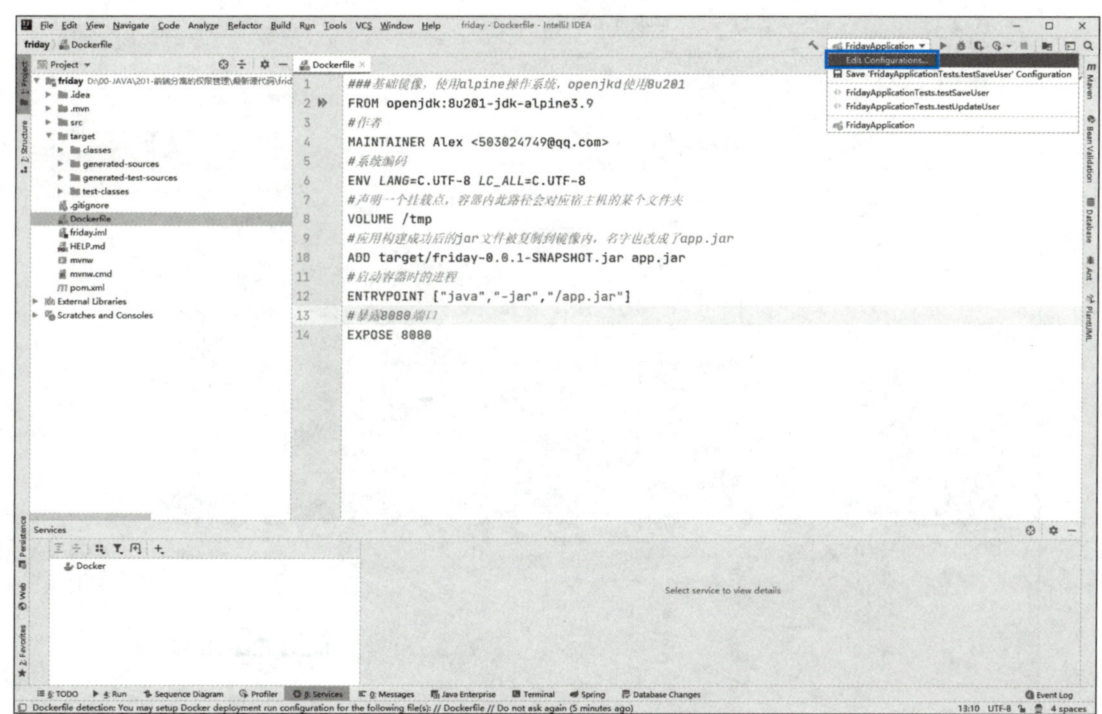

图 12-7　配置 Dockerfile 1

2）单击左上角的“+”按钮，选择左侧菜单栏的“Docker”选项，在弹出菜单里选择“Dockerfile”选项，如图 12-8 所示。

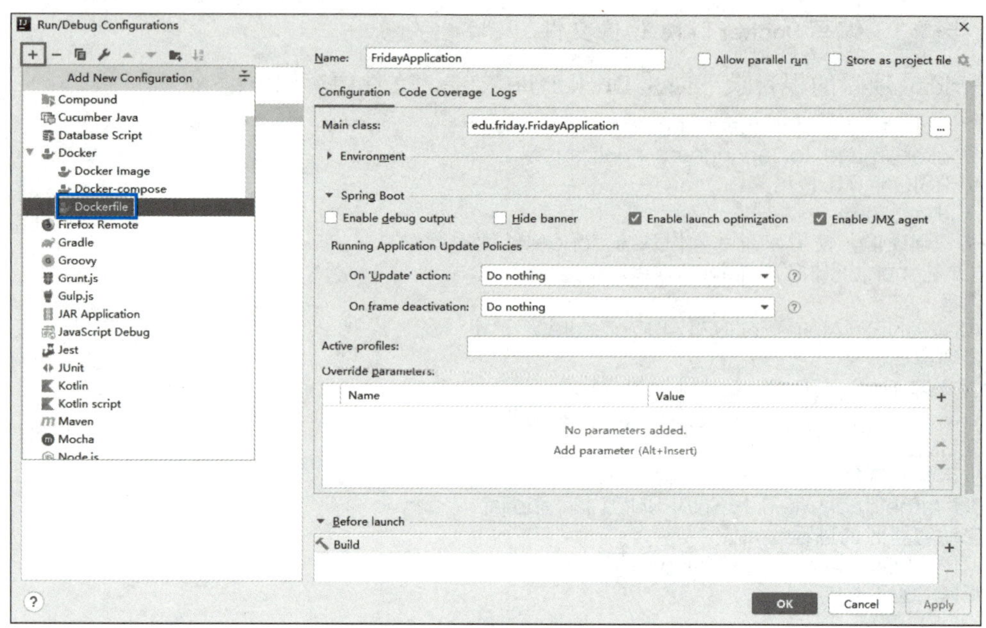

图 12-8　配置 Dockerfile 2

3）在 Docker 配置页面中配置下面的信息，如图 12-9 所示。

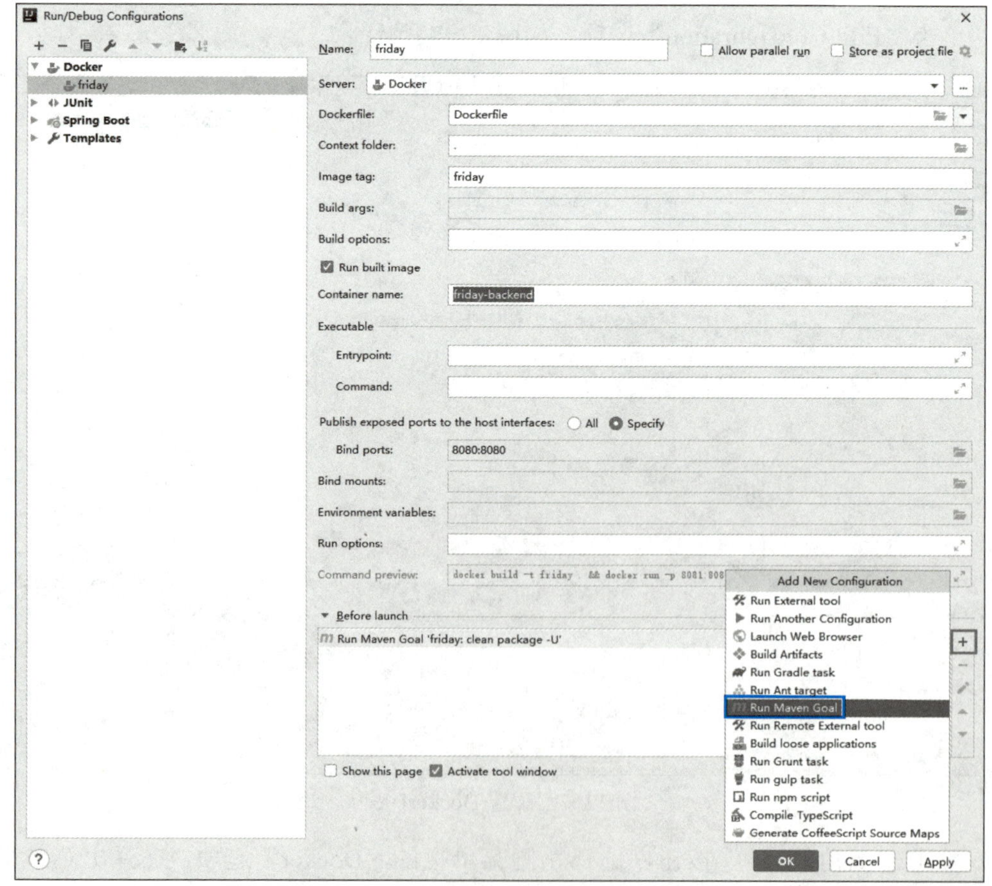

图 12-9　配置 Dockerfile 3

➢ Name：输入 friday。
➢ Server：选择前面配置的 Docker 服务。
➢ Dockerfile：选择前面增加的 Dockerfile 文件。
➢ Image tag：输入 friday。
➢ Container name：输入 friday-backend。
➢ Bind ports：配置 8080:8080，分别映射宿主机端口和容器端口。

然后单击“Before launch”下的“+”按钮，选择“Run Maven Goal”命令配置 Maven。

4）单击“Run Maven Goal”后，输入要执行的 Maven 命令“clean package -U”，表示每次在构建镜像之前，都会将当前工程清理掉并且重新编译构建，然后单击“OK”按钮，如图 12-10 所示。

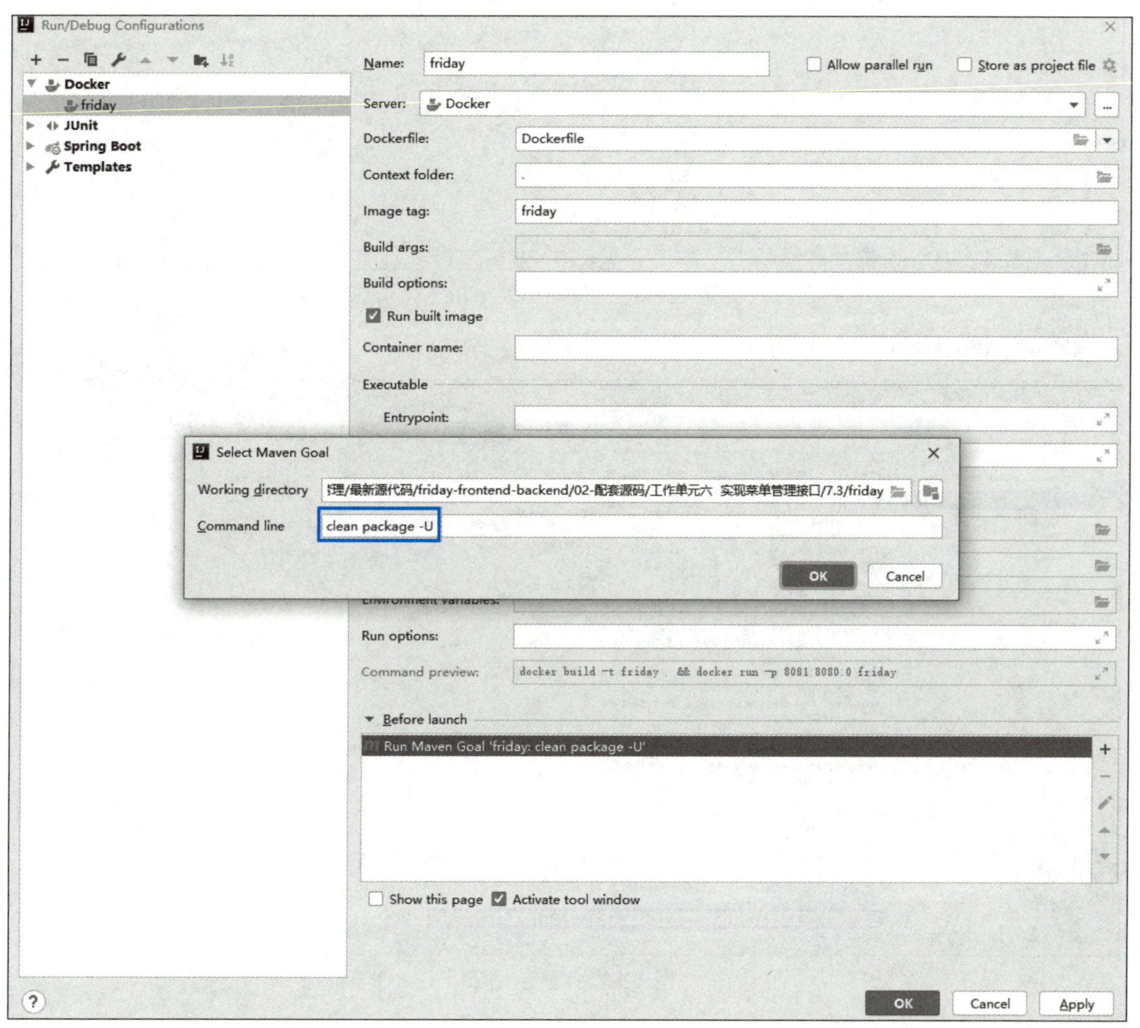

图 12-10　配置 Dockerfile 4

5）选择配置好的 Docker 运行方式，单击“▶”按钮运行验证，如图 12-11 所示。

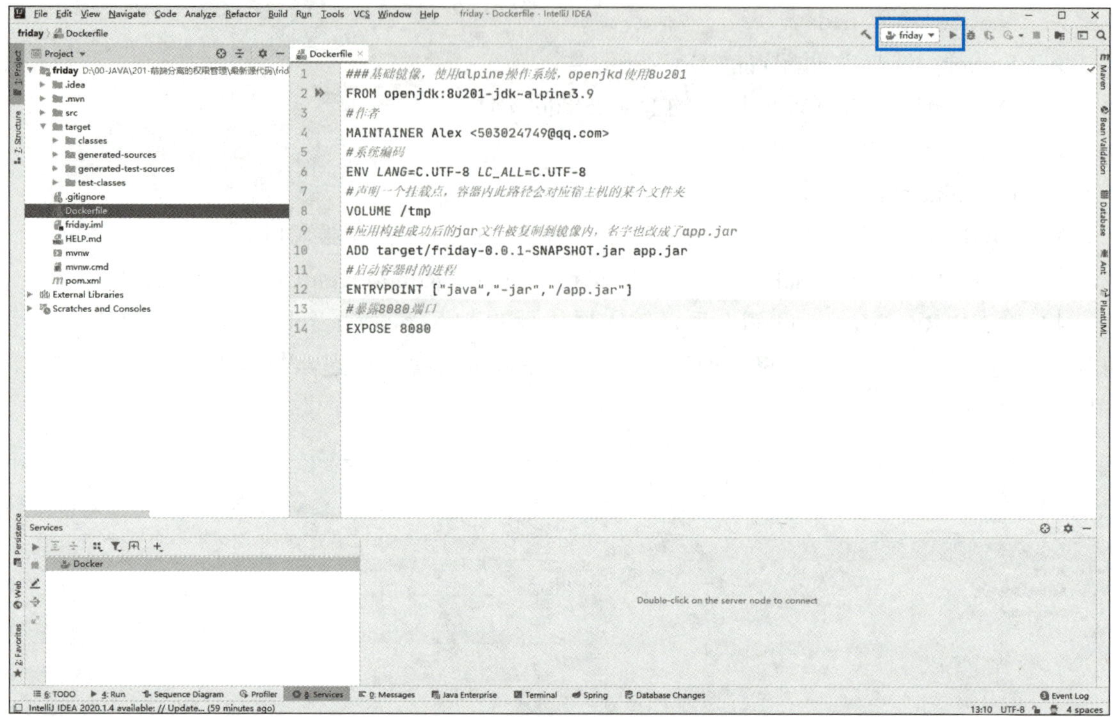

图 12-11　配置 Dockerfile 5

6）显示结果如图 12-12 所示，即为发布成功。

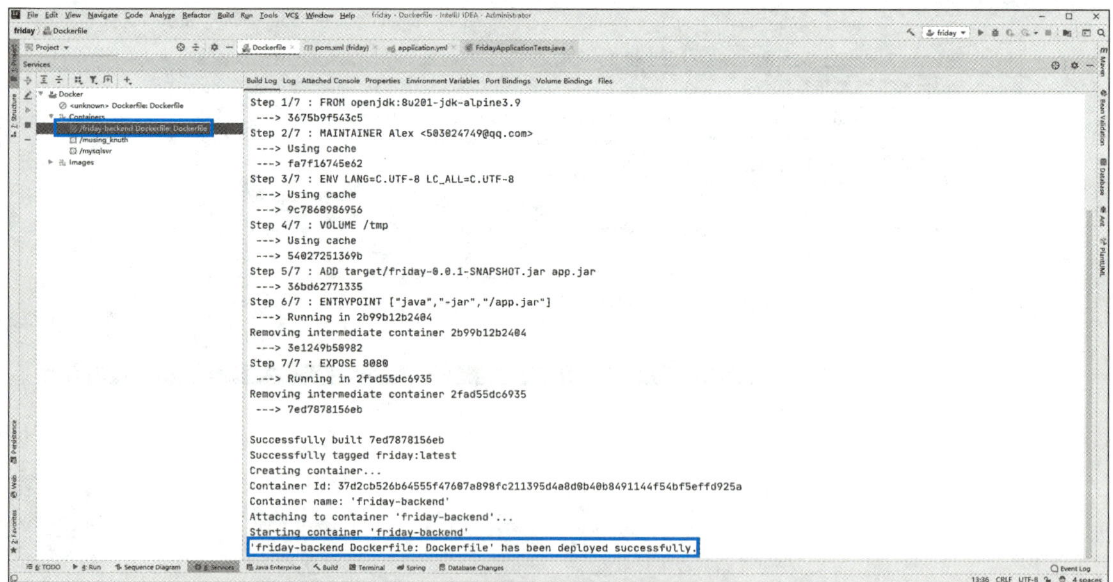

图 12-12　显示结果

知识小结

Dockerfile

Dockerfile 是一个用来构建镜像的文本文件，文件里面是构建镜像所需的指令和说明。

下面是 Dockerfile 常用指令的解释。

- FROM：指定基础镜像。
- VOLUME：声明一个挂载点，容器内此路径会对应宿主机的某个文件夹。
- ADD：将构建成功后的 jar 文件复制到镜像，与 COPY 命令类似。
- EXPOSE：暴露容器运行时的监听端口给外部。
- ENV：设置环境变量。
- MAINTAINER：指定作者。
- ENTRYPOINT：启动命令，与 CMD 命令类似。

任务评价

技 能 点	知 识 点	自我评价（不熟悉/基本掌握/熟练掌握/灵活运用）
编写 Dockerfile 构建镜像	Dockerfile 语法知识	
使用 IDEA 工具实现自动发布后端项目		

任务 3 自动发布前端项目到 Docker 容器

微课 12-3 自动发布前端项目到 Docker 容器

任务分析

通过在本地安装 Docker 工具集，配合 VS CODE 插件，编译前端项目，并将打包好的前端项目实时部署到远程服务器上。

任务实施

步骤 1 安装 Docker 工具

VS CODE 没有内置 Docker 工具，所以需要单独安装 Docker 工具，这里我们使用 Windows 的 Docker 桌面版。由于 Windows 的 Docker 桌面版和 Docker Toolbox 已经包括 Compose 和其他 Docker 应用程序，因此用户不需要单独安装 Compose，仅需要安装 Windows 的 Docker 桌面版即可。

1）启用 CPU 虚拟化 Docker 桌面版的虚拟化技术在 Windows 10 系统中，需要 CPU 提供虚拟化支持，所以在安装前，需要先在 BIOS 中开启 CPU 虚拟化。首先打开任务管理器，切换到性能选项，如图 12-13 所示。

如果在虚拟化选项显示“已启用”，说明 CPU 虚拟化已经开启，可以进行下一步，如果显示“已禁用”，则需要在 BIOS 中开启；由于各品牌的 BIOS 设置不同，读者可自行搜索“Windows 10 开启 CPU 虚拟化”查找设置路径。

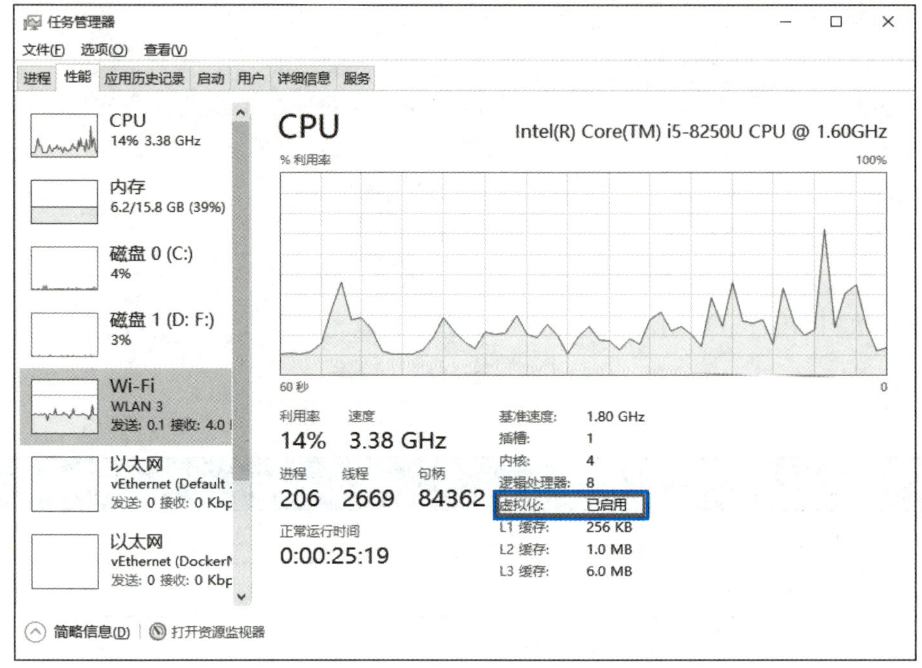

图 12-13　任务管理器的性能选项

2）安装 Hyper-V　Hyper-V 是微软开发的虚拟机，类似于 VMWare 或 VirtualBox，仅适用于 Windows 10 的企业版、专业版或教育版中的一个，是 Docker Desktop for Windows 所使用的虚拟机。首先打开 Windows 功能，开启 Hyper-V，如图 12-14 所示。

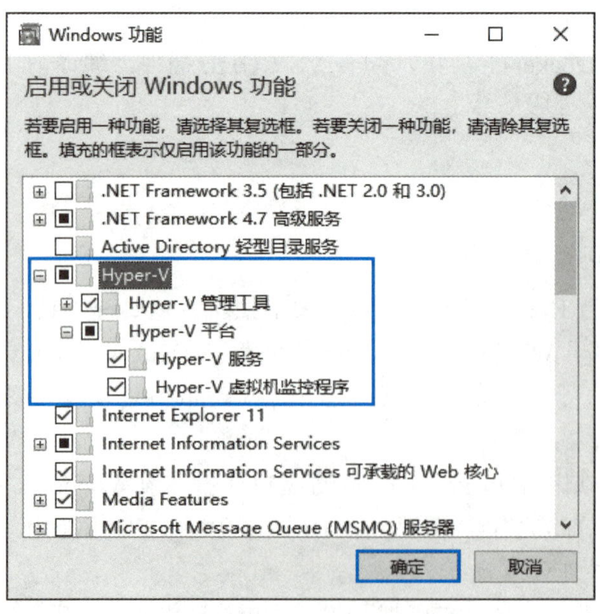

图 12-14　开启 Hyper-V

3）在 Docker 官网下载 Docker for Windows 的版本，双击下载的 Docker for Windows

Installer 安装文件，然后一直单击“Next”按钮，最后单击“Close”按钮完成安装。安装完成后，Docker 会自动启动。通知栏上出现小鲸鱼的图标，表示 Docker 正在运行。

4）安装完毕后，使用图 12-15 所示的命令查看使用的 Docker Compose 版本。

```
C:\Users\emoker>docker --version
Docker version 18.09.0, build 4d60db4

C:\Users\emoker>docker-compose --version
docker-compose version 1.23.2, build 1110ad01

C:\Users\emoker>
```

图 12-15　检查 Docker Compose 版本

步骤 2　在 VS CODE 中安装 Docker 插件

1）在左侧的菜单中找到并单击插件选项。在插件搜索栏中，输入“docker”并按 <Enter> 键，如图 12-16 所示。在搜索结果中找到并单击 Docker 插件打开插件主页，在插件主页中单击“Install”按钮安装该插件。

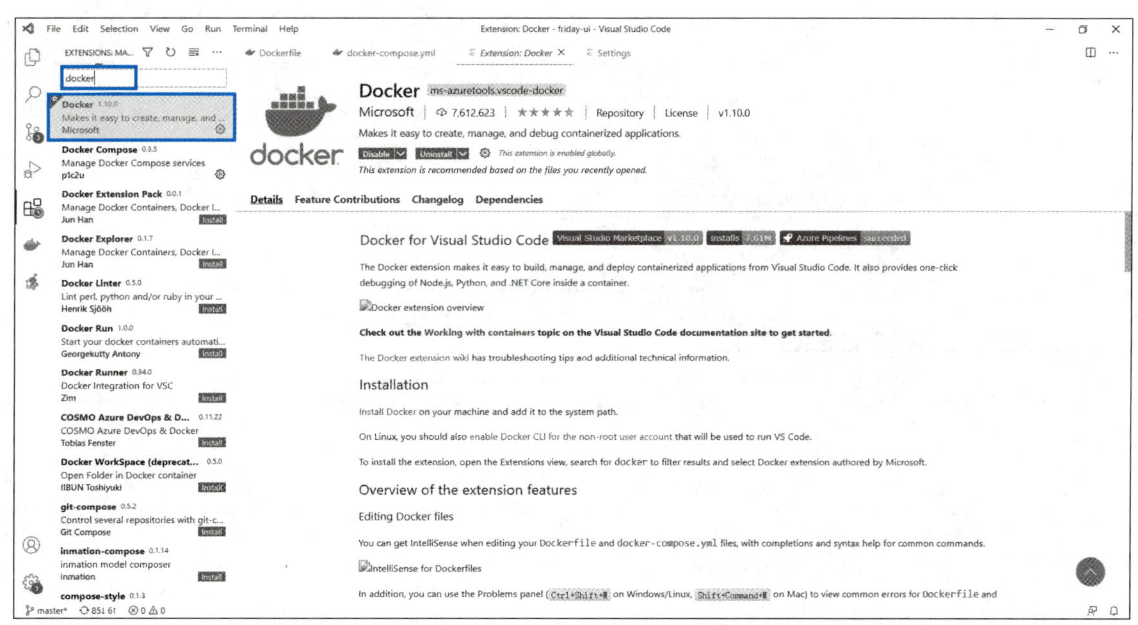

图 12-16　在 VS CODE 中安装 Docker 插件

2）在左侧的菜单中找到并单击插件选项。在插件搜索栏中，输入“docker compose”并按 <Enter> 键，如图 12-17 所示。在搜索结果中找到并单击“Docker Compose”插件打开插件主页，在插件主页中单击“Install”按钮安装该插件。

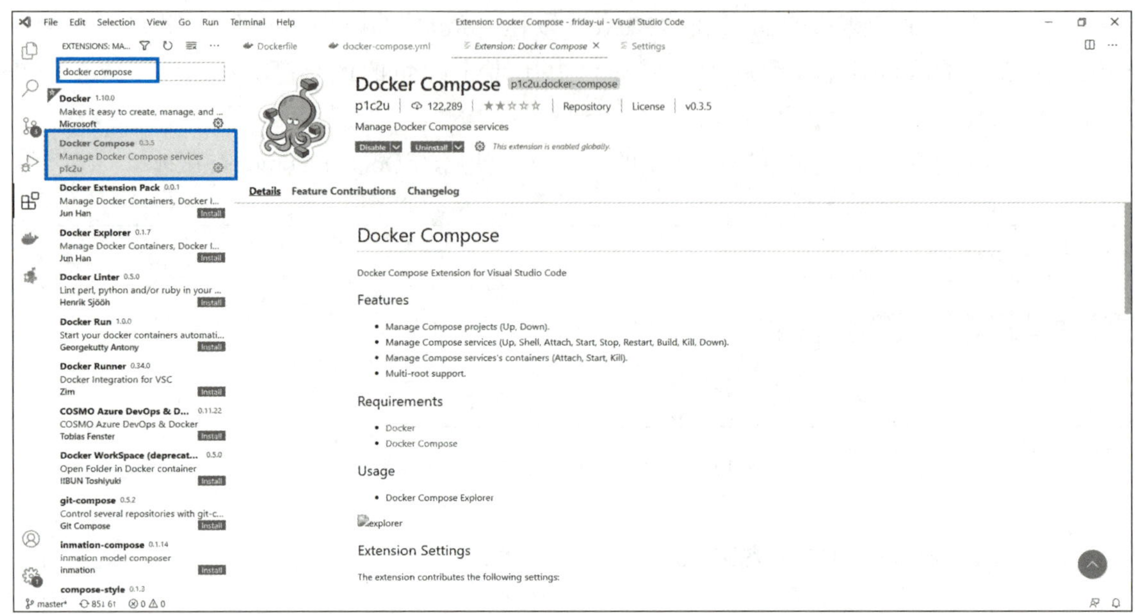

图 12-17　安装 Docker Compose 插件

步骤 3　在 VS CODE 中配置 Docker

1）打开 VS CODE 的设置，执行 File → Preferences → Settings 命令，如图 12-18 所示。

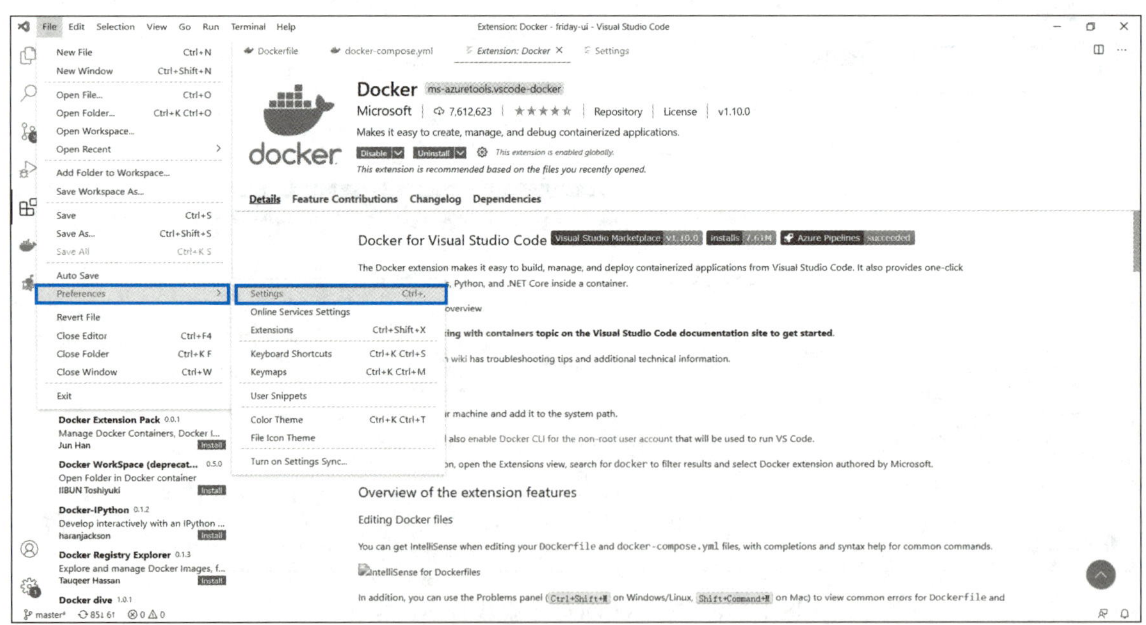

图 12-18　打开 VS CODE 的设置

2）在 VS CODE 设置页面的文本框中输入“docker host”后按 <Enter> 键，在出现的 Docker 设置中，参照图 12-19 的设置填好服务器 Docker 的远程控制地址，格式是“http://ip 地址：端口号”。

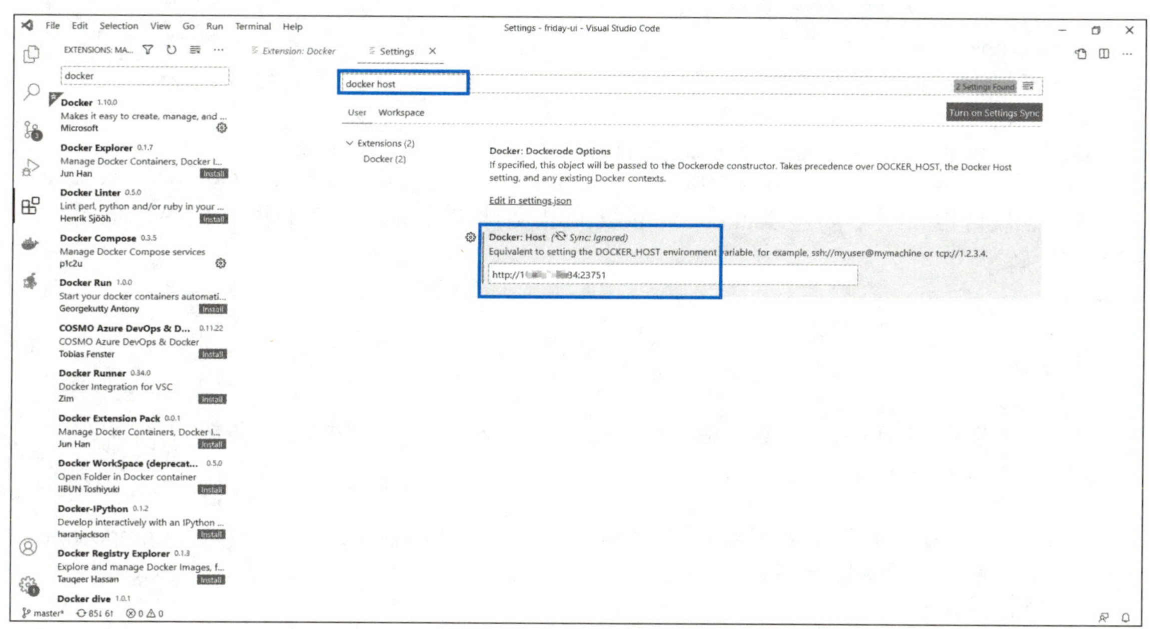

图 12-19　配置 Docker

步骤 4　编写配置文件

1）在项目根目录下创建 default.conf、docker-compose.yml、Dockerfile 等配置文件，如图 12-20 所示。

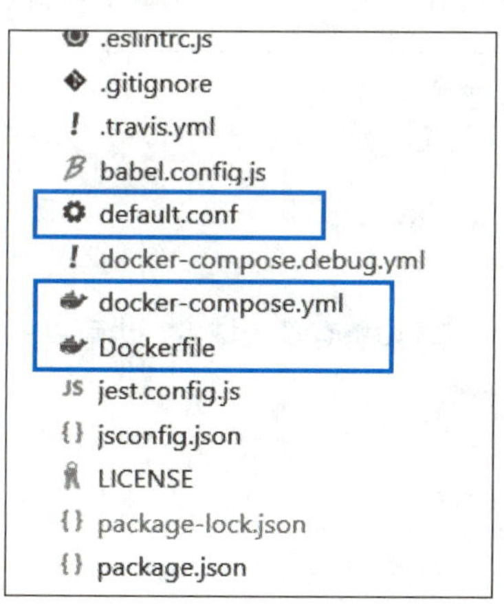

图 12-20　在 VS CODE 中创建 Docker 相关的配置文件

2）在项目根目录下创建 Dockerfile 文件，并编写下面的代码，指定 nginx 作为基础镜像，复制下面的目录和文件到镜像中。

```
# 指定 nginx 作为基础镜像构建
FROM nginx
# 执行文件复制命令复制文件
COPY dist/ /usr/share/nginx/html/
COPY default.conf /etc/nginx/conf.d/default.conf
```

3）在项目根目录下创建 docker-compose.yml 文件，并编写下面的指令代码，每条指令都配有注解。

```
version: "3.3"
services:
  fridyui:
# 指定镜像名或镜像 ID，如果镜像在本地不存在，Compose 将会尝试拉取这个镜像
    image: fridyui
build: .
# 设置环境变量。可以使用数组或字典两种格式。只给定名称的变量会自动获取运行 Compose 主机上对应变量的值，可以用来防止泄露不必要的数据
    environment:
# 指定 node 的环境是生产环境（production）
      NODE_ENV: production
# 暴露端口信息，使用宿主端口：容器端口（HOST:CONTAINER）的格式，也可以仅指定容器的端口（宿主将会随机选择端口）
    ports:
      #"host:contariner"
      - "80:80"
```

4）在项目根目录下创建 default.conf 文件，并编写下面的代码。

```
# 通过 nginx 配置文件配置一个虚拟主机
# 配置虚拟主机的相关参数
server {
# 指定这个主机监听 80 端口
    listen       80;
# 这个虚拟主机的名称是 localhost，如果有多个域名，也可以将 localhost 换成具体的域名
    server_name  localhost;
# 指定字符集
    #charset koi8-r;
# 设置日志保存路径
    access_log  /var/log/nginx/host.access.log  main;
# 配置请求的路由，以及各种页面的处理情况
    location / {
        root   /usr/share/nginx/html;
        index  index.html index.htm;
        try_files $uri $uri/ /index.html;
    }
```

```
    location /prod-api/ {
        #proxy_pass http://emok.utools.club/;
        proxy_pass http://119.29.79.234:8080/;
    }
    #error_page  404              /404.html;
    # redirect server error pages to the static page /50x.html
    #
    error_page   500 502 503 504  /50x.html;
    location = /50x.html {
        root   /usr/share/nginx/html;
    }
}
```

步骤 5　本地打包前端项目

1）运行下面的命令打包前端项目。

```
npm run build
```

2）单击“docker-compose.yml”文件，在弹出的菜单中单击“Compose Up”命令，如图 12-21 所示。

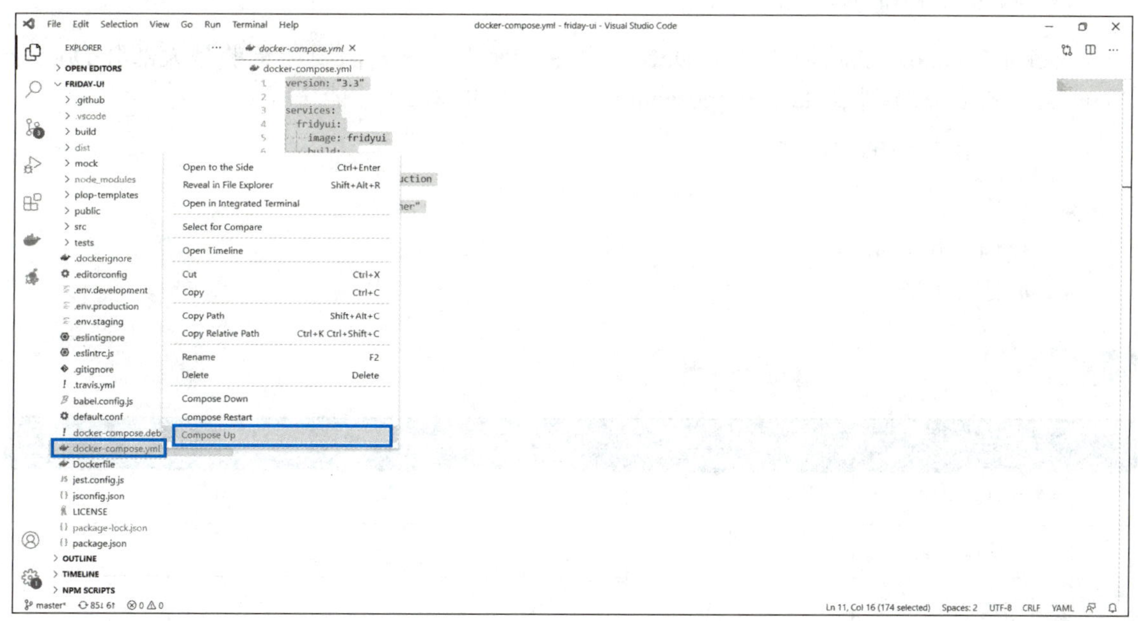

图 12-21　执行 Compose Up 命令

3）结果提示“friday-ui_fridyui_1 is up-to-date”，说明项目成功部署到 Docker 容器中，如图 12-22 所示。

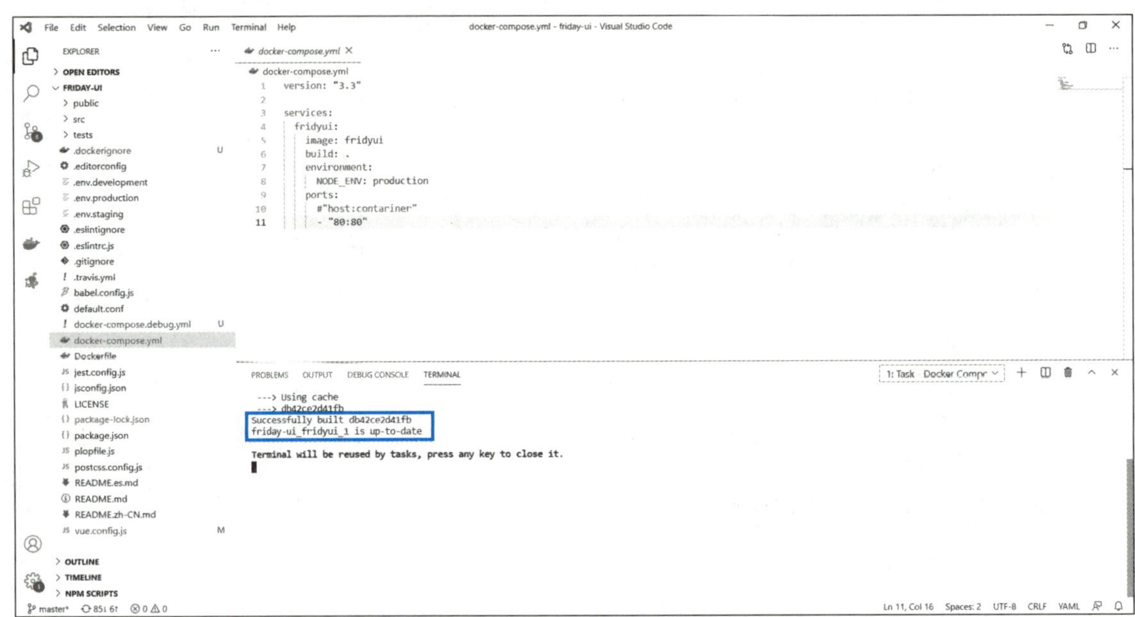

图 12-22　前端项目成功部署到 Docker 容器

知识小结

1. Docker Compose 配置文件

Docker Compose 的使用类似于 Docker 命令的使用，但是需要注意的是大部分 Docker Compose 命令都需要到 docker-compose.yml 文件所在的目录下才能执行。

2. YML 文件的编写规则

- 大小写敏感。
- 使用缩进表示层级关系。
- # 表示注释。

任务评价

技　能　点	知　识　点	自我评价（不熟悉 / 基本掌握 / 熟练掌握 / 灵活运用）
安装 Docker 工具	Dockerfile 指令 YML 文件	
安装配置 Docker 插件		
项目打包部署到 Docker 容器		

单元小结

本工作单元实现项目在生产环境 Docker 体系下的部署，由安装 Docker 服务、自动发布后端项目到 Docker 容器、自动发布前端项目到 Docker 容器三个任务组成。通过本单元的学习，掌握基于 Docker 技术的基本镜像操作、基于 Dockerfile 文件和 IDEA 插件的后端项目自动化部署以及基于 Dockerfile 文件和 Docker Compose 的前端项目自动化部署的职业技能，同时掌握了 1+X 职业技能等级证书（Java 应用开发）对容器镜像制作和系统部署与维护的职业技能等级要求。

实战强化

1．按照本工作单元任务 2 的内容，打包实训项目“诚品书城”的后端服务。

2．按照本工作单元任务 3 的内容，打包实训项目“诚品书城”的前端应用。

附录　实训项目（诚品书城）

一、用户故事列表

用户故事列表见附表 1。

附表 1　用户故事列表

序　号	故事名称	交互动作	交互信息
1	用户注册	注册分为三步： 1. 填写信息 2. 验证邮箱 3. 注册成功	1. 填写信息 需要填写邮箱地址、昵称、密码、确认密码以及验证码。 2. 填写信息 1）需要填写有效的 Email 地址； 2）昵称必须由小写英文字母、中文、数字组成，长度 4 ～ 20 个字符，一个汉字为两个字符； 3）密码由大小写英文字母、数字组成，长度 6 ～ 20 位； 4）生成验证码 3. 验证邮箱 向注册邮箱发送一个验证码，收到验证码后验证激活 4. 注册成功 提示用户注册成功并登录，提供跳转首页链接 5. 邮箱验证匹配，成功之后跳转到注册成功页面 6. 在注册成功页面，显示用户的 Email 信息，以及能够跳转到首页
2	用户登录	单击登录按钮，输入邮箱和密码，成功登录	1. 验证邮箱是否符合标准 2. 验证邮箱和密码是否可以匹配对应用户信息 3. 登录成功后，跳转到首页
3	首页－用户登出	单击登出按钮，跳转到用户登录页面	1. 单击登出按钮，跳转到用户登录页面 2. 保存在 session 中的用户信息被清除
4	首页－分类浏览模块	默认跳转到用户首页	1. 能够生成一级和二级菜单 2. 单击二级菜单的时候可以查询出相关的所有书籍
5	首页－编辑推荐模块	默认跳转到用户首页	1. 随机获取两本书的信息展示 2. 要求显示信息包括书封面图片、书名、作者、出版社、出版时间、简介、定价、诚品书城价（金额为小数点后保留一位） 3. 单击书封面图片、书名可跳转到该书的详细信息
6	首页－热销图书模块	默认跳转到用户首页	1. 要求显示信息包括书封面图片、书名、作者、出版社、出版时间、简介、定价、诚品书城价（金额为小数点后保留一位） 2. 单击书名可跳转到该书的详细信息 3. 根据销售（d_item 表）的数量来排序
7	首页－最新上架图书模块	默认跳转到用户首页	1. 要求显示信息包括书封面图片、书名、作者、出版社、出版时间、简介、定价、诚品书城价（金额为小数点后保留一位） 2. 单击书名可跳转到该书的详细信息 3. 根据产品（d_product 表）的 add_time 来排序
8	图书展示模块	在商品列表页面单击封面图片或者书名则跳转到图书展示模块	1. 展示图书的详细信息以及作者的相关介绍 2. 通过购买按钮可以将商品加入购物车中
9	购物车模块	1. 在图书展示模块单击购买后跳转到购物车页面 2. 在购物车页面进行增删改查的购物维护工作	1. 用户购买书之后，将购买的商品保存在购物车中 2. 用户可以修改购物车中购买的数量 3. 用户可以删除购物车中的商品 4. 系统能够自动计算购物车中商品的总价
10	订单生成模块	在购物车单击结算按钮，跳转到订单生成模块	确认订单，展示序号、商品名称、商品单价、商品数量、小计、总价
11	送货地址管理模块	在生成订单模块单击下一步按钮，跳转到送货地址管理模块，最后确认完成订单	1. 需要展示收件人姓名、收件人详细地址、邮政编码、联系电话 2. 提供一个地址管理的下拉列表，如果选择其中的下拉列表，数据自动添加到对应的控件

二、功能结构图

诚品书城功能结构图如附图 1 所示。

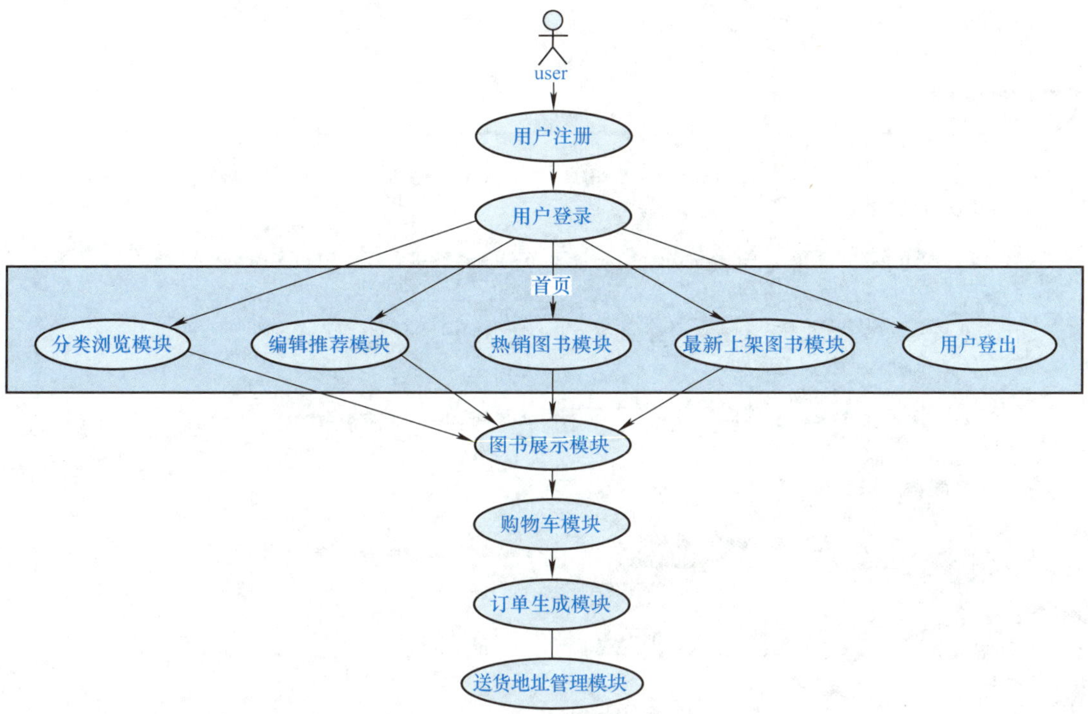

附图 1　诚品书城功能结构图

三、数据库设计

诚品书城数据库设计图如附图 2 所示。

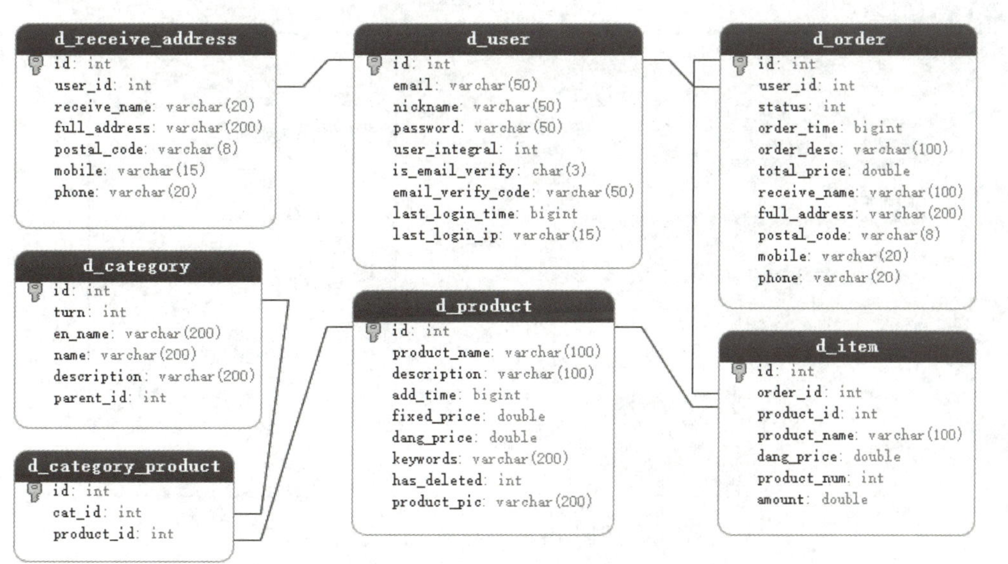

附图 2　诚品书城数据库设计图

四、主要的 UI 页面设计

1. 用户注册（见附图 3）

诚品书城

注册步骤：1.填写信息 > 2.验证邮箱 > 3.注册成功

以下均为必填项

请填写您的Email地址：	5030749@qq.com	请填写有效的Email地址，在下一步中您将用此邮箱接收验证邮件。 恭喜，该Email可用
设置您在诚品书城网的昵称：	chinaalex	您的昵称可以由小写英文字母、中文、数字组成， 长度4－20个字符，一个汉字为两个字符。
设置密码：	••••••	您的密码可以由大小写英文字母、数字组成，长度6－20位。
再次输入您设置的密码：	••••••	
验证码：	MA7LK ma7lk	请输入图片中的四个字母。 看不清楚？换个图片

注 册

附图 3　用户注册页面

2. 用户登录（见附图 4）

诚品书城

- 更多选择
  60万种图书音像，共计百万种商品。
- 更低价格
  电视购物的3-5折，传统书店的5-7折，其他网站的7-9折。
- 更方便、更安全
  全国超过300个城市送货上门、货到付款。零风险购物，便捷到家。

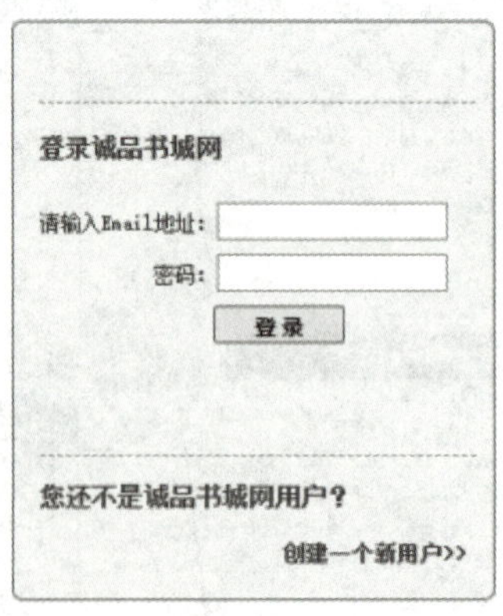

附图 4　用户登录页面

3. 首页 - 用户登出（见附图 5）

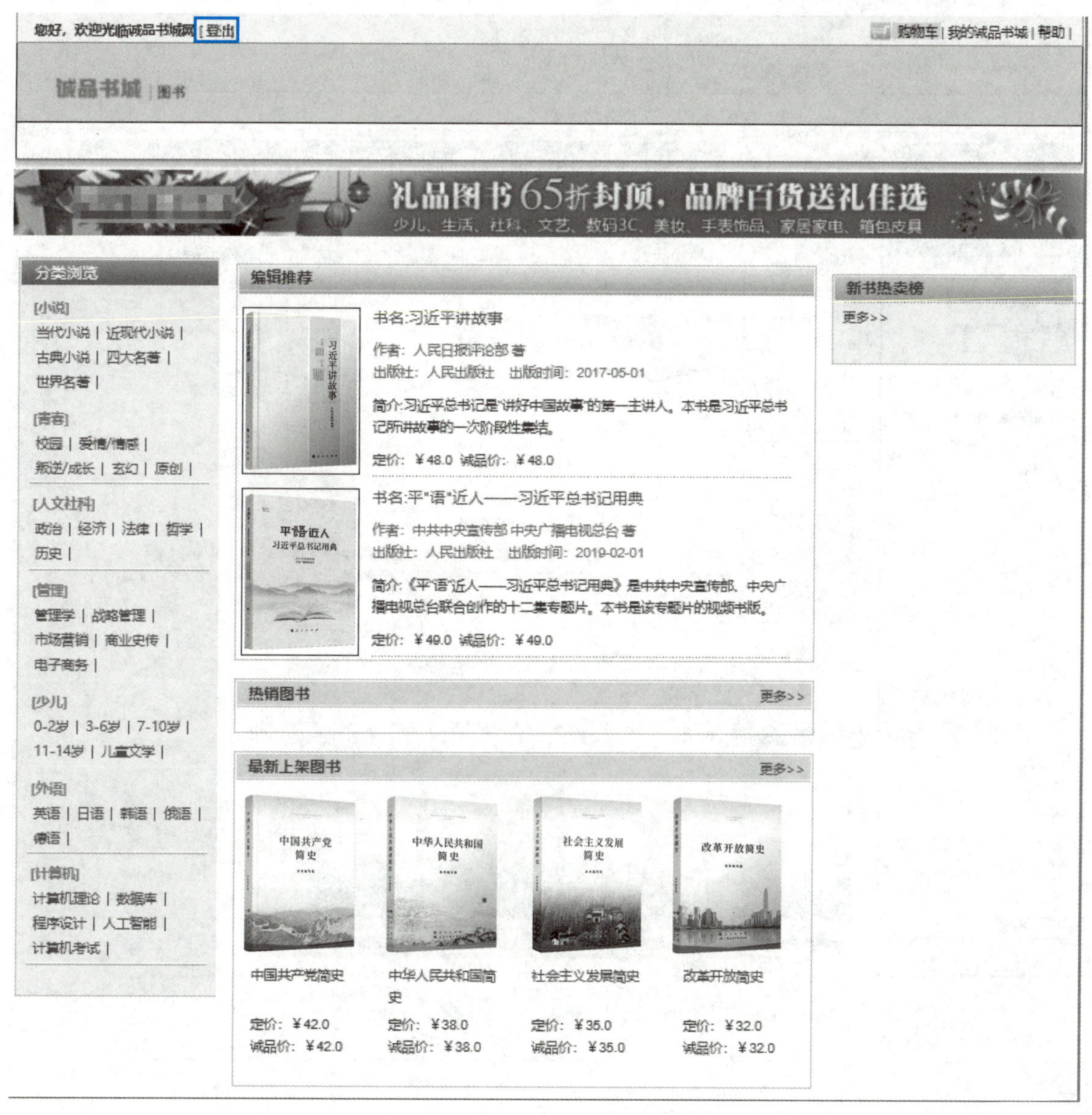

附图 5　首页 - 用户登出页面

4. 首页－分类浏览模块（见附图 6）

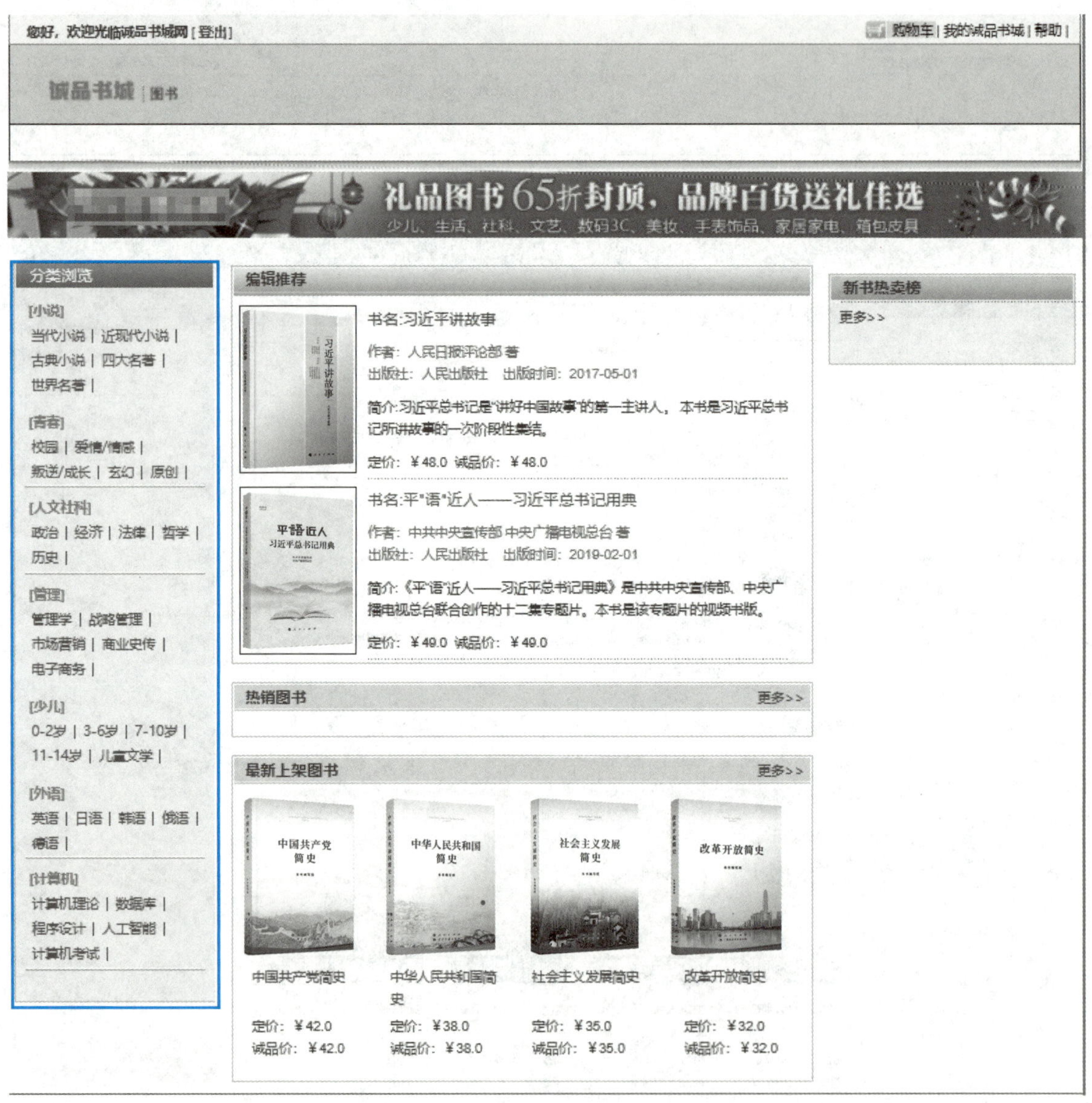

附图 6　首页－分类浏览模块

5. 首页 - 编辑推荐模块（见附图 7）

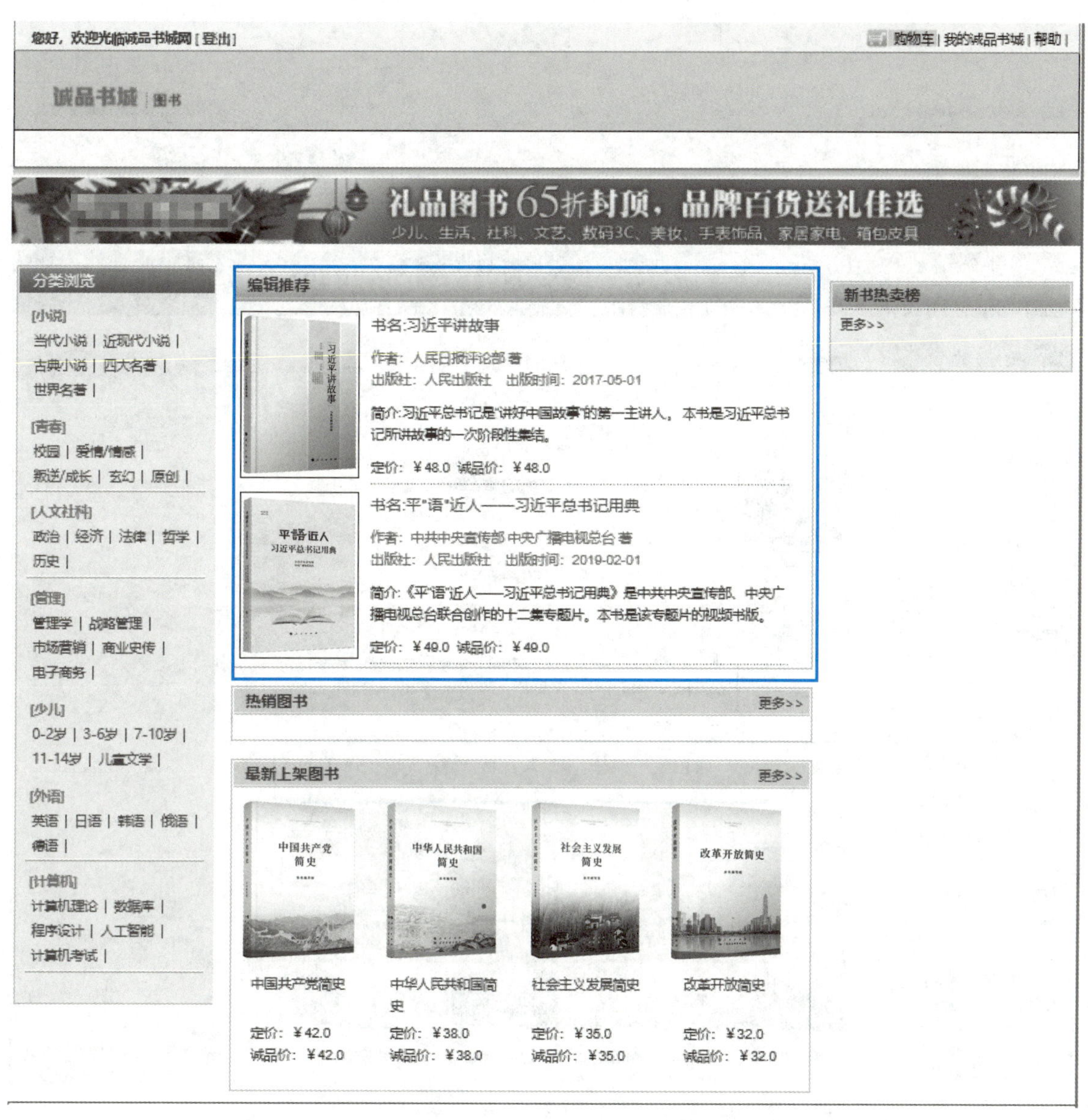

附图 7 首页 - 编辑推荐模块

6. 首页－热销图书模块（见附图 8）

附图 8　首页－热销图书模块

7. 首页－最新上架图书模块（见附图 9）

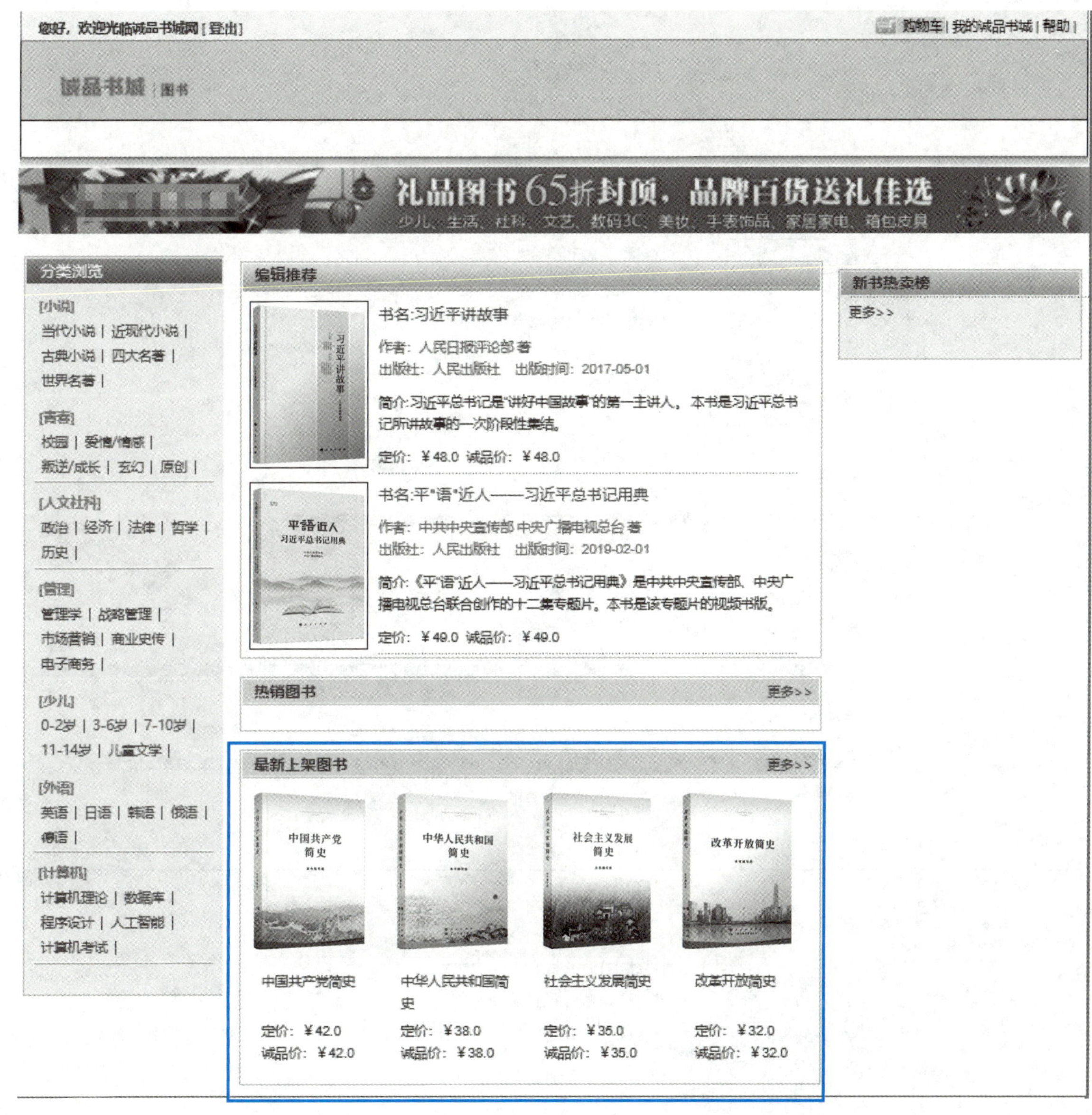

附图 9 首页－最新上架图书模块

8. 图书展示模块（见附图 10）

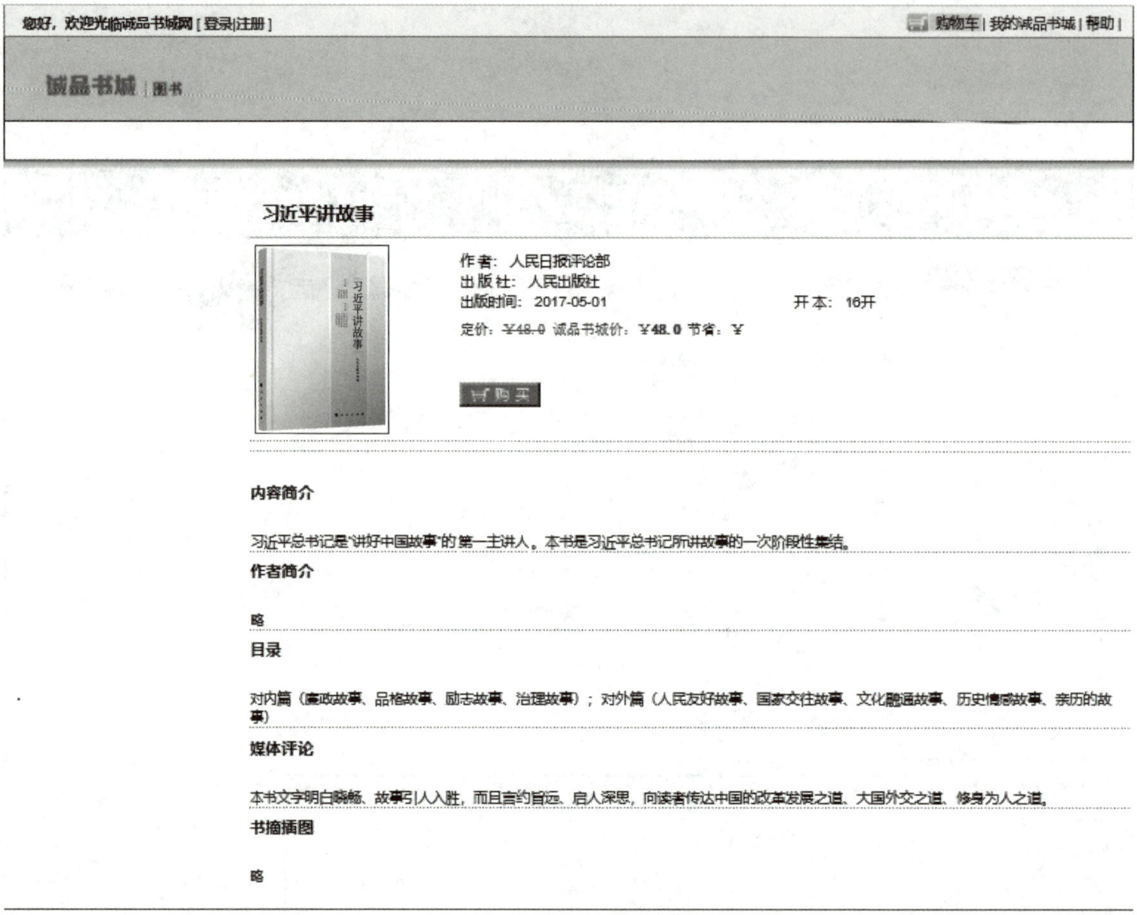

您好，欢迎光临诚品书城网 [登录|注册]　　购物车 | 我的诚品书城 | 帮助 |

诚品书城 | 图书

习近平讲故事

作者：人民日报评论部
出版社：人民出版社
出版时间：2017-05-01　　开本：16开
定价：~~¥48.0~~ 诚品书城价：¥48.0 节省：¥

购买

内容简介

习近平总书记是"讲好中国故事"的第一主讲人。本书是习近平总书记所讲故事的一次阶段性集结。

作者简介

略

目录

对内篇（廉政故事、品格故事、励志故事、治理故事）；对外篇（人民友好故事、国家交往故事、文化融通故事、历史情感故事、亲历的故事）

媒体评论

本书文字明白晓畅、故事引人入胜，而且言约旨远、启人深思，向读者传达中国的改革发展之道、大国外交之道、修身为人之道。

书摘插图

略

附图 10　图书展示模块

9. 购物车模块（见附图 11、附图 12）

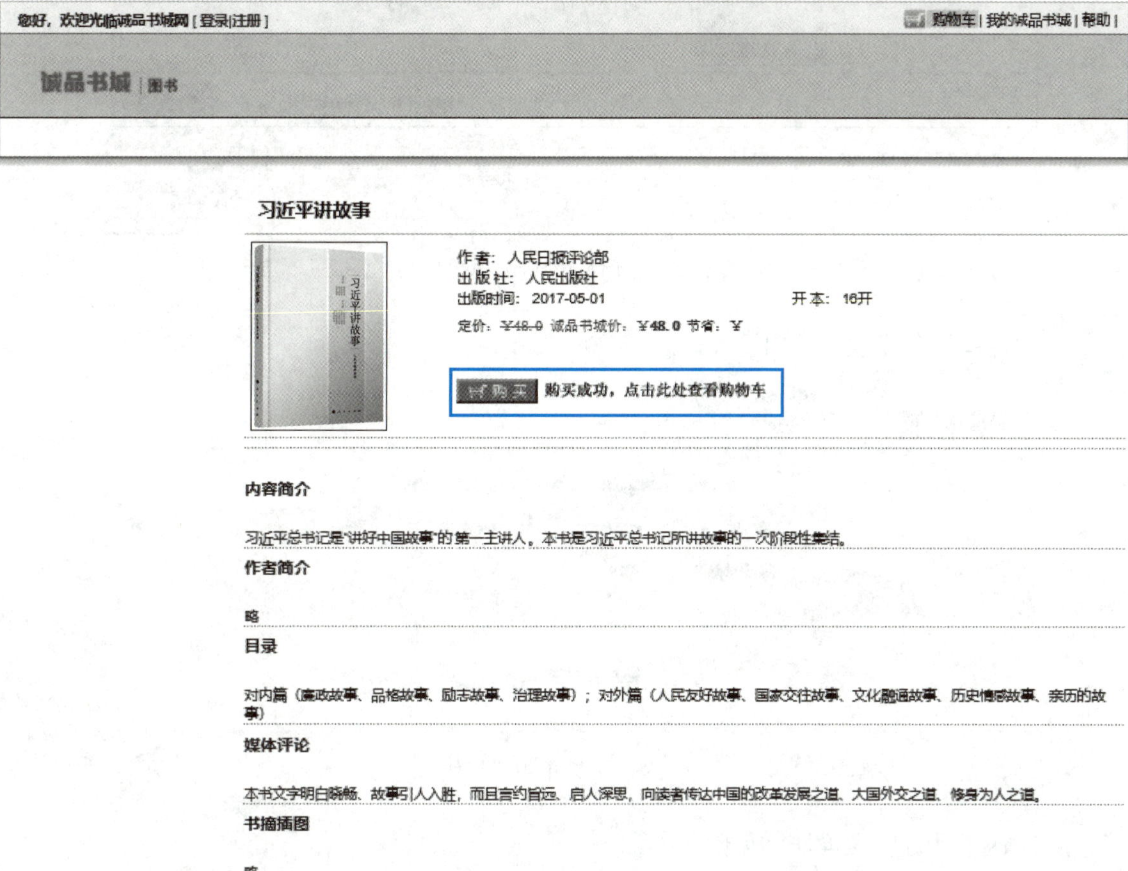

附图 11 购物车模块购买页面

我的购物车

您已选购以下商品　清空购物车

商品名	封面	市场价	诚品书城价	数量	变更数量	删除
习近平讲故事	习近平讲故事	48.0	¥48.0	2	变更	删除

继续挑选商品>>　您共节省：¥48 | 商品金额总计：¥48　结 算

您已删除以下商品，如果想重新购买，请点击"恢复"

¥　¥　恢复

附图 12　购物车模块：我的购物车

10. 订单生成模块（见附图 13）

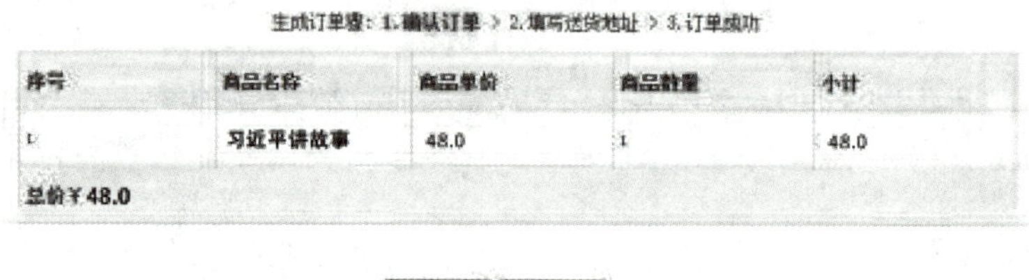

附图 13　订单生成模块

11. 送货地址管理模块（见附图 14）

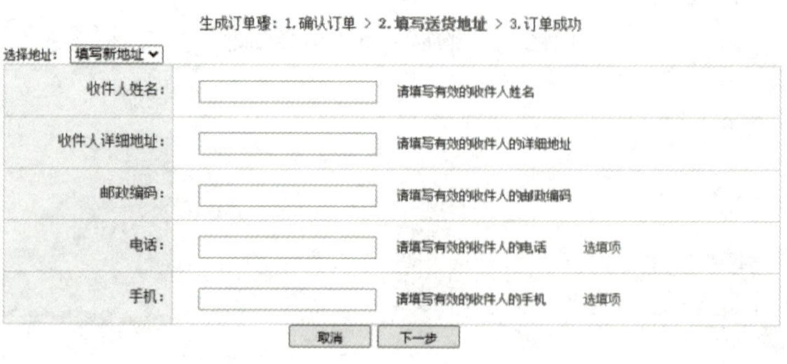

附图 14　送货地址管理模块

参 考 文 献

[1] 王松．SpringBoot+Vue 全栈开发实战 [M]．北京：清华大学出版社，2018．

[2] 陈学明．Spring+Spring MVC+MyBatis 整合开发实战 [M]．北京：机械工业出版社，2020．

[3] 贾志杰．Vue+SpringBoot 前后端分离开发实战 [M]．北京：清华大学出版社，2021．

[4] 李兴华．名师讲坛：Spring 实战开发（Redis+SpringDataJPA+SpringMVC+SpringSecurity）[M]．北京：清华大学出版社，2019．

[5] 徐丽健．SpringBoot+Spring Cloud+Vue+Element 项目实战：手把手教你开发权限管理系统 [M]．北京：清华大学出版社，2019．

[6] 王松．深入浅出 Spring Security[M]．北京：清华大学出版社，2021．

[7] 陈木鑫．Spring Security 实战 [M]．北京：电子工业出版社，2019．

[8] 宁海元，周振兴，彭立勋，等．高性能 MySQL[M]．3 版．北京：电子工业出版社，2013．

[9] 汪云飞．SpringBoot 实战 [M]．北京：电子工业出版社，2016．

[10] 梁灏．Vue js 实战 [M]．北京：清华大学出版社，2017．